Climate Change Adaptation Strategies – An Upstream-downstream Perspective

Nadine Salzmann • Christian Huggel
Samuel U. Nussbaumer • Gina Ziervogel
Editors

Climate Change Adaptation Strategies – An Upstream-downstream Perspective

Springer

Editors
Nadine Salzmann
Department of Geosciences
University of Fribourg
Fribourg, Switzerland

Samuel U. Nussbaumer
Department of Geosciences
University of Fribourg
Fribourg, Switzerland

Department of Geography
University of Zurich
Zurich, Switzerland

Christian Huggel
Department of Geography
University of Zurich
Zurich, Switzerland

Gina Ziervogel
Department of Environmental & Geographical Science and African Climate and Development Initiative
University of Cape Town
Cape Town, South Africa

ISBN 978-3-319-82179-5 ISBN 978-3-319-40773-9 (eBook)
DOI 10.1007/978-3-319-40773-9

Softcover reprint of the hardcover 1st edition 2016

Cover illustration: View towards the village of Kara-Jygach and the plains surrounding the inflow of the Naryn river into Toktogul reservoir, Jalal-Abad Province, Kyrgyzstan (photograph by Horst Machguth, 13 August 2013).

Printed on acid-free paper

This Springer imprint is published by Springer Nature
The registered company is Springer International Publishing AG Switzerland

Foreword

Climate change and the related adverse impacts are among the greatest challenges facing humankind in the twenty-first century. As a result of the significant increase of greenhouse gases in the atmosphere largely caused by human activities, the global phenomenon of climate change affects many different sectors including agriculture and water supply. Emitted greenhouse gases distribute homogeneously across the Earth's atmosphere, irrespectively of their source of emission, leading to global climate trends that do not recognize nor respect our man-made boundaries. Still, the impacts of climate variability or change become manifest on a local or regional level, asking for tailor-made solutions on the very same level, guided by national responsibility and global solidarity.

In such a framework, questions of how to tackle the challenge of adapting to climate change in upstream-downstream areas become increasingly important. Mountains as typical upstream areas are highly sensitive to global changes, as evidenced for instance by glacier retreat among the most obvious signs of climate change. At the same time, mountains are key contexts for sustainable development because of the indispensable goods and services they provide locally and to their adjacent downstream areas. Mountains are the world's water towers, providing freshwater to more than the half of humankind. They are centres of biological diversity, key sources of raw materials and important tourist destinations. Still, mountains are among the most disadvantaged regions in the world, with some of the highest poverty rates and greatest ecological vulnerability to global climatic, environmental and socio-economic change.

The obvious mismatch between the vulnerability and disregard of mountains at the one hand and their importance for the provision of key mountain ecosystem services on the other hand calls for urgent changes, which basically include four components, namely, (1) the recognition of mountain areas as key development contexts in global and national policy frames, (2) a scientifically sound information base related to mountains, (3) innovative approaches for action on the ground and (4) sufficient funding for (2) and (3). Fortunately, we thereby do not need to start from scratch, as evidenced, e.g., by the inclusion of mountains in three targets of the

Agenda 2030. But more is needed as obvious from the global climate change policy framework, where mountains only figure as a marginalia in the original UN Framework Convention on Climate Change (Art. 4.8.g) but not in the new Paris Agreement concluded at UNFCCC's COP 22 in Paris 2015.

The present book, prepared under the umbrella of the Swiss initiated and facilitated mountain programme SMD4GC[1], aims at creating a better understanding of how to tackle the four components mentioned above. It thereby provides an insight on how to reduce or avoid the adverse impacts and risks from climate change and to move towards a sustainable future in mountain regions. After an introductory part, which sets the scene on the current state of adaptation in mountainous regions and its challenges, the book highlights a dedicated number of selected case studies that introduce good adaptation practices from all over the world. The book concludes with some global considerations related to aspects of resilience building and science-policy dialogue for climate change adaptation in mountain regions, showing that important and encouraging inroads have been made.

We hope that this book will raise the awareness of the challenges of climate change adaptation in mountainous areas. At the same time, we expect the book to foster a comprehensive understanding of the role and importance of mountain ecosystem goods and services for global sustainable development. This, in turn, will hopefully contribute to trigger practical action to tackle climate change in the often neglected yet so important mountain regions of this world.

Mountain Desk André Wehrli
Swiss Agency for Development and Cooperation
Berne, Switzerland

[1]SMD4GC is the acronym for Sustainable Mountain Development for Global Change, a programme induced and supported by the Swiss Agency for Development and Cooperation, which aims at contributing to sustainable mountain development under uncertain changes in climatic, environmental and socio-economic conditions, focusing on poverty and risk reduction.

Acknowledgements and Disclaimer

This book is a contribution to the *Sustainable Mountain Development for Global Change (SMD4GC)* programme that is supported by the Swiss Agency for Development and Cooperation. Special thanks for providing reviews for the individual chapters are due to numerous anonymous referees.

The designations employed and the presentation of material in this publication do not imply the expression of any opinion whatsoever on the part of the publisher and partners concerning the legal or development status of any country, territory, city or area or of its authorities, the delimitation of its frontiers or boundaries. The mention of specific companies or products of manufacturers, whether or not these have been patented, does not imply that these have been endorsed or recommended by the institutions mentioned in preference to others of a similar nature that are not mentioned. The views expressed in this publication are those of the author(s) and do not necessarily reflect the views or policies of the institutions mentioned.

Contents

Part A

Setting the Scene: Adapting to Climate Change – A Large-Scale Challenge with Local-Scale Impacts 3
Nadine Salzmann, Christian Huggel, Samuel U. Nussbaumer, and Gina Ziervogel

Adaptation, Adaptation Science, and the Status of Adaptation in Mountain Regions 17
Graham McDowell, Eleanor Stephenson, and James Ford

Part B

Science in the Context of Climate Change Adaptation: Case Studies from the Peruvian Andes 41
Boris Orlowsky, Norina Andres, Nadine Salzmann, Christian Huggel, Christine Jurt, Luis Vicuña, Mario Rohrer, Pierluigi Calanca, Raphael Neukom, and Fabian Drenkhan

Managing Glacier Related Risks Disaster in the Chucchún Catchment, Cordillera Blanca, Peru 59
Randy Muñoz, César Gonzales, Karen Price, Ana Rosario, Christian Huggel, Holger Frey, Javier García, Alejo Cochachín, César Portocarrero, and Luis Mesa

Climate Change Adaptation in the Carpathian Mountain Region 79
Saskia Elisabeth Werners, Sándor Szalai, Henk Zingstra, Éva Kőpataki, Andreas Beckmann, Ernst Bos, Kristijan Civic, Tomas Hlásny, Orieta Hulea, Matthias Jurek, Hagen Koch, Attila Csaba Kondor, Aleksandra Kovbasko, M. Lakatos, Stijn Lambert, Richard Peters, Jiří Trombik, Ilse van de Velde, and István Zsuffa

Community Forest Management as Climate Change Adaptation Measure in Nepal's Himalaya 101
Rabin Raj Niraula and Bharat K. Pokharel

Ecosystem-based Adaptation (EbA) of African Mountain Ecosystems: Experiences from Mount Elgon, Uganda 121
Musonda Mumba, Sophie Kutegeka, Barbara Nakangu, Richard Munang, and Charles Sebukeera

Vulnerability Assessments for Ecosystem-based Adaptation: Lessons from the Nor Yauyos Cochas Landscape Reserve in Peru 141
Pablo Dourojeanni, Edith Fernandez-Baca, Silvia Giada, James Leslie, Karen Podvin, and Florencia Zapata

The Role of Ecosystem-based Adaptation in the Swiss Mountains 161
Veruska Muccione and Ben Daley

Community Perceptions and Responses to Climate Variability: Insights from the Himalayas 179
Anju Pandit, Anmol Jain, Randhir Singha, Augustus Suting, Senti Jamir, Neera Shresthra Pradhan, and Dhrupad Choudhury

Drought: In Search of Sustainable Solutions to a Persistent, 'Wicked' Problem in South Africa 195
Coleen Vogel and Koos van Zyl

Making Climate Resilience a Private Sector Business: Insights from the Agricultural Sector in Nepal 213
Chiara Trabacchi and Martin Stadelmann

Part C

Shaping Climate Resilient Development: Economics of Climate Adaptation 241
David N. Bresch

Building Resilience: World Bank Group Experience in Climate and Disaster Resilient Development 255
Daniel Kull, Habiba Gitay, Sofia Bettencourt, Robert Reid, Alanna Simpson, and Kevin McCall

The Science-Policy Dialogue for Climate Change Adaptation in Mountain Regions 271
Thomas Kohler, André Wehrli, Elbegzaya Batjargal, Sam Kanyamibwa, Daniel Maselli, and Urs Wiesmann

Part A

Setting the Scene: Adapting to Climate Change – A Large-Scale Challenge with Local-Scale Impacts

Nadine Salzmann, Christian Huggel, Samuel U. Nussbaumer, and Gina Ziervogel

Abstract This chapter's main objective is to provide the context of the book and to introduce the subsequent chapters.

The physical basis of the global climate change challenge is briefly outlined and the consequences for the societies primarily at the local scale are discussed. A short overview of how the international policy level responds to the challenge of global climate change impacts and risks is provided. Key terms related to different types of adaptation are also introduced and reasons for the complexity of climate change adaptation discussed. Then, the evidence for the importance of mountain ecosystems and adjacent downstream areas, which are critically linked through water, is briefly reviewed. Finally, each chapter of the book is introduced, followed by key conclusions we can draw from the book concerning the state and experiences of adaptation in upstream and downstream areas.

Keywords Mountain regions • Climate change • Vulnerability • Hazards and risks • Types of adaptation • Upstream and downstream areas • Global-local scale

N. Salzmann (✉)
Department of Geosciences, University of Fribourg, Fribourg, Switzerland
e-mail: nadine.salzmann@unifr.ch

C. Huggel
Department of Geography, University of Zurich, Zurich, Switzerland

S.U. Nussbaumer
Department of Geosciences, University of Fribourg, Fribourg, Switzerland

Department of Geography, University of Zurich, Zurich, Switzerland

G. Ziervogel
Department of Environmental & Geographical Science and African Climate and Development Initiative, University of Cape Town, Cape Town, South Africa

N. Salzmann et al. (eds.), *Climate Change Adaptation Strategies – An Upstream-downstream Perspective*, DOI 10.1007/978-3-319-40773-9_1

1 Introduction

Global climate change and the related adverse impacts and risks are among the greatest challenges facing humankind in the twenty-first century. This is recognized by all governments of the world, as clearly showcased through the UNFCCC (United Nations Framework Convention on Climate Change) Agreement of Paris in December 2015 signed by all 196 countries. Only a few months earlier, the UN's 193 Member States adopted the 2030 Agenda for Sustainable Development, with the broader aim of achieving sustainable development, of which climate change adaptation and mitigation is one part. Whether these agreements are worth the paper they are written on will depend on the actions that follow, both on emission reduction and adaptation. This book focuses on the adaptation part of the challenge, on how to reduce or avoid the adverse impacts and risks from climate change and to move towards a sustainable future. The focus is narrowed down further to examine mountain regions and their adjacent downstream areas. Upstream-downstream areas are among the ecosystems most vulnerable to climate change impacts and risks and essential for sustainable development. These areas are therefore also key regions for international cooperation and development. Switzerland as a small mountain country has a long-standing tradition, interest and expertise in sustainable development of upstream and downstream regions. In Switzerland as in other mountain and downstream countries, people have been living for centuries from the benefits, goods and services provided by mountain ecosystems and at the same time have been living with the constant risks inherent in these topographically complex and highly dynamic landscapes. With changing climatic conditions a key factor for sustainable development is altering and calls for adequate response. It is in this consequence, that the Swiss government has launched the programme 'Sustainable Mountain Development for Global Change' (SMD4GC) under the Global Programme Climate Change (GPCC) of the Swiss Agency for Development and Cooperation (SDC). This book, which is supported by SMD4GC is a contribution to the challenges arising from global climate change for societies.

This first chapter of the book is setting the scene by providing an introduction into the overall thematic and context of the book and thus for the chapters following.

2 Global-Scale Changes with Local-Scale Impacts

Global climate change as a result of the significant increase of greenhouse gases (GHGs) in the atmosphere since the pre-industrial times is largely caused by human activities (IPCC 2013). This includes how food is produced, how mobility, transportation and industry have developed, and the type of energy used to enable these activities. Because emitted GHGs distribute homogeneously across the Earth's atmosphere, irrespectively of where they are emitted, the warming trend is observed globally and varies only in its spatial-temporal magnitudes, which is caused by regional and local surface properties and meteorological characteristics. In addition to air temperature, precipitation is another key climate variable in the climate system with significant impacts to the environment and society when patterns change. For precipitation, however, there is no such clear global trend in one direction as for air temperature. Precipitation amounts and intensities nevertheless tend to more extremes values either towards dry periods in some regions or heavy precipitation events in others (Donat et al. 2016). Impacts of global climate change affect every place on Earth, independent on where and when the effective emissions are released (Knutti and Rogelj 2015). Accordingly, negative impacts of climate change on natural and human systems have been reported all over the world and evidence is provided by a great number of scientific assessment studies, including those assessed by the Intergovernmental Panel on Climate Change (IPCC). There is evidence for instance of reducing and/or altering of water resource regimes from snow and glaciers, changing characteristics of hazards and risks in high-mountain regions, loss in crop production and extremes such as heat waves or droughts in downstream regions and related impacts on socio-economic activities and life. The effective dimension of the impacts finally felt by a society at a specific location depends however not only on the magnitude of changed climate variables or how much of the changes are caused by anthropogenic GHG emissions; but importantly on a societies' vulnerability and exposure to climatic changes. Vulnerability moreover depends on the propensity to be adversely effected and the capacity of people, livelihoods, environmental services and resources, infrastructure, or economic, social, or cultural assets to respond to the risks faced. The core concept of integrated vulnerability, hazard and risks and the related terms as used by IPCC is illustrated by their respective figure (Fig. 1).

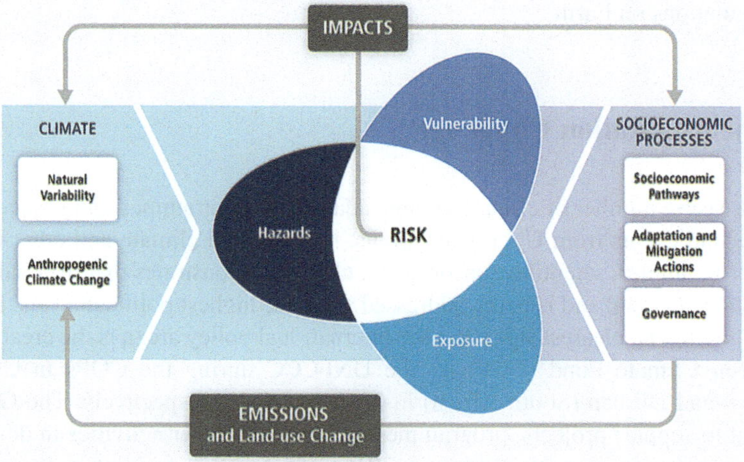

Fig. 1 The core concept and terms related to risks of climate related impacts and hazards as introduced by IPCC (2014) (Source: IPCC 2014)

Generally, increasing temperatures raise the risks for natural and human systems as often illustrated by using the so-called 'burning embers' diagram (IPCC 2014). Since projections from climate models suggest significant warming for the coming decades, irrespective of the emission scenario considered (RCP, Representative Concentration Pathways; van Vuuren et al. 2011), increases of related risks to ecosystems, economy, and daily life over the coming decades are hence very likely. The challenge is meanwhile recognized at the international level. Governments agreed in 2010 at the UNFCCC Conference of the Parties (COP16) in Cancún (Mexico) of the need to limit warming in order to keep climate change impacts at a manageable level. In December 2015 in Paris, all 196 countries finally adopted a new Climate Agreement as the follow-up agreement of the Kyoto Protocol. The Paris Agreement aims at a target of below 2 °C global average warming, and states at the same time that efforts should be pursued to limit warming to 1.5 °C. These targets refer about to a two in three probability of keeping warming to 2 °C or less. Or in other words, there is a remaining carbon budget of about 20 years' worth of current levels of emissions. Although these targets are not based on scientific evidence, it is assumed by the policy that limiting global warming to 2 °C compared to the pre-industrial level will keep the impacts of climate change at a manageable and thus safe level.

Regardless of the debate around the 'two-degree target', the most import part of the discussion is the recognition of the need to drastically decrease emissions to zero or even below by the end of the twenty-first century, and that this calls for tremendous efforts at all levels (international, regional and national) and by all players, including policy, public and private sectors, and the society. In addition to the mitigation efforts and liberated from any discussion whether limiting warming to agreed targets is sufficient and at all still reachable, it is also clear that along with decarbonization, there is an urgent need for strategies to adapt to impacts which already happened or are unavoidable. This is particularly true for the most vulnerable regions and populations on Earth.

3 The Adaptation Challenge

The increasing number of climate change adaptation programmes and related conferences and events from UN organizations, international climate and other monitoring programmes, scientific consortia etc. clearly demonstrates that the challenge has been understood and is being addressed up to the highest political levels. In this regard, among the clearest signals in the international policy arena is the creation of the Green Climate Fund (GCF) by the UNFCCC during the COPs in Cancún (Mexico) and Durban (South Africa) in 2010 and 2011, respectively. The GCF is intended to support projects, programmes, policies and other activities in developing countries (first projects started in 2015) and be the operating entity of the UNFCCC's financial mechanism for both mitigation and adaptation. Finally, the Paris Agreement emphasized again the urgency of adaptation efforts and support, particularly for the most vulnerable regions, which are often found in developing countries.

According to the definition by IPCC, climate change adaptation refers to the adjustment of natural or human systems as a response to actual or expected climatic stimuli or their effects, which moderates harms or exploits beneficial opportunities. Thereby adaptation can be 'anticipatory', 'autonomous' or 'planned'. Adaptation is anticipatory (or proactive) when it takes place before any impacts are observed. Autonomous adaptation means spontaneous adjustment to changes that does not necessarily constitute a conscious response to climatic stimuli but is triggered by ecological changes in natural systems and by market or welfare changes in human systems. Finally, planned adaptation is the result of a deliberate policy decision, based on the awareness that conditions have changed or are about to change and that action is required to return to, maintain, or achieve a desired state.

As outlined above, the need for adaptation actions (anticipatory and planned) is urgent and scientists, governments, non-governmental organizations and the private sector alike have been putting efforts into the development and implementation of adequate adaptation measures for several years. However, the challenge of adaptation is multidimensional and highly complex. There are still lots of remaining open questions and numerous aspects to be clarified around how to best respond and adapt to climate change impacts and risks, and also how to assess the effective potential and the limitation of adaptation measures to reduce adverse effects of climate change (Adger et al. 2009; Dow et al. 2013). The reasons for the adaptation's complexity are multifold as outlined in the following paragraphs (key issues in bold), and include the multidisciplinary nature of the scientific assessments required behind, future uncertainties, the close link with the policy level, and the need to embed responses into the local context of societies.

An important prerequisite for anticipatory or planned adaptation is the knowledge of the characteristics and magnitudes of changes and trends of key climate variables and the associated (potential) impacts, risks and vulnerabilities. In other words, to which changes, impacts and risks does a society need to adapt to. To provide this basic information, **reliable, long-term and continuous baseline data** of climatic and societal characteristics is necessary in order to derive trends and detect and determine changes. Ideally, such (climate) data sets are available over a time span of about 30 years, as recommended by the World Meteorological Organization (WMO) for climate-related assessments. As changes and impacts are often local in scale, respective data are needed on site. This is often a serious impediment for adaptation planning because of a general scarcity of reliable, spatially and temporally continuous data series, particularly in remote mountain regions (Salzmann et al. 2014). In addition, the effective risks for a specific location and its society are often not known at a level of details, as would ideally be required for appropriate adaptation planning. As a consequence, adaptation measures must often be developed based on incomplete or weak databases.

As climate will very likely continue to change, **modelled future projections of climate, impacts and risks** need to be taken into account when aiming at sustainable adaptation measures. Outputs from models, and in particular model output of climate projections, have by definition a certain range of **uncertainty** inherent, which typically increases from global to local scales and the further they project into the future (Hawkins and Sutton 2009). Adaptation measures must therefore be flexible and adjustable as time evolves and incorporate different strategies for handling uncertainties from the various sources including those from limited physical knowledge and uncertain behavior of societies.

These broad sources of uncertainties call for **inter-, multi- and transdisciplinary** investigations and integrated approaches. While the investigations of changes and impacts mainly demand scientific approaches, the development of adaptation measures requires a participative approach that engages the affected local communities and the decision makers within their respective **socio-economic context**. Only this way, any measures have a chance to be accepted, to be effective and thus to be sustainable. This means that the process of adaptation is also strongly **linked to, and needs to be mainstreamed into the policy and decision-making level**, which requires a high and mature level of science-policy dialogues and openness, confidence and trust between the different players. As such, adaptation efforts are naturally long-term processes, where measures must periodically be evaluated and adjusted to new climatic patterns, societal changes or socio-economic realities.

This is however in contrast to the often applied **'project'-approaches**, where adaptation measures are developed and implemented within a specific funded project and thus within a limited time period of typically about 3–4 years. Consequences of project-based approaches that are not embedded in long-term processes mean that there is often not enough time available to implement the measures. Or, priority is given to **'robust' and 'no-regret' (standard) measures** that can be implemented relatively easily and quickly. Such measures may have been applied in other contexts before, and are then simply transferred to other regions. However, can measures simply be transferred from one context to another without risks of maladaptation or benefit to some but harm to others? There are questions on the extent to which adaptation measures are able to neutralize adverse impacts of climate change. In general, there is a lack of monitoring and evaluation of the success (or failure) of measures and their appropriateness for other regions. This is certainly also due to the relatively short period of available experience in adaptation to climate change impacts and risks. Nevertheless, it is obvious that important open questions remain about the effective potentials and limits of adaptation measures. Therefore, a systematic documentation, analyses and synthesis of experiences from the past years are urgently needed in order to share the gathered know-how so far and to derive best practices and guidelines. Such analyses will help to support decision-making processes and maximize possible co-benefits and minimize potential adverse side-effects of adaptation measures. As such, this know-how will facilitate, and enable efficient use of resources that are going to be released for instance through the aforementioned GCF by UNFCCC and moreover be an important step towards sustainable development.

4 Mountain Regions and Adjacent Downstream Areas: Vulnerable Systems

Only a few months before the Paris Agreement, the UN's 193 Member States adopted the 2030 Agenda for Sustainable Development. These new Sustainable Development Goals (SDGs) build on the eight Millennium Development Goals (MDGs) that had come to conclusion by end of 2015, and address and emphasize the needs of people in both developed and developing countries in 17 goals and 169 targets. They center all on the three fundamental dimensions of sustainable development (social, economic and environmental) as well as on important aspects related to peace, justice and effective institutions. The SDGs highlight mountain environments as among the ecosystems most essential for sustainable development. Mountains, including the indigenous people and local communities that live there, are considered as vulnerable and fragile ecosystems to the adverse impacts of climate change. Mountain regions and their embedded socio-economic systems are in addition typically remote, not only by means of spatial distances but also in the sense of poor transport, infrastructure and connections to larger cities or modern infrastructure, and national decision makers, and making them potentially even more vulnerable. High-mountain environments are also naturally extreme in terms of climatic and environmental conditions caused by steep and complex topography. As a result, highly specialized and adapted natural ecosystems and societies have been evolving over long time scales. Despite their remoteness and their often marginalized status, mountain regions are of high importance for many of the Earth's populations as recognized by the aforementioned SDGs. They provide many important goods and service directly and indirectly to the highly populated downstream areas, in particular water, for consumption, irrigation or power production. As hotspots of change, high-mountain ecosystems and their services are becoming increasingly important, in upstream and downstream areas alike. Mountain ecosystems have significant capacity to support adaptation to global change for instance through ecosystem-based adaptation. Upstream and downstream areas are closely linked to and depend on each other, which become obvious when looking through the lens of water – a globally and fundamentally essential resource.

5 Water: Or the Link Between Upstream and Downstream Areas

Water is central to ecosystems, people and economies in both upstream and downstream areas. Changes in climate, including temporal and spatial variability and magnitudes of precipitation, have impacts directly on water quality and quantity and hence on the impacts of climate change across scales.

The general characteristic pattern of precipitation at specific locations is primary determined by large-scale climate circulation. Mountain topography plays an important role in the regional and local spatial-temporal distribution of precipitation and determines whether it hits the ground as snow or rain. Mountain ranges can block moisture transport, which results in dry areas in the rain shadow, cause heavy precipitation systems in front of mountain ranges, or provoke precipitation through convection potentially leading to thunderstorms. Heavy precipitation events can provoke floods or trigger debris flows, and may interrupt and destroy routes of transportation or infrastructure. On the other side, droughts can cause significant reduction or damage to food production, hydropower generation or lead to contamination of fresh water in both upstream and downstream areas. In high-elevation regions, precipitation often falls in solid form (snow), and mountain areas can thus dump the impact of heavy precipitation events through delayed and decelerated runoff. Glaciated high-mountain environments can store precipitation in frozen form and as such act as important natural temporal water storage system, which store water during the wet seasons and release the water only during the dry season. In such regions, melt water from the cryosphere is often the only water resource available to cover the demand for fresh water to be used for irrigation in upstream and downstream regions. Sectors like agricultural or energy production have adapted to these naturally-driven timing of a reliable water source in many upstream and downstream regions.

The observed and projected increases in air temperatures cause on the one hand further significant glacier recession and on the other hand lead to more frequent rain than snow events. Both effects result in a reduction of the cryospheric storing capacity. Recent studies furthermore provided evidence that warming is enhanced with increasing elevation, yet not uniformly around the globe (Mountain Research Initiative EDW Working Group 2015). Changing precipitation pattern moreover influence also the amount and timing of precipitation. Consequently, changed runoff and extreme events such as heavy precipitation events or dry periods can pose risks to societies in upstream and downstream areas by floods, landslides, debris flows or droughts. Due to the close linkage of upstream and downstream areas through water, adaptation measures might be particularly complex in these regions, as measures taken at one site might impact other areas. For instance, if a dam is built to artificially store water in the upstream area as a response to reduced precipitation and/or loss of natural water storage capacities (e.g. glaciers), downstream areas are then likely to have even higher water scarcity, which can potentially lead to conflict situations, in particular in transboundary settings (Bichsel 2011). This example shows an additional challenge of climate change impacts and adaptation. As climate change impacts in mountain regions can affect downstream areas far away through the interlinkage of water, likewise can local adaptation measures implemented in upstream areas have (positive or negative) effects on downstream areas. This furthermore highlights again the importance of interdisciplinary approaches, the importance of considering local and regional aspects when taking action and hence the importance of engaging with policy and decision makers at different levels. And finally, it shows the importance to document and synthesize experiences in order to derive and share guidelines for good practices.

6 Organization of this Book

This book is a contribution to the important challenge of adapting to adverse impacts and risks of climate change in mountain ecosystems and the adjacent downstream areas, and as such supports the overarching goal of sustainable development. As adaptation measures must be streamlined with and mainstreamed into overall development and planning within the local and regional contexts, there are typically no standard solutions available for successful adaptation. Moreover, societies are currently only about to learn from and evaluate first experiences with the development and implementation of adaptation measures as there are only few long-term examples available. Therefore, and in view of the emerging needs of adaptation, it is time to pause for a moment and to summarize and synthesize the experiences and lessons learnt so far. To serve this purpose, this book provides insights from several case studies from different upstream and downstream regions of the world (Fig. 2) and syntheses from global perspectives. The book offers a range of rich and intriguing perspectives from experts with diverse background and context, including science, governmental and non-governmental organizations, private sector or international organizations.

The book is organized in three main parts (A, B, C). Part A provides an introduction into the main thematic of the book followed by ten case studies in Part B from different upstream and downstream regions (cf. Fig. 2) of the world, which report on actions taken as well as experiences and lessons learnt. In Part C the global perspectives enter the discussion and synthesize the adaptation efforts by including aspects of resilience, economy and the science-policy dialogue.

In Part A, McDowell et al. (chapter "Adaptation, Adaptation Science, and the Status of Adaptation in Mountain Regions") provide an overview on the scientific

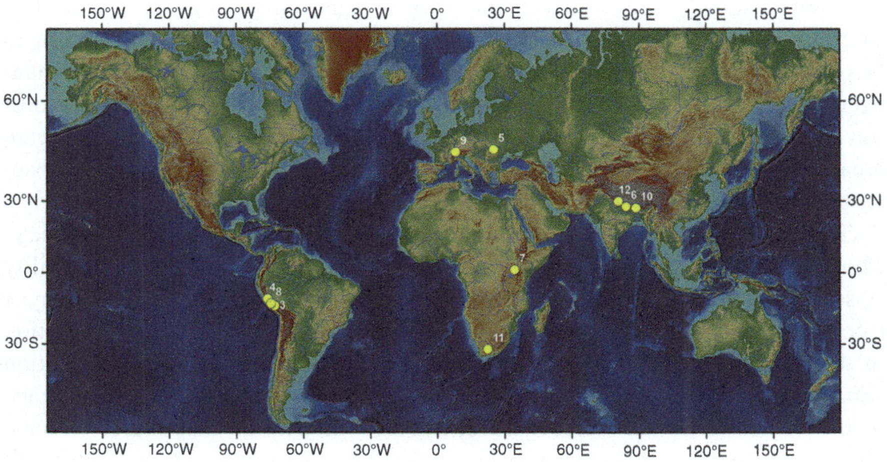

Fig. 2 Upstream and downstream regions indicated by topography and river systems. The *yellow dots* and *numbers* indicate the case study sites of the chapters in this book

status of climate change adaptation in upstream and downstream areas and draw out the linkages between adaptation scholarship and specific socio-economic and environmental conditions. As such, the chapter also points to the challenge of moving from the theory to the practice of adaptation.

The first case study (chapter "Science in the Context of Climate Change Adaptation: Case Studies from the Peruvian Andes") of Part B centers on the role of scientific studies in the process of assessing the needs for adaptation. In this chapter presented by Orlowksy et al. it is shown how scientific assessments are motivated by the local community's perception on climate change impacts. In this context the authors also stress that they found an important lack of social sciences studies compared to those from natural science. Chapters "Science in the Context of Climate Change Adaptation: Case Studies from the Peruvian Andes" and "Managing Glacier Related Risks Disaster in the Chucchún Catchment, Cordillera Blanca, Peru" focus both on the Andes. In chapter "Managing Glacier Related Risks Disaster in the Chucchún Catchment, Cordillera Blanca, Peru", Muñoz et al. report on a case from the Cordillera Blanca, Peru, where a village is in risk of glacier lake outburst floods (GLOFs) and document how the affected community manages these risks. Jointly with an international multidisciplinary team of experts, the community has planned and implemented a comprehensive early warning system, where particular emphasis was given to technical and social aspects.

A comprehensive analysis of climatic trends, impacts and adaptation options follows in chapter "Climate Change Adaptation in the Carpathian Mountain Region" for the case of the Carpathian mountain region, which represents a significant natural refuge on the European continent and stretches over several countries. The assessments undertaken by Werners et al. had inspired a strategic agenda on adaptation to be implemented under the Carpathian Convention. The chapter thus illustrates the important and clear link needed between science and policy and across borders. Chapter "Community Forest Management as Climate Change Adaptation Measure in Nepal's Himalaya" centers on forest management as a measure to adapt to climate change. Raj and Bharat show how community forest management in Nepal's Himalayas has led to conservation of biological diversity, ecosystem management etc. which ultimately turned into positive effects on overall livelihood of the local people depending of forest resources in both upstream and downstream areas. In particular, these measures have increased the population's adaptive capacity and the resilience of forest to changing climatic conditions.

The following chapters "Ecosystem-based Adaptation (EbA) of African Mountain Ecosystems: Experiences from Mount Elgon, Uganda", "Vulnerability Assessments for Ecosystem-based Adaptation: Lessons from the Nor Yauyos Cochas Landscape Reserve in Peru" and "The Role of Ecosystem-based Adaptation in the Swiss Mountains" focus on the concept of Ecosystem-based Adaptation (EbA). Defined by the Convention on Biological Diversity (CBD) 10th Conference of the Parties (COP) in October 2010, EbA is one way of adapting to climatic changes by using biodiversity and ecosystem services to help people to adapt to the adverse effects of climate change and thus links ecosystem services, sustainable resource management and climate change adaptation. In chapter "Ecosystem-based

Adaptation (EbA) of African Mountain Ecosystems: Experiences from Mount Elgon, Uganda", Musonda et al. discus in-depth an EbA approach which was implemented in the Mount Elgon region (Africa). They document all steps from initial vulnerability analyses to the implementation of an EbA method, here a gravity water scheme. Dourojeanni et al. in chapter "Vulnerability Assessments for Ecosystem-based Adaptation: Lessons from the Nor Yauyos Cochas Landscape Reserve in Peru" focus particularly on the vulnerability assessment for EbA. They analyze the results from a unique opportunity given at Nor Yauyos Cochas Landscape Reserve in Peru where three different vulnerability assessment approaches had been applied in the same location. Although all three approaches resulted in similar recommendation for adaptation measures, it also becomes obvious from the comparative study that only the application of a participatory approach did not require additional studies to implement measures following the vulnerability assessment. In chapter "The Role of Ecosystem-based Adaptation in the Swiss Mountains", Muccione and Daley explore the role of EbA in a highly developed region as for instance the Swiss Alps are. They identified EbA interventions taken particularly in the field of risk management, water management and agriculture and conclude that challenges and opportunities of EbA in Switzerland can mostly be attributed to knowledge, acceptance and other socio-economic factors.

The topic of community perception is tackled in chapter "Community Perceptions and Responses to Climate Variability: Insights from the Himalayas". Pandit et al. present the results of a survey on community perception and response to climate variability in the Himalayas (Bhutan, India, Nepal). Their findings show that the communities' challenges are complex and seldom only related to climate variability and change. However, climate change importantly exacerbates their basic challenges. The chapter further points towards the needs of the communities to enhance adaptive capacity and resilience. Chapter "Drought: In Search of Sustainable Solutions to a Persistent, 'Wicked' Problem in South Africa" focuses on particular climate-related stress inherent for many countries in South Africa. Vogel and van Zyl provide a historical, comparative assessment of the role of the State and other institutional settings for the reduction of drought risks with particular focus on sustainable solutions.

In the last contribution of the book's case study part, Trabacchi and Stadelmann raise in chapter "Making Climate Resilience a Private Sector Business: Insights from the Agricultural Sector in Nepal" the importance of including the private sector as a complement to the limited and often not sufficient public finances in fighting against climate change impacts in developing countries. At the example of the agricultural sector in Nepal the authors investigate leverages, which international public finance can use to involve domestic private actors. Based on semi-structured interviews with project stakeholders, finance modeling, risk assessment and a cost-benefit analysis they found key elements that help to involve and raise interest of the private sector.

Part C of the book finally takes a global view on the thematic presented in the case studies in Part B. The first chapter of Part C (chapter "Shaping Climate Resilient Development: Economics of Climate Adaptation") takes up the issue of the previous chapter and focuses on the private sector and its role for shaping climate resilient development. Bresch introduces the methodology of Economics of Adaptation (ECA) and synthesizes the results from 20 ECA studies. As an overall result of these studies, it was found that the key drivers are mostly today's weather and climate risks and economic development. In chapter "Building Resilience: World Bank Group Experience in Climate and Disaster Resilient Development", Kull et al. provide the perspective from several years of experiences gathered by the World Bank Group. They conclude that integrating climate and disaster resilience into the development planning processes leads to immediate and sustainable development gains, particularly in the most vulnerable countries. Along with these findings from Parts B and C of this book, it becomes obvious that communication plays a critical role for climate change adaptation and thus sustainable development. In the final chapter of the book, Kohler et al. take up the issue of science-policy dialogue and show how these processes have been institutionalized in both developed and developing countries. In their chapter "The Science-Policy Dialogue for Climate Change Adaptation in Mountain Regions", they propose to strengthen the science-policy dialogue by closing the data and information gap relating to mountain climates and existing adaptive actions.

In summary, the knowledge gathered in this book allows us to draw a number of important conclusions concerning the state and experiences of adaptation in upstream and downstream areas. The first set of the following conclusions does not necessarily imply upstream and downstream specific issues and therefore shows that many mechanisms of adaptation operate along the same or similar lines independent on the regions:

- Climate change, variability and extremes represent the immediate threats and needs and are therefore generally at the forefront of current adaptation practices.
- Well-designed adaptation comes with several important co-gains, in particular for sustainable development and disaster risk reduction. Adaptation thus contributes to, and needs to be seen in the context of the Sustainable Development Goals (Agenda 2030) and the Sendai Framework for Disaster Risk Reduction.
- Non-climatic factors often outweigh the climatic factors in terms of contributing to, or enhancing risks and negative effects of climate change, and therefore need to be considered for adaptation.
- Adaptation implies an iterative learning process.

In addition to the more generally applicable points above we can draw a number of specific issues for mountain regions and adjacent downstream areas from this book:

- A major challenge for adaptation in mountains is due to limited availability of data, both for the natural as well as social environment. This limitation is even more pronounced in developing countries, especially in terms of long-term observation series that are important to design reasonably robust and sustainable adaptation strategies and measures.

- There is now a considerable number and diversity of adaptation experiences in upstream and downstream regions available. Adaptation experiences in mountains primarily relate to water resources, forest and ecosystem management, and risk reduction of extreme events, in particular mass flows and movements.
- The involvement of the private sector is minimal so far in adaptation planning in mountain regions, which contrasts with some of the current strategies of international climate policy. There are also only very few experiences concerning economic evaluation of adaptation in mountains.
- The participatory work with local mountain communities is an essential element for success in adaptation efforts.
- The different types of remoteness of mountain communities need to be considered in the adaptation planning.

References

Adger, W. N., Dessai, S., Goulden, M., Hulme, M., Lorenzoni, I., Nelson, D. R., Naess, N. O., Wolf, J., & Wreford, A. (2009). Are there social limits to adaptation to climate change? *Climatic Change, 93*(3), 335–354.

Bichsel, C. (2011). Liquid challenges. Contested water in Central Asia. *Sustainable Development Law and Policy, 12*(1), 24–30.

Donat, M. G., Lowry, A. L., Alexander, L. A., O'Gorman, P. A., & Maher, N. (2016). More extreme precipitation in the world's dry and wet regions. *Nature Climate Change, 6*(5), 508–513.

Dow, K., Berkhout, F., Preston, B. L., Klein, R. J. T., Midgley, G., & Shaw, M. R. (2013). Limits to adaptation. *Nature Climate Change, 3*(4), 305–307.

Mountain Research Initiative EDW Working Group (2015). Elevation-dependent warming in mountain regions of the world. *Nature Climate Change, 5*(5), 424–430.

Hawkins, E., & Sutton, R. (2009). The potential to narrow uncertainty in regional climate predictions. *Bulletin of the American Meteorological Society, 90*, 1095–1107.

IPCC (2013). Climate change 2013: The physical science basis. In T. F. Stocker, D. Qin, G.-K. Plattner, M. Tignor, S. K. Allen, J. Boschung, A. Nauels, Y. Xia, V. Bex, & P. M. Midgley (Eds.), *Contribution of Working Group I to the Fifth Assessment Report of the Intergovernmental Panel on Climate Change* (p. 1535). Cambridge: Cambridge University Press.

IPCC (2014). Climate change 2014: Impacts, adaptation, and vulnerability. Part B: Regional aspects. In V. R. Barros, C. B. Field, D. J. Dokken, M. D. Mastrandrea, K. J. Mach, T. E. Bilir, M. Chatterjee, K. L. Ebi, Y. O. Estrada, R. C. Genova, B. Girma, E. S. Kissel, A. N. Levy, S. MacCracken, P. R. Mastrandrea, & L. L. White (Eds.), *Contribution of Working Group II to the Fifth Assessment Report of the Intergovernmental Panel on Climate Change* (p. 688). Cambridge: Cambridge University Press.

Knutti, R., & Rogelj, J. (2015). The legacy of our CO_2 emissions: A clash of scientific facts, politics and ethics. *Climatic Change, 133*(3), 361–373.

Salzmann, N., Huggel, C., Rohrer, M., & Stoffel, M. (2014). Data gaps and research needs on glacier and snow cover change and related runoff – A climate change adaptation perspective. *Journal of Hydrology, 518, Part B*, 225–234.

van Vuuren, D. P., Edmonds, J., Kainuma, M., Riahi, K., Thomson, A., Hibbard, K., et al. (2011). The representative concentration pathways: An overview. *Climatic Change, 109*(1), 5–31.

Adaptation, Adaptation Science, and the Status of Adaptation in Mountain Regions

Graham McDowell, Eleanor Stephenson, and James Ford

Abstract This chapter introduces the conceptual foundations of, and core themes within, the climate change adaptation scholarship; outlines common approaches to adaptation science; presents key critiques of how adaptation is conceptualized and examined; and discusses the status of adaptation in upstream-downstream environments. The chapter draws out linkages between adaptation scholarship and mountain-specific socio-economic and environmental conditions. It also addressed an important gap in the broader adaptation scholarship, where there have been few studies characterizing and examining adaptation in mountain regions. Topics covered clarify key conceptual and analytical aspects of climate change adaptation and strengthen rationale for efforts to increase understanding of adaptation in upstream-downstream systems. The chapter also facilitates more informed engagement with subsequent chapters in *Climate Change Adaptation Strategies – An Upstream-downstream Perspective*.

Keywords Adaptation • Adaptation science • Climate change • Mountains • Upstream-downstream

1 Introduction

Interest in how society can respond to the effects of climate change in upstream-downstream systems is growing as the fingerprint of environmental change becomes more conspicuous in mountain regions. Compounding present-day concerns, IPCC AR5 projections indicate further reductions in global glacier volumes of 15–85 % by 2100, with northern hemisphere snowcover contracting by 7–25 % over the same period (IPCC 2013). Because mountain regions and immediately proximate

G. McDowell (✉)
Institute for Resources, Environment, and Sustainability, University of British Columbia, 429 2202 Main Mall, Vancouver, BC V6T 1Z4, Canada
e-mail: grahammcdowell@gmail.com

E. Stephenson • J. Ford
Department of Geography, McGill University, 805 Sherbrooke Street West, Montréal, QC H3A 0B9, Canada

N. Salzmann et al. (eds.), *Climate Change Adaptation Strategies – An Upstream-downstream Perspective*, DOI 10.1007/978-3-319-40773-9_2

lowlands are home to 26 % of the global population and contain some of the most biodiverse ecosystems on the planet, climate-related changes in upstream-downstream systems portend potentially significant and far-reaching implications (Huddleston et al. 2003; Ariza et al. 2013; Price and Weingartner 2012). However, caution is warranted in presuming that such figures signify unavoidable and necessarily adverse impacts; the pathways through which climate change affects society are rarely straightforward, inevitable, or immutable (Turner et al. 2003). Accordingly, efforts to identify and address the effects of climate change in upstream-downstream systems must recognize the unique circumstances that affect how people and institutions navigate the challenges and opportunities of environmental change; that is, there is a need to understand adaptation.

This chapter is concerned primarily with adaptation in humans systems, and engages with three overarching questions: What is adaptation, how is adaptation studied, and what do we know about adaptation in upstream-downstream systems? The chapter aims to draw out linkages between the adaptation scholarship and mountain-specific socio-economic and environmental conditions in an effort to support more advanced work on the socio-ecological dimensions of climate change in mountain regions. In pursuit of this objective, the chapter: (i) introduces the conceptual foundations of, and core themes within, the climate change adaptation scholarship; (ii) outlines common approaches to adaptation science; (iii) presents key critiques of how adaptation is conceptualized and examined; (iv) and discusses the status of adaptation in upstream-downstream environments. The chapter concludes by setting out some of the current challenges in adaptation scholarship and science, as well as the particular contributions of this book to adaptation science in mountain regions. More broadly, the chapter addresses an important gap in the adaptation scholarship, where there have been few studies characterizing and examining approaches to adaptation, and their implications for adaptation research, in mountain regions.

2 Conceptual Foundations

The term *adaptation* has become part of the lexicon of climate change scholarship and policy, but its origins predate the climate change field by many decades. Adaptation was first theorized in early evolutionary biology scholarship, where it refers to genetic or behavioral characteristics that support fitness and survival in the face of environmental pressures (Smit and Wandel 2006). Mid-twentieth century anthropologists were the first to extend the concept to social spheres, where they argued that humans are inherently adaptive in the face of social and environmental stressors (Engle 2011; Smit and Wandel 2006); sociologists, psychologists, and geographers also played a role in advancing early theories of adaptation in human systems (Simonet 2010). This work laid the foundation for problematizing environmentally deterministic conceptualizations of human-environment relations, and informed how natural hazards, political ecology, and development studies scholars' conceptualized the impacts of climate variability and change on people and social systems (Bassett and Fogelman 2013). These fields, in turn, underpin much of the contemporary work on adaptation to climate change.

Today, the IPCC's definition of climate change adaptation—"the adjustment in natural or human systems in response to actual or expected climatic stimuli or their effects, which moderates harm or exploits beneficial opportunities"—provides the benchmark conceptualization for academic and policy work in this area (IPCC 2013). However, notwithstanding this important reference point, the proliferation of scholarly work on climate change adaptation is driving conceptual thinking, and aspects of climate change adaptation theory continue to evolve as new ideas are proposed and debated. Scholars working on environmental change in mountains are contributing to this evolution of thinking, although their contributions are still limited compared to those working in climate-affected regions such as the Arctic. Herein, key concepts in climate change adaptation scholarship, including some emerging conceptual trajectories, are introduced and discussed.

2.1 Adaptation

Adaptation is a response to phenomena that stress a system or present new opportunities. In the context of climate change, these phenomena can be broken down into several types of stimuli: inter-annual or decadal variability such as changing precipitation patters, long-term changes in climate norms like shifting discharge regimes, and isolated extreme events like glacial lake outburst floods (GLOFs). Adaptation to such stimuli is undertaken in the pursuit of several broadly identifiable goals, including *preventing loss*, *spreading or sharing loss*, *diversifying* to moderate harm and/or take advantage of new opportunities, *migrating* to reduce exposure to climate stimuli, and *restoring* climate-affected systems (Smit et al. 1999). More recently theorized goals of adaptation include *transforming* affected systems and *preserving ecosystem services* (Jones et al. 2012; Folke et al. 2010).

Adaptations can be pursued individually or collectively, and can be classified in terms of their timing, intent, scope, and form (Smit and Wandel 2006). *Timing* is an indication of when adaptations occur relative to climatic stressors. Adaptations can be reactive (occurred after climatic stress) or anticipatory (occurring in preparation for climatic stress). Adaptations that occur in preparation for climatic stress are often viewed as preferable, as they are thought to have the greatest potential for reducing harm (Ebi and Burton 2008). *Intent* describes the extent to which adaptations are conceived and implemented through formalized processes. Autonomous adaptations are viewed as less formal in their development and implementation whereas planned adaptations are viewed as representing a more formalized approach to adaptation. In areas where access to formal decision-making processes is limited or local customs and livelihoods conflict with state led planning (e.g. some mountain communities), autonomous adaptation may be more appropriate (Thornton and Manasfi 2010; McDowell et al. 2014). However, autonomous adaptation may also represent exclusion from relevant services in geopolitically marginalized regions, a situation common to many highland communities (McDowell et al. 2014). Mainstreaming—embedding climate change adaptation efforts within existing management and development processes—is a frequently referenced objective of planned adaptations (Berrang-Ford et al. 2011; Lesnikowski et al. 2011), although this presupposes the existence of institutions capable of responding to adaptation needs. *Scope* refers to the spatial and temporal extent of adaptation efforts. Adaptations may be conceived and enacted at various spatial scales: individual, household, community, regional, or national levels. Similarly, adaptations can range from short-term interventions (e.g. short-term investments) to longer-term system adjustments (e.g. institutional changes), with the appropriate scale of adaptation depending on variables such as the goal of the adaptation, the level of knowledge about the system adapting, and the resources available. *Form* describes the adaptation approach, with commonly employed strategies being technological, behavioral, financial, institutional, regulatory, and informational. There is no theoretically best form of adaptation. Rather the most appropriate form is contingent on the goal of adaptation and a range of contextual factors.

Adaptations can also be thought of as either 'soft' or 'hard' (Sovacool 2011). These implementation approaches reflect differing ideas about how adaptation is best carried out and, to a lesser extent, efforts to deal with different climatic stimuli. *Hard adaptations* rely primarily on infrastructure development, are complex and capital intensive, and often lack community consultation. Sea walls that protect against sea level rise and storm surges are the archetypal hard adaptation example. Although hard adaptations have been critiqued for being socially and ecologically disruptive and inflexible to uncertain future climate conditions (Sovacool 2011), they remain a commonly advocated adaptation approach. This is especially true of adaptations supported by international funding mechanisms such as the UNFCCC Adaptation Fund, which favor measureable adaptation activities vis-à-vis climate change (Khan and Roberts 2013). *Soft adaptations* are often less tangible, focusing on empowering communities, drawing on local skills and knowledge, utilizing locally appropriate technologies, and promoting flexibility to changing climatic conditions. A soft adaptation might focus on identifying intrinsic adaptive abilities within a community, then working to promote and enhance these abilities, for example by promoting the transmission of local ecological knowledge and insights about effective responses to past experiences of climatic change. Soft adaptations may also involve working with existing natural capital to moderate the effects of climate change, as is the case with ecosystem-based adaptation (discussed in Sect. 3). Although many scholars posit that soft adaptation approaches are preferable in many cases, such adaptations can be limited in their scope (e.g. community focused), are difficult to evaluate, and are subject to elite capture (Khan and Roberts 2013). Hard and soft adaptations are different approaches to implementing adaptation, but they can be complimentary (Carey et al. 2014). For example, in a highland agricultural community exposed to GLOFs and changing precipitation regimes, installing infrastructure to protect against floods (hard adaptation) while also supporting food sharing networks (soft adaptation) could be appropriate.

Adaptation is thought to be successful when it is effective, efficient, equitable, and legitimate (Adger et al. 2005). *Effectiveness* refers to an adaptation action achieving its goals. *Efficiency* can be broadly defined as a situation where the benefits of an adaptation outweigh the cost of its implementation, including the opportunity cost of not using resources for other socially beneficial ends. *Equity* is greater when the benefits of adaptation reduce or do not worsen underlying disparities within a system. That is, adaptation is more equitable when the distributional consequences of adaptation tend to benefit the most vulnerable. *Legitimacy* concerns the process of adaptation planning and implementation, with legitimacy increasing when inclusive decision-making processes underpin adaptation. Defining success in the context of mountain regions can be especially complex due to the interlinked nature of ecosystem services in upstream-downstream systems as well as the often complex socio-economic and political circumstances within alpine catchments (e.g. marginalization of highland communities).

Long-term trajectories of adaptation are the subject of much contemporary work on adaptation, with mal-adaptation, sustainable adaptation, and transformative adaptation emerging as important themes. *Mal-adaptation* refers to situations in which adaptation increases emissions of greenhouse gases; disproportionately burdens the most vulnerable, other groups/systems, or time periods; has high opportunity costs; reduces incentives to adapt; or promotes inflexibility and path dependency (Barnett and O'Neill 2010). Empirical work has begun to revel the existence of mal-adaptation in the context of climate change adaptation (Adger and Barnett 2009; Ford et al. 2013b), leading to efforts to identify why mal-adaptation emerges and how it can be avoided. This work has promoted a corresponding line of questioning about *sustainable adaptation* among development scholars, where mal-adaptation is conceptualized as a problem for sustainable development (Eriksen and Brown 2011). Kates et al. (2012) describes *transformative adaptation* as adaptations that "are adopted at a much larger scale, that are truly new to a particular region or resource system, and that transform places and shift locations" (p. 7165). Others such as Folke et al. (2010) go beyond this definition, suggesting that transformational change implies as a system scale step change from a less desirable state to a more desirable state. In all cases, it is argued that climate change poses risks in some systems that are so significant they will require novel or dramatically different approaches in order to be managed effectively. For example, the dramatic changes in snow and ice environments projected over the coming century could present unprecedented water and hazard management challenges. Implementing and sustaining transformational adaptations, however, is expected to be very difficult, especially in light of institutional and behavioral inertia that favors the persistence of existing socio-economic conditions (Pelling 2011; O'Brien 2012). See Table 1 for a summary of key adaptation themes.

Finally, a now extensive body of empirical work has established that adaptations rarely occur in response to climate change alone (Berrang-Ford et al. 2011; Ford et al. in press; Lesnikowski et al. 2015); adaptation to multiple, often interacting stressors is the norm. Thus, although it is tempting to assume that adaptation is a direct product of climatic change, it is important to appreciate that adaptation often

Table 1 Adaptation goals, classifications, implementation, success, and long-term outlook

Theme	Key themes
Goals	Prevention of loss
	Spreading or sharing loss
	Diversification
	Migration
	Restoration
	Transformation
	Preservation of ecosystem services
Classifications	Timing
	Intent
	Scope
	Form
Implementation	Hard adaptation
	Soft adaptation
Success	Effective
	Efficient
	Equitable
	Legitimate
Long-term outlook	Mal-adaptation
	Sustainable adaptation
	Transformative adaptation

addresses interlinked environmental and socio-economic opportunities and constraints. For example, a highland farmer's decision to switch to a more drought resistant crop may be driven both by changing precipitation dynamics and better market prices for the new crop.

2.2 *Adaptive Capacity*

Adaptive capacity refers to the ability to devise and implement adaptations (Engle 2011). High adaptive capacity can enable stressors to be addressed without incurring harm, whereas low adaptive capacity can engender vulnerability to seemingly minor perturbations. Importantly, adaptive capacity is not a homogenous feature; it is differentiated within and between systems due to a range of socio-economic and political factors (Adger et al. 2007; Ford and Smit 2004). As such, adaptive capacity provides insights about who can adapt, how they adapt, and what effect their adaptations have on reducing harm and/or accessing new opportunities (Smit et al. 2000).

Adaptive capacity can be evaluated by exploring the factors that support (determinants) and constrain (barriers) adaptation. Frequently identified determinants include access to economic resources, technology, and information; high levels of social and cultural capital, equitable socio-economic conditions, and supportive social networks; and well functioning governance arrangements (Adger et al. 2007). When such conditions exist, higher levels of adaptive capacity are often observed. Notwithstanding, barriers can undermine adaptive capacity, often leading to uneven distributions of adaptability (Moser and Ekstrom 2010; Biesbroek et al. 2013). IPCC AR5 defines barriers as "factors that make it harder to plan and implement adaptation actions or that restrict options" (IPCC 2014). Poverty, social and political marginalization, and attendant challenges are among the most commonly documented barriers to adaptation at the community level, whereas institutional fragmentation, conflicting timescales, and lack of resources are commonly reported institutional barriers (Eisenack et al. 2014; Ford and King 2015; Gupta et al. 2010). Such barriers lead to complex mosaics of adaptive capacity across scales (e.g. within communities, between upstream and downstream areas, between nations) and through time (as socio-economic and political factors change so too does the nature and distribution of adaptive capacity).

Increasing adaptive capacity is a critical aspect of policies aimed at reducing vulnerability to climate change. Enhancing adaptive capacity is often, although not always, analogous to human development (e.g. poverty reduction, access to education), pointing up synergistic opportunities to improve well-being while also reducing the adverse impacts of a changing climate (Ayers and Huq 2009; Conway and Mustelin 2014; Lemos et al. 2007). This realization should be encouraging for mountain scholars, given the human development work happening in many mountain ranges. However, climate change may produce novel challenges requiring targeted capacity-building endeavors above and beyond human development initiatives (Füssel 2007). Regardless of how adaptive capacity is strengthened, knowledge of how determinants and barriers are distributed within and between systems can help target the allocation of scarce resources.

3 Approaches to Adaptation Science

Adaptation science refers to research efforts aimed at understanding the factors driving adaptation (stimuli), who or what adapts (systems), how they adapt (processes), and the effects of their adaptation (outcomes) (Smit et al. 1999; Mustelin et al. 2013). It is a form of practice-oriented scholarship, where providing insights relevant to the development of adaptation policies guides many research projects (Swart et al. 2014). Although assessments typically represent transdiciplinary endeavors, particular disciplines (e.g. geography, natural hazards) have tended to focus on specific research foci, leading to several distinct approaches to adaptation science. Although each approach has distinct strengths and weaknesses, all provide worthwhile analytical, methodological, and practical considerations for the study of adaptation in upstream-downstream systems. A selection of common adaptation science approaches is presented below.

3.1 Community-Based Adaptation

Community-based adaptation (CBA) "combines information from both local, participatory forms of assessment, and wider scale assessments of risks from climate change scientists and modelers" (Forsyth 2013a, p. 441; Ayers and Forsyth 2009). The approach has its roots in natural hazards, political ecology, and development scholarship, where identifying and investigating the context specific circumstances that condition how people experience and respond to past, present, and projected climatic variability and change is central to analysis (Dodman and Mitlin 2013). This work is focused primarily on socio-economic and political dimensions of determinants and barriers of adaptive capacity (also conceptualized as underlying drivers of vulnerability), and the evaluation of existing or planned adaptation actions undertaken at the community level (Dodman and Mitlin 2013). Proponents posit that consulting community members about their experiences of and responses to environmental change provides the type of grounded information needed to develop effective capacity building endeavors, to devise locally accepted adaptation initiatives, and to identify and support existing autonomous adaptations that appear to be effective, efficient, equitable, and legitimate (Burton et al. 2005). In CBA, enhancing the ability of communities to manage the effects of climate change is conceptualized as closely related to human development (soft adaptation), where poverty reduction and increasing social and cultural capital are central objectives (Huq and Reid 2007). Examples of CBA assessments in upstream-downstream contexts can be found in chapters "Managing Glacier Related Risks Disaster in the Chucchún Catchment, Cordillera Blanca, Peru", "Climate Change Adaptation in the Carpathian Mountain Region", "Community Forest Management as Climate Change Adaptation Measure in Nepal's Himalaya", "Ecosystem-based Adaptation (EbA) of African Mountain Ecosystems: Experiences from Mount Elgon, Uganda", and "Community Perceptions and Responses to Climate Variability: Insights from the Himalayas" of this book.

3.2 Ecosystem-Based Adaptation

Whereas CBA assessments are primarily concerned with understanding the socio-economic and political dimensions of adaptation and adaptive capacity, ecosystem-based adaptation (EBA) assessments are concerned with understanding the role natural capital can play in enabling 'win-win' outcomes for climate change adaptation and ecosystem conservation (Colls et al. 2009). EBA perspectives and assessment protocols have emerged from collaboration between conservation and development communities, where efforts to use adaptation as a tool for reducing vulnerability and improving conservation outcomes are advocated. The focus is on identifying how ecosystem services—short hand for the multitude of benefits humanity receives from the environment—can be managed in ways that protect human systems from harm (e.g. flood protection, enhancing food security) and provide new opportunities (e.g. livelihood diversification) while also supplying co-benefits such as biodiversity conservation and carbon sequestration (Jones et al. 2012). This is an important conceptual and analytical development, especially in view of awareness that human adaptation could become a major driver of environmental degradation without greater attention to the potential for unintended ecological side effects (Turner et al. 2010). For example, new alpine water management schemes could adversely affect aquatic ecology if environmental impacts are not treated with due concern in adaptation plans. EBA assessments are cross-scale in nature, focusing on how local-level circumstances and adaptation needs can be attended to through linkages to development and conservation objectives and funding programs at higher levels (Vignola et al. 2009). In practice, many EBA assessments and projects are being carried out by development and conservation organizations (e.g. UNEP; UNDP; IUCN) (see UNEP 2010 for examples of mountain specific EBA activities). This book profiles EBA programs in chapters "Ecosystem-based Adaptation (EbA) of African Mountain Ecosystems: Experiences from Mount Elgon, Uganda", "Vulnerability Assessments for Ecosystem-based Adaptation: Lessons from the Nor Yauyos Cochas Landscape Reserve in Peru", and "The Role of Ecosystem-based Adaptation in the Swiss Mountains".

3.3 Resilience as Adaptation

Resilience perspectives are based on explicit recognition of feedbacks and interdependencies between social and ecological systems. These functional dynamics were first recognized by ecologists, but became more fully articulated through interdisciplinary collaboration across social and physical sciences (Folke 2006). Conceptually and analytically, coupled 'socio-ecological systems' are the focus of resilience scholarship, with attention paid to how these systems are affected by and respond to stress (Folke et al. 2010). In this work, the term resilience is generally synonymous with adaptive capacity, providing an indication of the ability of systems to absorb, respond to, or capitalize on the effects of stimuli (Engle 2011). However, whereas CBA focuses on the human dimensions of adaptation and EBA focuses on the role of natural capital in adaptation, resilience perspectives focus squarely on the interplay of multiple climatic and non-climatic stimuli. This integrative approach requires information about social dynamics and institutions, ecosystem characteristics, and functional linkages thereof. Resilience scholars have also been at the forefront of theories of transformational change (e.g. Walker et al. 2004; Gunderson and Holling 2001), prompting much of the discussion about transformational adaptation introduced above. Resilience scholars' caution against oversimplifying socio-ecological dynamics, leading to assessment approaches embracing complexity and uncertainty (Nelson et al. 2007). These assessments tend to be based on empirical evidence, but conducted through a more theoretical lens; insights about systems' sources of resilience and responses to stressors take center stage. Resilience-based approaches to adaptation science often provide insights about broad-level system dynamics, and are gaining increasing purchase in discussion about adaptation to global environmental change. Resilience approaches are discussed in chapters "Making Climate Resilience a Private Sector Business: Insights from the Agricultural Sector in Nepal", "Shaping Climate Resilient Development: Economics of Climate Adaptation", and "Building Resilience: World Bank Group Experience in Climate and Disaster Resilient Development".

3.4 Governance as Adaptation

Governance as a form of adaptation science deals most deeply with questions of process and outcome; namely, how do people and institutions navigate the challenges and opportunities of climate change, and what are the effects of these procedures on effectiveness, efficiency, equity, and legitimacy. In the context of environmental change, much governance work draws on insights from resilience, political ecology, and political science scholarship (e.g. Benson and Garmestani 2013; Forsyth 2013b; Vink et al. 2013), where socio-ecological interdependencies and/or questions of power are prominent. Analytically, this form of adaptation science evaluates governance both as a *facilitator of adaptation* and as a *form of adaptation*. That is, governance represents both the process through which decisions about adaptations are negotiated, and the arrangements through which adaptations are implemented. Evaluations vary in scale from international climate negotiations such as UNFCCC COP meetings (Biermann et al. 2010) to decision-making processes in small subsistence-based communities (Armitage et al. 2011). Regardless of analytical scale, questions of who is involved or not in negotiations about adaptation and why, who benefits or not from proposed or actual adaptations and why, the efficiency of adaptive actions, and the degree to which adaptations reduce vulnerability are central. Accordingly, this form of adaptation science provides critical information about the nature of social processes and arrangements that translate the effects of climatic stimuli into society's adaptive actions. Chapters "Building Resilience: World Bank Group Experience in Climate and Disaster Resilient Development" and "The Science-Policy Dialogue for Climate Change Adaptation in Mountain Regions" expand on governance in the context of climate change adaptation. Table 2 provides a summary of the adaptation science approaches discussed above.

Table 2 Summary of adaptation science foci

	Stimuli focus	System focus	Process focus	Outcome focus
Community-based adaptation	Social/non-climatic>climatic	Social	Socio-economic and political	Human well-being improvements, reduced vulnerability to climate change
Ecosystem-based adaptation	Social/non-climatic<climatic	Social and environmental	Ecological; socio-economic and political	Conservation, reduced exposure to climatic stimuli, livelihood improvements
Resilience	Social/non-climatic=climatic	Socio-ecological	Socio-ecological	Ability of systems to absorb, respond, or capitalize on effects of stimuli
Governance	Social/non-climatic=climatic	Social	Socio-economic and political	More effective, efficient, equitable, and legitimate adaptations

4 Critiques of Adaptation and Adaptation Science

Herein, some key critiques of adaptation are introduced in an effort to better understand perceived and de facto shortcomings of adaptation scholarship and science, to support more informed engagement with adaptation in mountain contexts, and to stimulate thinking about opportunities to engage productively with criticisms.

4.1 Critiques of Adaptation

Perhaps the most well-known critique of adaptation is that it distracts from mitigation—efforts to reduce, prevent, or counteract the emission of greenhouse gases. However, the veracity of this critique has waned as awareness of society's commitment to climate-related changes has become more certain. According, to IPCC AR5 "A large fraction of anthropogenic climate change resulting from CO_2 emissions is irreversible on a multi-century to millennial time scale" (IPCC 2013). Thus, although mitigation is essential to avoid further climatic perturbation, work on adaptation is now widely viewed as appropriate and complementary.

Some of today's most substantive critiques of adaptation come from development scholars, political ecologists, conservation biologists, and resilience scholars. For example, some development scholars have questioned whether or not adaptation is distinct from human development, and suggested that the focus on climate change adaptation may be directing resources away from human development initiatives already working to enhance the underlying determinants adaptive capacity (Gupta 2009). More starkly, political ecologists such as Ribot (2010) have suggested that "The term 'adaptation', although common in climate discussions, is highly problematic. It naturalizes the vulnerable populations, implying that, like plants, they should adjust to stimuli. The term implicitly places the burden of change on the affected unit—rather than on those causing vulnerability or with responsibility (e.g. government) to help with coping and enable wellbeing. 'Adaptation' also suggests 'survival of the fittest', which is not a desirable ethic for society." (p. 3). These critiques highlight the need to recognize and resolve the unintended consequences of an adaptation framing on other efforts to improve human well-being. Consideration of these points is especially important in the mountain regions, where marginalization is already a problem in many ranges, and where adaptation and development communities will be working together more closely and frequently in the future.

Another critique is that existing approaches to human adaptation have paid too little attention to unintended ecological side effects (Turner et al. 2010). Human adaptations can add to the pressure placed on already stressed ecosystems, threatening biological diversity and ultimately the flow of ecosystem services that underpin human well-being. Although some scholars acknowledge such interdependencies and feedbacks, avoiding unintended ecosystem impacts it is not yet a core tenet of most work on human adaptation. Given the sensitivity of alpine environments, this should become a core focus for those working on adaptation in mountain contexts. Finally, a growing number of scholars suggest that adaptation may be an insufficient or inappropriate way to conceptualize responding to contemporary socio-ecological changes and challenges (O'Brien 2012). Here, adaptation is seen as a conservative approach that can serve to reinforce and perpetuate the dynamics of untenable systems; that is, it can occlude the need for transformational socio-economic and political changes. This critique has initiated an important discussion among climate change adaptation scholars about when, where, and why pursuing transformational changes might be more appropriate that adaptation (e.g. Pelling 2011).

Operationalizing adaptation concepts is another area where critiques have emerged, with concerns about the vagueness of ideas such as hard and soft adaptation, planned and autonomous adaptation, and mal-adaptation raising question about the usefulness of adaptation scholarship from a practical perspective. This suggests that much work remains to be done with regards to linking adaptation theory to the practice of adaptation. Similarly, it has proven challenging to evaluate the successfulness of adaptations in practice (Ford et al. 2013a), leading to difficulty in obtaining institutional support and funding for adaptation projects (especially soft adaptations).

4.2 Critiques of Adaptation Science

Adaptation science has also been the subject of numerous critiques. Perhaps most broadly, Swart et al. (2014) have suggested that there is not yet a discernable 'science of adaptation', with issues such as untested heuristics and confusing terminology undermining the emergence of adaptation science. However, although such observation encourage critical reflection about the nature of adaptation science, it is apparent that the scientific investigation of adaptation is well established in the peer-reviewed literature (Berrang-Ford et al. 2011; Bassett and Fogelman 2013).

Despite the growing number of adaptation science publications, it is sometimes argued that the context specific nature of adaptation assessments limits the elicitation of transferable insights about adaptation (e.g. between study sites and across scales) (Adger and Barnett 2009); however, new methodological approaches are helping to resolve this limitation (e.g. Berrang-Ford et al. 2015). Others (especially anthropologists) have suggested that adaptation science approaches are too coarse conceptually and analytically, leading to consequential omissions about aspects of how people manage and adapt to change, and constraints likely to affect local adaptation options (e.g. Wenzel 2009). While some CBA work is quite sophisticated in terms of social engagement, more focused attention on socio-cultural dynamics and human agency in EBA, resilience, and governance work may be beneficial. On the other hand, CBA rarely critically evaluates socio-ecological dynamics and feedbacks (Dodman and Mitlin 2013), where resilience and EBA approaches have made substantial progress in assessing the ecological implications of human adaptation. And while CBA and EBA assessments are viewed as more likely to obtain practical suggestions, governance and resilience approaches are sometimes criticized for lacking actionable recommendations (Fabinyi et al. 2014). Given the relative strengths and weaknesses of adaptation science approaches discussed herein, enhanced communication and collaboration across existing approaches to adaptation science may be a logical first step to addressing critiques.

Notwithstanding some existing conceptual and analytical shortcoming, adaptation research has made important contributions to our understanding of the unique circumstances that condition how people and institutions navigate the challenges and opportunities of climate change. A recent synthesis of climate change adaptation research focused on mountain regions helps to elucidate what is known about adaptation in upstream-downstream systems.

5 The State of Knowledge on Adaptation in Mountain Regions

5.1 Review of Adaptation Research in Upstream-Downstream Systems

This section situates adaptation and adaptation science in the context of upstream-downstream systems by profiling results from a recent systematic review of peer-reviewed literature on adaptation in glaciated mountain regions (see McDowell et al. 2014 for full research protocol and results), and interpreting these findings in relation to the themes, approaches, and critiques outlined above. Systematic review methods are used to synthesize the state of knowledge and elucidate key

Table 3 Adaptation research in glaciated mountain regions

Studies on human adaptation were identified for only 40 % of countries with glaciated mountain regions
Certain countries dominate the literature (Peru, Nepal, and India accounted for 67 % of adaptation plans or actions documented)
Certain mountain regions have received more research attention (44 % of adaptation initiatives documented were in the Himalaya, 27 % in the Andes, 24 % in the European Alps, 0 % in the North American Rockies)
Most studies were undertaken by researchers in geography or environmental studies departments (combined 64 %), with notable absence of research undertaken within health fields
Most studies were undertaken by western academics (with 72 % of first authors based in North America or Europe), although the majority of adaptation initiatives documented occurred in developing nations

Based on McDowell et al. (2014)

characteristics of adaptation emerging from specific ecosystems or regions, and to track changes in the state of knowledge about adaptation over time (Berrang-Ford et al. 2015). Given its glaciated mountain regions focus, the review only speaks to a subset of adaptations occurring in upstream-downstream environments more broadly. However, as the only systematic review of adaptation in mountain regions to date, it provides a useful context for unpacking the topics outlined above, and situating chapters to come.

A key finding of the review was that globally, scholarship on human adaptation to climate change in upstream regions remains limited: only 36 studies explicitly addressing human adaptation in glaciated mountain regions appeared in peer-reviewed English-language literature published within the past decade. In total, only 74 discrete adaptation plans or actions were identified. Consequently, large geographical gaps in the scholarship remain, with no studies reported for 60 % of countries with alpine glaciation; the geographical distribution of research is also highly uneven, with certain countries and regions receiving the majority of research attention (Table 3).

Because certain regions, populations, and topics have received comparatively little attention in the peer-reviewed literature, the review does not provide firm conclusions about the status of adaptation in upstream environments. However, it does identify several characteristics of adaptation occurring in mountain regions (Table 4), and is consistent with the findings of other more general regional reviews of the status of adaptation (e.g. Ford et al. 2014, Press, Sud et al. 2015).

5.2 *The Heterogeneity of Adaptation in Mountain Regions*

Although the review profiled above provides an overview of key trends in research and some commonalities in adaptation needs, such aggregated results risk obscuring the diversity of adaptation needs and actions underway in upstream-downstream

Table 4 Characteristics of adaptation in mountain regions

The majority of adaptation initiatives were undertaken in sectors affected by climate-related cryospheric/hydrological changes, including agriculture, hazard management, tourism, water management, and health
In terms of *climate stimuli* motivating adaptation, changing hydrological regimes (including changes in precipitation and geohazards such as landslides and floods) were reported in 37 % of adaptations reviewed, followed by changes in temperature and seasonality. However, non-climatic stressors such as livelihood transitions, new demands on water systems, non-climatic environmental changes, and poverty played a role in motivating more than half of adaptations
In terms of *timing*, most adaptations reviewed (65 %) were reactive rather than anticipatory
In terms of *intent*, nearly half (46 %) of adaptations described in the literature were autonomous adaptations undertaken by upland populations, while mainstreamed adaptation initiatives made up only 20 % of adaptations documented
In terms of *scope*, the majority (58 %) of adaptation initiatives documented were undertaken at the individual, household, or community scale. Adaptations spanning several scales were also documented, including soft adaptation initiatives to improve local adaptive capacity and empower communities supported by outside development institutions
In terms of *equity*, many of the adaptation initiatives reviewed made note of vulnerable populations (38 %), but fewer (16 %) contained concrete provisions to ensure adaptation was indeed equitable and did not reinforce or worsen underlying inequalities, suggesting *legitimacy* may be in question
In terms of *effectiveness*, only 4 % of the adaptations reviewed had been evaluated, although approximately half were completed

Based on McDowell et al. (2014)

systems. Adaptation is undeniably heterogeneous, as an overview of representative cases within the mountain adaptation literature attests.

In low income upland regions where development needs dominate, stressors like increasing temperature, changing seasonality, and decreased snow cover are documented as motivating changes in livelihood practices. For example, upland agriculturalist communities in the Nepal Himalayas are adapting by growing new crops suitable for warmer conditions to moderate the livelihood impacts of changing seasonality, temperature, and snowpack (Chaudhary et al. 2011), and seeking more diversified live hoods through trade and day labor (Onta and Resurreccion 2011). Similarly, Chinese farmers in the Urumqi River Basin report seeding crops earlier and conserving water in response to changing seasonality and hydrological regimes, (Deng et al. 2011, 2012). On the other side of the world, Andean pastoralist communities are also contending with decreased water availability, and are modifying livestock grazing patterns and increasing water storage to help reduce livelihood impacts (Young et al. 2010).

At the national and regional scale, new disaster risk reduction initiatives are underway to address climate-related hazards. For example, the United Nations Development Program is coordinating disaster-warning systems with national governments in India and Bhutan to address glacial lake outburst flood risks in the Himalayas (i.e. UNDP GLOF program), but faces challenges of low institutional capacity, lack of data, and inadequate infrastructure (see Meenawat and Sovacool

2011; Moors et al. 2011; Sovacool et al. 2012). Many of these challenges translate to the Cordillera Blanca in the Andes, where threats posed by climate-sensitive hazards (e.g. GLOFs, ice-rock avalanches) are compounded by lack of information and funding (Carey 2005). Here, hard adaptation solutions such as draining GLOF hazards have achieved mixed results, for example resulting in unintended displacement of vulnerable populations (Carey 2005; Carey et al. 2012). High-income nations such as Austria, Switzerland, and Norway, are also adapting to upland climate hazards, but here the academic literature to date focuses on adaptation initiatives to mitigate recreational and economic impacts, such as mapping mountaineering hazards in the Alps (Rixen et al. 2011).

Though upland hazards represent a key category of adaptation initiatives at the national and regional scale, initiatives to address slow-onset climate impacts are also underway. For example, new watershed councils for integrated water management in Peru have been created to improve coordination and governance of water resources (Bury et al. 2013). Water governance marks an important adaptation priority globally, affecting high-income nations also. For example, in the Rhone Catchment in Switzerland, the national government is supporting new integrated watershed management policies that take into account climate-related water scarcity and existing rivalries for water resources between sectors, and in some cases even create public infrastructure for controlled flooding (Beniston et al. 2011).

The above cases cover a limited subset of adaptation research, but exemplify the range of adaptation actions undertaken by people and institutions to adapt to climate change in mountainous regions. As adaptation in upstream-downstream systems transects this wide range of sectors and both spatial and temporal scales, different scholarly approaches will profoundly shape research outcomes, drawing attention to particular adaptation features, roles, and outcomes. The following section, provides insights about the interplay between upland adaptation characteristics and adaptation science, and situates chapters to come within wider scholarly context.

5.3 *Synthesis*

Read against the preceding overview of adaptation and adaptation science, this review helps clarify the utility and limitations of the particular approaches to adaptation research in upstream-downstream environments. For example, given the prevalence of literature documenting adaptations at the local scale and the focus on autonomous adaptations, such as agricultural and livelihood diversification, CBA approaches will continue to be important in understanding local experiences and supporting local adaptive capacity. However, the prevalence of localized non-climatic environmental changes—such as water rivalries and scarcity—as a secondary motivator of adaptation suggests the importance and merits of EBA approaches. The wider array of non-climatic stressors documented also points to the utility of resilience approaches that attend to multiple intersecting stressors, such as lack of data, low institutional capacity, and infrastructural barriers. Finally, the fact that many upland adaptation studies note the presence of vulnerable populations with specific adaptation needs—while also documenting the failure of many adaptation actions to provide for these vulnerable groups—suggests the importance of understanding adaptation as a governance issue, where who is involved in and served by adaptation decisions will determine the legitimacy and equity of adaptation initiatives. These approaches can help prevent unintended and maladaptive consequences, such as displacement of vulnerable populations, during a planned adaptation action.

While the review results can validate the utility of these particular approaches to adaptation science, they can also be situated within the critiques of adaptation and adaptation scholarship outlined above. For example, critics who argue that adaptation framing displaces responsibility onto those most vulnerable by locating adaptation at a local scale may find evidence in the dominance of local and regional scale studies in the literature, and relative absence of national and international adaptation measures. Meanwhile, critics who argue that adaptation approaches are insufficient for conceptualizing the ecological challenges we face from fundamentally unsustainable socio-ecological systems may find evidence in the prevalence of non-climatic environmental degradation as a secondary motivator for adaptation. As well, those development scholars who contest the usefulness of overlaying an adaptation lens on existing efforts to improve human well-being may find credence in the prevalence of non-climatic factors motivating adaptation, which suggest the difficulty of disaggregating adaptation from development needs. Finally, the very low number of adaptation studies including an evaluation component suggests that critiques about operationalizing adaptation remain unresolved.

Responding to these critiques and gaps in knowledge marks a crucial agenda for adaptation research focused on upstream-downstream systems. A number of chapters in this book engage with issues noted here and in Sect. 4 of this chapter, including adaptation vs. development, the political ecology of adaptation, adaptation and ecosystem impacts and feedbacks, adaptation and transformation, and operationalizing and evaluating adaptation.

6 Conclusion

This chapter provided an overview of adaptation, adaptation science, and the status of adaptation in upstream-downstream systems. Specifically, it introduced the conceptual foundations of, and core themes within, the climate change adaptation scholarship; outlined common approaches to adaptation science; presented key critiques of how adaptation is conceptualized and examined; and discussed the status of adaptation in upstream-downstream environments. In so doing, the chapter drew out linkages between adaptation scholarship and mountain-specific socio-economic and environmental conditions. It also addressed an important gap in the broader adaptation scholarship, where there have been few studies characterizing and examining approaches to adaptation in mountain regions. Topics covered in the first four sections of the chapter facilitate thoughtful engagement with the conceptual and analytical aspects of subsequent case studies, while the review profiled in Sect. 5 strengthens rationale for this book and related efforts to increase understanding of adaptation in climate-affected upstream-downstream systems.

Despite significant work on climate change adaptation, there are still major challenges in terms of moving from theories of adaptation to the practice of adaptation (Berrang-Ford et al. 2011). For example, efforts to develop, fund, and implement adaptations are often mired in disagreement; monitoring and evaluation protocols are nascent and largely untested; and arrangements for addressing socio-ecological interactions and cross-scale dynamics remain largely hypothetical. Moreover, in view of twenty-first century warming that may dramatically exceed the UNFCCC definition of "dangerous anthropogenic interference with the climate system" (UNFCCC 1992), limited success in initiating and sustaining transformational adaptations is concerning. There are also a number of challenges specific to adaptation research and practice in upstream-downstream systems. For example, given the diverse geographical contexts of these systems, linguistic, cultural, political, and institutional differences compound challenges of research coordination and results dissemination, with attendant effects on adaptation planning and action.

Notwithstanding these challenges, much progress is being made. Adaptation scholars and practitioners are making significant contributions to our ability to understand and implement planned adaptations, to identify and support effective autonomous adaptations, and to respond to climatic stimuli in ways that are attentive to human well-being and ecosystem integrity. In the following chapters, such contributions are highlighted in the context of mountain regions, providing a timely vignette of how people and institutions are responding to climate change in upstream-downstream systems.

References

Adger, W. N., & Barnett, J. (2009). Four reasons for concern about adaptation to climate change. *Environment and Planning A, 41*, 2800–2805.

Adger, W. N., Arnell, W. N., & Tompkins, L. E. (2005). Successful adaptation to climate change across scales. *Global Environmental Change, 15*, 77–86.

Adger, W. N., Agrawala, S., Mirza, M. M. Q., Conde, C., O'Brien, K., Pulhin, J., Pulwarty, R., Smit, B., Takahashi, K. (2007). Assessment of adaptation practices, options, constraints and capacity. In *Chapter 17: IPCC AR4 Working Group 2* (pp. 717–743). Cambridge: Cambridge University Press.

Ariza, C., Maselli, D., & Kohler, T. (2013). *Mountains: Our life, our future progress and perspectives on sustainable mountain development*. Bern: Swiss Agency for Development and Cooperation (SDC), Centre for Development and Environment (CDE).

Armitage, D., Berkes, F., Dale, A., Kocho-Schellenberg, E., & Patton, E. (2011). Co-management and the co-production of knowledge: Learning to adapt in Canada's Arctic. *Global Environmental Change, 21*, 995–1004.

Ayers, J., & Forsyth, T. (2009). Community-based adaptatoin to climate change: Strengthening resilience through development. *Environment, 51*, 22–31.

Ayers, J. M., & Huq, S. (2009). Supporting adaptation to climate change: What role for official development assistance? *Development Policy Review, 27*, 675–692.

Barnett, J., & O'Neill, S. (2010). Maladaptation. *Global Environmental Change, 20*, 211–213.
Bassett, T. J., & Fogelman, C. (2013). Déjà vu or something new? The adaptation concept in the climate change literature. *Geoforum, 48*, 42–53.
Beniston, M., Stoffel, M., & Hill, M. (2011). Impacts of climatic change on water and natural hazards in the Alps: Can current water governance cope with future challenges? Examples from the European "ACQWA" project. *Environmental Science & Policy, 14*, 734–743.
Benson, M. H., & Garmestani, A. S. (2013). A framework for resilience-based governance of social-ecological systems. *Ecology and Society, 18*, 9.
Berrang-Ford, L., Ford, J. D., & Paterson, J. (2011). Are we adapting to climate change? *Global Environmental Change, 21*, 25–33.
Berrang-Ford, L., Pearce, T., & Ford, J. (2015). Systematic review approaches for climate change adaptation research. *Regional Environmental Change, 15*, 755–769.
Biermann, F., Pattberg, P., & Zelli, F. (2010). *Global climate governance beyond 2012: Architecture, agency and adaptation*. Cambridge: Cambridge University Press.
Biesbroek, G. R., Klostermann, J. E., Termeer, C. J., & Kabat, P. (2013). On the nature of barriers to climate change adaptation. *Regional Environmental Change, 13*, 1119–1129.
Burton, I., M. E, H. S, Lim, B. (2005). *Adaptation policy frameworks for climate change: Developing strategies, policies and measures*. Cambridge: Cambridge University Press.
Bury, J., Mark, B. G., Carey, M., Young, K. R., McKenzie, J. M., Baraer, M., French, A., & Polk, M. H. (2013). New geographies of water and climate change in Peru: Coupled natural and social transformations in the Santa River watershed. *Annals of the Association of American Geographers, 103*, 363–374.
Carey, M. (2005). Living and dying with glaciers: People's historical vulnerability to avalanches and outburst floods in Peru. *Global and Planetary Change, 47*, 122–134.
Carey, M., French, A., & O'Brien, E. (2012). Unintended effects of technology on climate change adaptation: An historical analysis of water conflicts below Andean Glaciers. *Journal of Historical Geography, 38*, 181–191.
Carey, M., McDowell, G., Huggel, C., Jackson, J., Portocarrero, C., Reynolds, J. M., Vicuna, L. (2014). Integrated approaches to adaptation and disaster risk reduction in dynamic socio-cryospheric systems. In W. Haeberli & C. Whiteman (Eds.), *Snow and ice-related hazards, risks and disasters* (pp. 219–261). Oxford: Elsevier.
Chaudhary, P., Rai, S., Wangdi, S., Mao, A., Rehman, N., Chettri, S., & Bawa, K. S. (2011). Consistency of local perceptions of climate change in the Kangchenjunga Himalaya landscape. *Current Science, 101*, 504–513.
Colls, A., Ash, N., Ikkala, N., & Conservancy, N. (2009). *Ecosystem-based adaptation: A natural response to climate change*. Gland: IUCN.
Conway, D., & Mustelin, J. (2014). Strategies for improving adaptation practice in developing countries. *Nature Climate Change, 4*, 339–342.
Deng, M., Zhang, H., Mao, W., & Wang, Y. (2011). Public perceptions of cryosphere change and the selection of adaptation measures in the Urumqi River Basin. *Advances in Climate Change Research, 2*, 149–158.
Deng, M., Qin, D., & Zhang, H. (2012). Public perceptions of climate and cryosphere change in typical arid inland river areas of China: Facts, impacts and selections of adaptation measures. *Quaternary International, 282*, 48–57.
Dodman, D., & Mitlin, D. (2013). Challenges for community-based adaptation: Discovering the potential for transformation. *Journal of International Development, 25*, 640–659.
Ebi, K. L., & Burton, I. (2008). Identifying practical adaptation options: An approach to address climate change-related health risks. *Environmental Science & Policy, 11*, 359–369.
Eisenack, K., Moser, S. C., Hoffmann, E., Klein, R. J., Oberlack, C., Pechan, A., Rotter, M., & Termeer, C. J. (2014). Explaining and overcoming barriers to climate change adaptation. *Nature Climate Change, 4*, 867–872.
Engle, N. L. (2011). Adaptive capacity and its assessment. *Global Environmental Change, 21*, 647–656.

Eriksen, S., & Brown, K. (2011). Sustainable adaptation to climate change. *Climate and Development, 3*, 3–6.
Fabinyi, M., Evans, L., & Foale, S. J. (2014). Social-ecological systems, social diversity, and power: Insights from anthropology and political ecology. *Ecology and Society, 19*, 28.
Folke, C. (2006). Resilience: The emergence of a perspective for social–ecological systems analyses. *Global Environmental Change, 16*, 253–267.
Folke, C., Carpenter, S. R., Walker, B., Scheffer, M., Chapin, T., & Rockström, J. (2010). Resilience thinking: Integrating resilience, adaptability and transformability. *Ecology and Society, 15*, 20.
Ford, J., & King, D. (2015). A framework for examining adaptation readiness. *Mitigation and Adaptation Strategies for Global Change, 20*, 505–526.
Ford, J. D., & Smit, B. (2004). A framework for assessing the vulnerability of communities in the Canadian arctic to risks associated with climate change. *Arctic, 57*, 389–400.
Ford, J. D., Berrang-Ford, L., Lesnikowski, A., Barrera, M., & Heymann, S. J. (2013a). How to track adaptation to climate change: A typology of approaches for national-level application. *Ecology and Society, 18*, 40.
Ford, J. D., McDowell, G., Shirley, J., Pitre, M., Siewierski, R., Gough, W., Duerden, F., Pearce, T., Adams, P., & Statham, S. (2013b). The dynamic multiscale nature of climate change vulnerability: An Inuit harvesting example. *Annals of the Association ofAmerican Geographers, 103*, 1193–1211.
Ford, J. D., McDowell, G., & Jones, J. (2014). The state of climate change adaptation in the Arctic. *Environmental Research Letters, 9*, 104005–104014.
Ford, J. D., Berrang-Ford, L., Bunce, A., Mckay, C., Irwin, M., Pearce, T., Irwin, (in press). The current status of climate change adaptation in Africa and Asia. In *Regional environmental change*.
Forsyth, T. (2013a). Community-based adaptation: A review of past and future challenges. *Wiley Interdisciplinary Reviews: Climate Change, 4*, 439–446.
Forsyth, T. (2013b). *Critical political ecology: The politics of environmental science*. Abingdon: Routledge.
Füssel, H.-M. (2007). Adaptation planning for climate change: Concepts, assessment approaches, and key lessons. *Sustainability Science, 2*, 265–275.
Gunderson, L., & Holling, C. (2001). *Panarchy: Understanding transformations in human and natural systems*. Washington, DC: Island press.
Gupta, J. (2009). Climate change and development cooperation: Trends and questions. *Current Opinion in Environmental Sustainability, 1*, 207–213.
Gupta, J., Termeer, C., Klostermann, J., Meijerink, S., van den Brink, M., Jong, P., Nooteboom, S., & Bergsma, E. (2010). The adaptive capacity wheel: A method to assess the inherent characteristics of institutions to enable the adaptive capacity of society. *Environmental Science & Policy, 13*, 459–471.
Huddleston, B., Ataman, E., De Salvo, P., Zanetti, M., Bloise, M., Bel, J., Francheschini, G., & Fed'Ostiani, L. (2003). *Towards a GIS-based analysis of mountain environments and populations*. Rome: FAO.
Huq, S., & Reid, H. (2007). Community-based adaptation: A vital approach to the threat climate change poses to the poor. *International Institute for Environment and Development (IIED). Briefing Paper.* London: International Institute for.Environment and Development (IIED).
IPCC. (2013). Climate change 2013: The physical science basis. In *Contribution of Working Group I to the Fifth Assessment Report of the Intergovernmental Panel on Climate Change*. Cambridge: Cambridge University Press.
IPCC. (2014). Climate change 2014: Impacts, adaptation, and vulnerability. In *Contribution of Working Group II to the Fifth Assessment Report of the Intergovernmental Panel on Climate Change*. Cambridge: Cambridge University Press.
Jones, H. P., Hole, D. G., & Zavaleta, E. S. (2012). Harnessing nature to help people adapt to climate change. *Nature Climate Change, 2*, 504–509.

Kates, R. W., Travis, W. R., & Wilbanks, T. J. (2012). Transformational adaptation when incremental adaptations to climate change are insufficient. *Proceedings of the National Academy of Sciences, 109*, 7156–7161.

Khan, M. R., & Roberts, J. T. (2013). Adaptation and international climate policy. *Wiley Interdisciplinary Reviews: Climate Change, 4*, 171–189.

Lemos, M. C., Boyd, E., Tompkins, E. L., Osbahr, H., & Liverman, D. (2007). Developing adaptation and adapting development. *Ecology and Society, 12*, 26.

Lesnikowski, A. C., Ford, J. D., Berrang-Ford, L., Paterson, J. A., Barrera, M., & Heymann, S. J. (2011). Adapting to health impacts of climate change: A study of UNFCCC annex I parties. *Environmental Research Letters, 6*, 1–9.

Lesnikowski, A., Ford, J. D., Berrang-Ford, L., Barrera, M., & Heymann, J. (2015). How are we adapting to climate change? A global assessment. *Mitigation and Adaptation Strategies for Global Change, 20*, 277–293.

McDowell, G., Stephenson, E., & Ford, J. (2014). Adaptation to climate change in glaciated mountain regions. *Climatic Change, 126*, 77–91.

Meenawat, H., & Sovacool, B. K. (2011). Improving adaptive capacity and resilience in Bhutan. *Mitigation and Adaptation Strategies for Global Change, 16*, 515–533.

Moors, E. J., Groot, A., Biemans, H., Van Scheltinga, C. T., Siderius, C., Stoffel, M., Huggel, C., Wiltshire, A., Mathison, C., Ridley, J., Jacob, D., Kumar, P., Bhadwal, S., Gosain, A., & Collins, D. N. (2011). Adaptation to changing water resources in the Ganges basin, northern India. *Environmental Science and Policy, 14*, 758–769.

Moser, S. C., & Ekstrom, J. A. (2010). A framework to diagnose barriers to climate change adaptation. *Proceedings of the National Academy of Sciences, 107*, 22026–22031.

Mustelin, J., Kuruppu, N., Matus Kramer, A., Daron, J., de Bruin, K., & Guerra Noriega, A. (2013). Climate adaptation research for the next generation. *Climate and Development, 5*, 189–193.

Nelson, D. R., Adger, W. N., & Brown, K. (2007). Adaptation to environmental change: Contributions of a resilience framework. *Annual Review of Environment and Resources, 32*, 395.

O'Brien, K. (2012). Global environmental change II from adaptation to deliberate transformation. *Progress in Human Geography, 36*, 667–676.

Onta, N., & Resurreccion, B. P. (2011). The role of gender and caste in climate adaptation strategies in Nepal emerging change and persistent inequalities in the far-western region. *Mountain Research and Development, 31*, 351–356.

Pelling, M. (2011). *Adaptation to climate change: From resilience to transformation*. Adingdon: Routledge.

Price, M. F., & Weingartner, R. (2012). Introduction: Global change and the world's mountains-Perth 2010. *Mountain Research and Development, 32*, S3–S6.

Ribot, J. (2010). Vulnerability does not fall from the sky: Toward multiscale, pro-poor climate policy. In R. Mearns & A. Norton (Eds.), *Social dimensions of climate change: Equity and vulnerability in a warming world* (pp. 47–74). Washington, DC: The World Bank.

Rixen, C., Teich, M., Lardelli, C., Gallati, D., Pohl, M., Putz, M., & Bebi, P. (2011). Winter tourism and climate change in the alps: An assessment of resource consumption, snow reliability, and future snowmaking potential. *Mountain Research and Development, 31*, 229–236.

Simonet, G. (2010). The concept of adaptation: Interdisciplinary scope and involvement in climate change. *Surveys and Perspectives Integrating Environment and Society, 3*, 1–9.

Smit, B., & Wandel, J. (2006). Adaptation, adaptive capacity and vulnerability. *Global Environmental Change, 16*, 282–292.

Smit, B., Burton, I., Klein, R. J., & Street, R. (1999). The science of adaptation: A framework for assessment. *Mitigation and Adaptation Strategies for Global Change, 4*, 199–213.

Smit, B., Burton, I., Klein, R. J., & Wandel, J. (2000). An anatomy of adaptation to climate change and variability. *Climatic Change, 45*, 223–251.

Sovacool, B. K. (2011). Hard and soft paths for climate change adaptation. *Climate Policy, 11*, 1177–1183.

Sovacool, B. K., D'Agostino, A. L., Rawlani, A., & Meenawat, H. (2012). Improving climate change adaptation in least developed Asia. *Environmental Science & Policy, 21*, 112–125.

Sud, R., Mishra, A., Varma, N., & Bhadwal, S. (2015). Adaptation policy and practice in densely populated glacier-fed river basins of South Asia: A systematic review. *Regional Environmental Change, 15*, 825–836.

Swart, R., Biesbroek, R., & Lourenço, T. C. (2014). Science of adaptation to climate change and science for adaptation. *Interdisciplinary Climate Studies, 2*, 29.

Thornton, T. F., & Manasfi, N. (2010). Adaptation—genuine and spurious: Demystifying adaptation processes in relation to climate change. *Environment and Society: Advances in Research, 1*, 132–155.

Turner, B. L., Kasperson, R. E., Matson, P. A., McCarthy, J. J., Corell, R. W., Christensen, L., Eckley, N., Kasperson, J. X., Luers, A., & Martello, M. L. (2003). A framework for vulnerability analysis in sustainability science. *Proceedings of the National Academy of Sciences, 100*, 8074–8079.

Turner, W. R., Bradley, B. A., Estes, L. D., Hole, D. G., Oppenheimer, M., & Wilcove, D. S. (2010). Climate change: Helping nature survive the human response. *Conservation Letters, 3*, 304–312.

UNEP. (2010). *Ecosystem based adaptation in mountain ecosystems*. http://www.unep.org/climatechange/adaptation/EcosystemBasedAdaptation/EcosystemBasedAdaptationinMountainEcosystems/tabid/51980/Default.aspx. Retrieved 9.2.2015).

UNFCCC. (1992). United Nations framework convention on climate change – Article 2.

Vignola, R., Locatelli, B., Martinez, C., & Imbach, P. (2009). Ecosystem-based adaptation to climate change: What role for policy-makers, society and scientists? *Mitigation and Adaptation Strategies for Global Change, 14*, 691–696.

Vink, M. J., Dewulf, A., & Termeer, C. (2013). The role of knowledge and power in climate change adaptation governance: A systematic literature review. *Ecology and Society, 18*, 46.

Walker, B., Holling, C. S., Carpenter, S. R., & Kinzig, A. (2004). Resilience, adaptability and transformability in social-ecological systems. *Ecology and Society, 9*, 5.

Wenzel, G. W. (2009). Canadian Inuit subsistence and ecological instability—if the climate changes, must the Inuit? *Polar Research, 28*, 89–99.

Young, G., Zavala, H., Wandel, J., Smit, B., Salas, S., Jimenez, E., Fiebig, M., Espinoza, R., Diaz, H., & Cepeda, J. (2010). Vulnerability and adaptation in a dryland community of the Elqui Valley, Chile. *Climatic Change, 98*, 245–276.

Part B

Science in the Context of Climate Change Adaptation: Case Studies from the Peruvian Andes

Boris Orlowsky, Norina Andres, Nadine Salzmann, Christian Huggel, Christine Jurt, Luis Vicuña, Mario Rohrer, Pierluigi Calanca, Raphael Neukom, and Fabian Drenkhan

Abstract Within the context of the Climate Change Adaptation Program (PACC), a number of scientific investigations on water resources, natural disasters and perceptions by local people highlight adaptation needs in the regions of Cusco and Apurímac in Peru, considering past, present-day and future climate conditions. This chapter compiles their findings and attempts a systematic evaluation with respect to their contributions to climate change adaptation. The studies consistently find aggravating water scarcity during the dry season (April to September) due to projected precipitation decreases and reduced storage capacity of shrinking glaciers. Impacts include below-capacity hydropower generation and increased crop failure risks. For natural disasters, database inconsistencies prevent a detection of trends. While the natural science studies have produced a new and more comprehensive

B. Orlowsky (✉) • C. Huggel • C. Jurt • L. Vicuña
Department of Geography, University of Zurich, Zurich, Switzerland
e-mail: boris.orlowsky@geo.uzh.ch

N. Andres
Swiss Federal Institute for Forest, Snow and Landscape Research WSL, Birmensdorf, Switzerland

N. Salzmann
Department of Geography, University of Zurich, Zurich, Switzerland

Department of Geosciences, University of Fribourg, Fribourg, Switzerland

M. Rohrer
Meteodat GmbH, Zurich, Switzerland

P. Calanca
Agroscope, Zurich, Switzerland

R. Neukom
Oeschger Centre for Climate Change Research and Institute of Geography, University of Bern, Bern, Switzerland

F. Drenkhan
Department of Geography, University of Zurich, Zurich, Switzerland

Department of Sciences, Pontificia Universidad Católica del Perú, Lima, Peru

N. Salzmann et al. (eds.), *Climate Change Adaptation Strategies – An Upstream-downstream Perspective*, DOI 10.1007/978-3-319-40773-9_3

understanding of the target regions, their implications for society have hardly been investigated anthropologically. One of the few social science studies emphasizes that climate change is only one out of many determinants of rural livelihoods in the target regions, which have not been addressed scientifically yet. We thereby find an imbalance of available scientific knowledge regarding natural vs. social sciences. Overcoming such imbalance would allow for a more comprehensive integration of scientific findings into design and implementation of adaptation measures within the local context.

Keywords Science-practice interface • Climate change • Adaptation • Andes • Peru

1 Introduction

The administrative regions of Cusco and Apurímac in the Central Andes of Peru are among the most biodiverse of the world, reaching from high mountain environments (above 6000 m a.s.l., see Fig. 1) to the Amazon. They feature unique ecosystems such as cloud and dry forests, high mountain wetlands and grass and agricultural lands. Most of their population lives in the semi-arid Andean valleys in and around

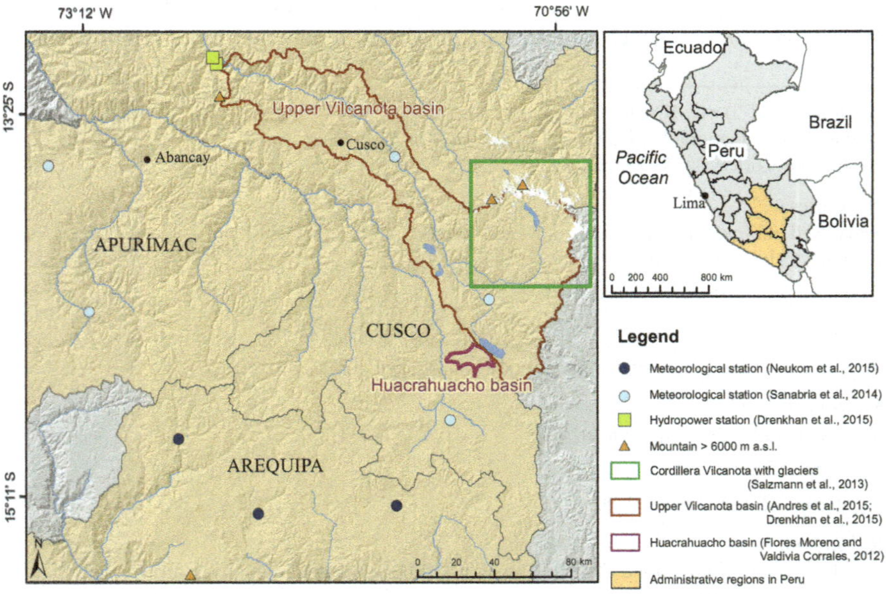

Fig. 1 PACC study areas in the regions Cusco and Apurímac, highlighting locations of the discussed studies. Meteorological stations (Neukom et al. 2015; Sanabria et al. 2014, *round dots*); hydropower stations (Drenkhan et al. 2015, *triangles*); Cordillera Vilcanota (Salzmann et al. 2013, *green rectangle*); upper Vilcanota basin (Andres et al. 2014); Huacrahuacho basin (Flores Moreno and Valdivia Corrales 2012)

the two capitals, Cusco and Abancay, respectively, with an overall rural population fraction of approximately 50 % (INEI 2008).

Observed precipitation shows the marked seasonality of a wet season between October and March and only very little precipitation during the rest of the year (Garreaud et al. 2009). The many dry months impose challenges for agriculture, which is often rainfed, although the fraction of irrigated land has increased substantially over the past 20 years (see Sect. 3). Also other water dependent activities such as hydropower generation have to cope with water scarcity during the dry season, which has additionally triggered conflicts about water resources (Lynch 2012). Over the past decades, the region has experienced several natural disasters (floods, debris and mud flows) with impacts on society and economy.

Rural farming is mostly for subsistence, requiring the farmers to generate additional income. Expanding mining activities in the two regions (the Peruvian Ministry of Energy and Mining reports one production unit and 23 exploration projects in Apurímac, and five production units and 21 exploration projects in Cusco, MINEM 2015) provide attractive opportunities for additional income and cause temporal migration. In consequence and due to general socioeconomic development, poverty rate has decreased from 41.5 to 33.8 % in 2011–2014 (INEI 2014).

The combination of a demanding climatological setting, frequent natural disasters and the economic situation makes these regions particularly vulnerable to impacts of climate change (e.g. on food security, Kaser et al. 2010; Baraer et al. 2012; Sietz et al. 2012; Bury et al. 2013). In response, the Peruvian Ministry of the Environment (MINAM) in collaboration with the Swiss Agency for Development and Cooperation (SDC) established in 2008 the Climate Change Adaptation Program (PACC) in the Peruvian Andes, aiming at an in-depth understanding of the present-day relations between society and natural resources and disasters, vulnerabilities to impacts of future climate change and the evaluation and implementation of suitable adaptation options. The program envisages making scientific understanding useable for adaptation practice in the thematic areas of water resources, food security and natural disasters. At the time of writing, the program approaches its last year.

The ambition of actually establishing this science-practice link is reflected in the main partners of the program, which include a practice side (responsible for implementation activities and knowledge transfer to the policy arena, carried out by different Swiss and Peruvian Non-Governmental Organisations) and a science side (responsible for performing and supporting local research and capacity building for Peruvian researchers, carried out by different Swiss and Peruvian universities and research institutions).

In our chapter we report on science contributions to such a science-practice link, drawing mainly on examples from PACC, and discuss their merits and shortcomings in the context of adaptation.

Past experience shows that linking science and practice in the context of international cooperation comes with many challenges, and e.g. that communicating science in an understandable language and iterative communication between science and policy are indispensable for a successful joint production of useable knowledge (Lemos and Morehouse 2005; Dilling and Lemos 2011). One framework about how to make science useable for climate change adaptation, which is the result of numerous consultations with various stakeholder in the Andes, is presented in Huggel et al. (2015b) and will guide the presentation and discussion of science contributions in this chapter.

We introduce this framework in the following Sect. 2 and use it as a lens through which to assess selected science contributions (Sect. 3). Discussion of their strengths and weaknesses from an adaptation perspective is given in Sect. 4.

2 A Framework for the Science Contribution in Climate Adaptation

The framework proposed in Huggel et al. (2015, b) consists of the three interconnected phases framing and scoping, scientific assessment and evaluation of adaptation options and their implementation (see Fig. 2). In the following we briefly summarize the three phases.

The first phase compiles the context. Ideally, the framing and scoping is a joint process, involving different stakeholders. Considering the stakeholders' social, cultural, political and environmental contexts in this process is essential, since objectives and expectations can diverge among stakeholders (Weichselgartner and Kasperson 2010; Hegger et al. 2012) This phase serves to identify pressing present and expected future risks together with the related processes and to recognize the involved stakeholders and their relation to each other.

In the second phase, this rather coarse sketch of the context is refined through scientific assessment. On the climate side, analysis of observational data provides a deeper understanding of the relevant systems and processes at local to regional scales, while future scenarios (e.g. climate projections) allow for assessing the identified vulnerabilities under future conditions. These analyses also help to identify knowledge and data gaps. Often present-day climate variability and extremes are considered a more immediate priority by authorities as compared to slow onset climate change, which is much less perceived and addressed through concrete measures. Any comprehensive vulnerability assessment needs to consider both (Wilby et al. 2009). Studies from the social sciences' side investigate the wider context (e.g., stakeholders' perceptions, their interests, power relations and connectedness to environment and natural resources). Bringing the perspectives from all stakeholders together is a necessary condition for joint production of effectively useable knowledge and sustainable development.

The third phase in the framework consists of the evaluation of adaptation options and their implementation, which requires a close collaboration of climate (impact) scientists, practitioners, decision-makers and other stakeholders (Fussel 2007). Adaptation to climate change or present climate variability is highly (local) context

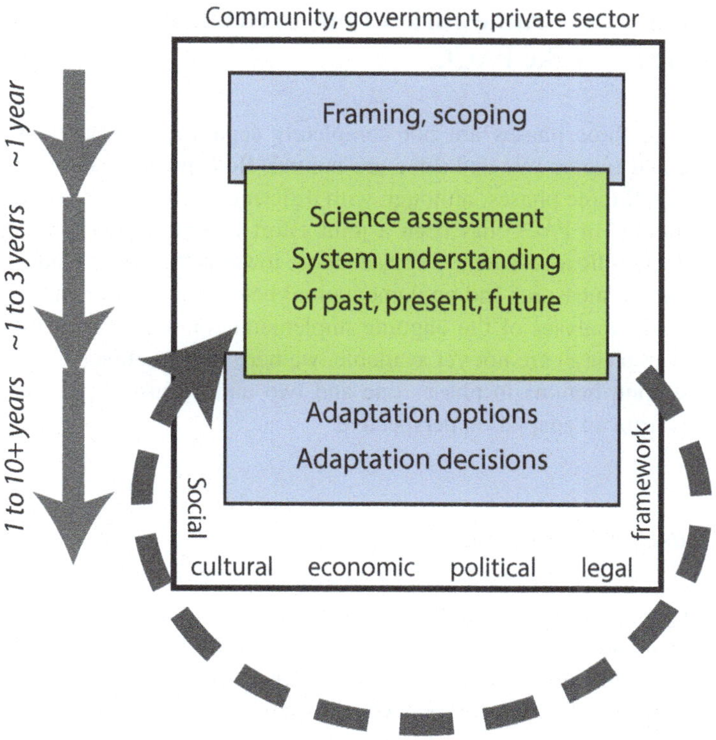

Fig. 2 Framework for a climate adaptation process with science contribution. Three main stages are distinguished and indicated as *boxes*, where *blue* denotes processes with a strong policy and stakeholder component, and *green* with a stronger focus on science. The *dashed arrow* indicates the iterative nature of the adaptation process. *Arrows at the left* provide an approximate duration of the respective stage of the framework (Reproduced from Huggel et al. 2015b)

specific. Therefore, scientific results and models as well as the adaptation decisions have to be placed in the respective social, cultural, economic, legal and political dynamic setting.

Experience shows that allocated time frames in many projects are too short, since the mutual trust, which is essential for collaboration yet often hampered by differences in culture, interests and perceptions, needs time to build up. Time scales and response times of different stakeholders differ as well, potentially slowing down progress and requiring constant revision of schedule and log-frame.

3 Applying the Framework's Lens to Science Contributions in Contexts Like PACC

Generally, the three phases are not completely separate, and overlaps between phases one and two or two and three are natural. Both practice and science sides contribute to all three phases, although with differing intensities. Phase one (framing and scoping) in PACC has been a joint effort of science and practice, while phase two (scientific assessment) is biased more towards the science side and phase three (adaptation measures and implementation) belongs to the domain of practice. Since scientific analyses of the ongoing implementation activities (see COSUDE 2013 for an overview) are not yet available, we here restrict ourselves to report on the science contributions in phases one and two and discuss their strengths and weaknesses from an adaptation perspective.

3.1 Phase 1: Framing and Scoping

The framing and scoping within PACC was brought together by experts from international cooperation, the Peruvian partner organizations, representatives of the local governments and Peruvian and international scientists. The different backgrounds of these professionals produced a broad and diverse range of contributions on the three main thematic lines of PACC (water resources, food security and natural disasters). From these many inputs and consultations we conceived a first synopsis of the situation, as summarized in the following.

Climatologically, the Central Andes are characterized by semi-arid conditions, with only small variations of monthly temperatures throughout the year (due to the tropical location), but with a pronounced seasonality of precipitation. Precipitation falls mainly between October and March, the remaining months are mostly dry (Garreaud et al. 2009). Especially during the dry season, water-dependent activities such as agriculture and hydropower can substantially rely on glacial melt water and storage in (artificial) lakes, depending on their location in the catchment and relative to glaciated head waters (Casassa et al. 2009; Drenkhan et al. 2015). Dry years have substantially cut agricultural yields. Heavy precipitation events have repeatedly caused mud and debris flows (e.g. Cusco/Machu Picchu in January 2010), destroying roads in the region with impacts on local population and the sensitive touristic sector.

Rising temperatures due to climate change have caused (and are expected to continuously cause) significant losses of glacier ice (see Fig. 3), decreasing on the one hand the glacial storage capacity and on the other hand increasing the risk of disasters such as from Glacial Lake Outburst Floods (GLOFs; Haeberli et al. in press). The decreased storage capacity will lead to a reduced stream water availability in the near future, especially during the dry season (Chevallier et al. 2011; Baraer et al. 2012). Vulnerabilities arising from this climatological situation concern

Fig. 3 View of the Suyuparina glacier margin in the Cordillera Vilcanota, where glaciers have lost 40–45 % of their volume since the mid 1980s (Salzmann et al. 2013). Note the extent of the previously glaciated (*dark*) bedrock (Photograph by F. Drenkhan, 2015)

domestic supply, agriculture irrigated from river water and hydropower generation during the dry season as well as settlements in the downstream basin of glaciers with an increased exposure to disasters such as GLOFs.

Besides climatological factors, these vulnerabilities relate to their specific social, cultural, political and economic context. For example, according to the last available census of 2012 (INEI 2013), in the Cusco region 32 % of agricultural land is irrigated, while for Apurímac the fraction is 48 %. Since the preceding census of 1994 (INEI 1995) agricultural area has increased significantly by 12 % (118 %) in Cusco (Apurímac), with an even stronger increase of the fraction of irrigated land of 83 % (164 %) in Cusco (Apurímac). The availability of irrigation water depends on local hydrology but also on access to these recourses, potentially competed for (under unequal conditions) by the water demand of hydropower and mining entities, both in terms of quantity and quality (pollution).

Pollution of water and land is a common cause of social conflicts, despite a recently adopted law on Integrated Water Resources Management (IWRM) promoting an effective, participatory and equitative water management at catchment scale (Drenkhan et al. 2015). Water management, (irrigation) use and demand, control and rights, and ecological impacts are further reasons for the 215 social conflicts in Peru, of which 23 take place in Apurímac and 16 in Cusco. One hundred seventeen of these are active socio-ecological conflicts, mostly related to mining and energy such as hydropower (DDP 2015). That farmers out of a need for extra income sometimes work for mining companies produces the paradoxical situation that farmers economically depend on a company which at the same time compromises the water and environmental resources they rely on.

3.2 Phase 2: Scientific Assessment

Scientific assessments of the framework's second phase (Huggel et al. 2015b) should therefore on the one hand address climate conditions and trends, for example with respect to the relevant components of the water balance (input from precipitation, outputs through river runoff and evapotranspiration and storage components such as glaciers and lakes), both in observations and future projections by climate models. On the other hand, the views of the involved stakeholders (farmers, mining companies, hydropower producers, policy makers) within their context (ecological, cultural, social, legal, political and economic) together with their power relations need to be assessed in order to provide a solid basis for the development and evaluation of adaptation options.

The following section introduces a number of studies (mostly originating from PACC), which address different aspects of these requirements. We present them along the hydrological cycle, beginning in the atmosphere with future precipitation. We continue with receding glaciers and investigate water resources and their future changes, discussing impacts of these changes for agriculture and (outside PACC) hydropower generation. Complementing these slow-onset changes, another study presents an inventory of recent natural disasters as a base for the design of Disaster Risk Reduction (DRR) measures. Finally, linking climate change and society, the last study investigates perceptions of rural farmers in the small basin of Huacrahuacho concerning the impact of climate change on their way of life. The map in Fig. 1 highlights the locations of these studies.

3.2.1 Neukom et al. (2015): Future Precipitation Over the Central Andes of Peru

The instrumental climate record of the Central Andes is short, most of the available station data starting only in about 1965 or even later, thereby precluding the estimation of long-term precipitation changes. Using precipitation projections from Global Climate Models (GCMs) is problematic as well, since precipitation depends on complex processes occurring on spatial and temporal scales which are much finer than the GCMs are able to resolve. This holds in particular for regions with complex topography such as the Andes. Using characteristics of large scale circulation, which is more reliably represented by the GCMs' dynamics, as predictors of precipitation, can result for some regions in more robust projections. For the Central Andes, the upper tropospheric Westerly flow is a good predictor for austral summer precipitation (Minvielle and Garreaud 2011). Neukom et al. (2015) analyze projections of these flows in combination with instrumental records of precipitation (see Fig. 1 for their locations), reanalysis data and paleoclimate proxy data to assess precipitation variability in the Central Andes over the period AD 1000–2100.

Neukom et al. (2015) find that current precipitation is still within the range of natural pre-industrial variability. However, total summer precipitation is projected to leave this range within the next few decades and by the end of the century, the GCM projections indicate a decrease of up to 30 %. Additionally the fraction of dry years is projected to increase by a factor of four from 5 % during pre-industrial times to 20 % during 2071–2100.

These results impose several scientific constraints for policy and adaptation planning in the area. First, they define a time frame: Adaptation measures addressing decreased precipitation of unprecedented degree should become effective within the next few decades. Second, they provide a quantitative estimate of the precipitation decrease, which is an important input for the scientific quantification of climate change impacts (e.g. for crop modelling, see below) and for long-term planning of water storage and allocation. Third, the projected increase in dry years urges policy makers to prepare for potential multi-year droughts rather than single drought years.

3.2.2 Salzmann et al. (2013): Glacier Changes and Climate Trends in the Cordillera Vilcanota

Compared to other high mountain regions worldwide, Peru has a relatively dense climate station network with several stations at high elevations (above 4000 m a.s.l.) and records dating back to 1965 (a few have even longer records), although many stations have considerable data gaps (Schwarb et al. 2011). Peru has also a relatively long history of glacier observation, particularly in the Cordillera Blanca (Kaser et al. 2003). It is therefore remarkable, that for Cordillera Vilcanota, the second largest glacier-covered Cordillera in Peru and an important water resource for the densely populated Cusco region, coordinated measurement programs had never been initiated. Consequently, within PACC dedicated studies and a mass balance measurement program combined with local capacity building courses have been set up. In the study of Salzmann et al. (2013), available glacier and climate data from multiple sources (satellite images, meteorological station data and climate reanalysis), have been collected and assimilated to generate a comprehensive multi-decadal analysis of respective changes in glacier area and volume and related trends in air temperature, precipitation and in a more general manner for specific humidity.

Salzmann et al. (2013) found only marginal glacier changes between 1962 and 1985, yet revealed a massive ice loss since 1985 (about 30 % of area and about 45 % of volume). These high numbers corroborate with studies from other glacierized cordilleras in Peru (see Fig. 3 for an example of a retreating glacier). The analyses of the climate data showed an overall moderate increase in air temperature, and mostly weak and non significant trends for precipitation sums, which thus probably do not fully explain the observed substantial ice loss. The study suggests that the likely increase of specific humidity and thereby increased longwave radiation in the upper troposphere, where the glaciers are located, has played a major role in the observed massive ice loss of the Cordillera Vilcanota over the past decades. In view of projected global temperature increase in future and suggested decrease in precipitation for the Central Andes (cf. section above), loss of snow and ice will be very likely, substantially affecting water resources.

3.2.3 Andres et al. (2014): Future Changes in Water Resources of the Vilcanota Basin

Given the strong seasonality of precipitation and runoff in the high Central Andes of Peru and its immediate importance for the different water user communities, estimates of future availability and seasonality of water resources in this region is crucial. After careful calibration of a hydrological model to reproduce the observed runoff at the gauge of a hydropower station at the Vilcanota river (see Fig. 1), Andres et al. (2014) investigate impacts of future climate change on Vilcanota runoff and further components of the hydrological system, for different time slices of the twenty-first century. To this end they use future projections from three GCMs to drive the calibrated local hydrological model.

The models investigated in Andres et al. (2014) project a systematic increase in temperature, which is consistent with other studies in the Andes (Bradley et al. 2006; Urrutia and Vuille 2009). Future monthly precipitation sums decrease, in particular from May to August, when precipitation totals are already low. Naturally, some part of the rainfall is stored as ice, snow, soil moisture, in lakes or underground reservoirs. The study shows that the overall storage capacity decreases – e.g. snowmelt reduces in future, most likely because there will be less snow to melt. Since less wet season (January–May) precipitation remains in the storage until the dry season, that water needs to go into and thereby increases runoff during the rainy season, potentially accompanied by higher flood peaks. In the dry season, on the other hand, decreased storage means water scarcity, since less precipitation can be stored from the rainy to the dry season. Water dependent activities such as agriculture or hydropower generation will be affected by these changes especially in the dry season when discharge is already low, even more so since their water demand is expected to further increase.

3.2.4 Sanabria et al. (2014): Rainfed Agriculture Under Future Climates in Cusco and Apurímac

There is little doubt that without adaptation agriculture will, in the long term, suffer from decreased water availability. Considering the possible future changes in precipitation indicated by the analysis by Neukom et al. (2015) three problems are likely to emerge in the context of crop production: (1) a decrease in annual precipitation amounts; (2) a shift in the onset of the rainy season; and, (3) an increase in inter-annual variability and the occurrence of dry or very dry years.

Sanabria et al. (2014) have studied the implications of these changes for rainfed annual crops, using a new set of early twenty-first century climate scenarios along with the crop model developed by Sanabria and Lhomme (2013). Crops targeted by this study were potato, maize, wheat, barley and broad bean, i.e. the five annual crops that account for 50 % of the cultivated area in the Department of Cusco and Apurímac (see Fig. 1) and deliver the dietary basis for a large share of the peasant population.

Results of Sanabria et al. (2014) suggest that the first of the three problems associated with changes in annual precipitation is probably not a major concern in the immediate future, as average yield deficits relative to today's mean productivity are not likely to exceed 5 % by 2030. However, given the extremely small number of climate change scenarios used by Sanabria et al. (2014) further analysis is necessary, in particular when taking the results by Neukom et al. (2015) into account, which highlight substantial precipitation decreases in projections of the coming decades.

More attention is required by the second of the three problems associated with changes in precipitation, i.e. a possible shift in the onset of the rains. In fact, the data used by Sanabria et al. (2014) to setup their modelling exercise suggest that the precipitation requirements for ensuring crop establishment after planting (which is scheduled by farmers to coincide with the onset of the rains) are quite stringent. For Cusco and Apurímac the simulations conducted by Sanabria et al. (2014) indicate a locally higher risk of planting failure by 2030. Although minimal, the increase in the risk of crop failure at planting can be considered as significant, since it already accounts for the possibility of a later planting entailed by increasing temperatures. This suggests that spontaneous adaptation is probably not sufficient to counterbalance the negative impacts of climate change.

Due to the lack of specific scenarios, the implications for crop cultivation of the third problem associated with changes in precipitation, i.e. a possible increase in inter-annual variability and the occurrence of dry years, has not been studied in detail up to now. In the framework of Phase 1 (framing and scoping) it nevertheless can be said that the problem is likely to represent a major challenge for adaptation, as year-to-year variations in crop production already limit food security.

3.2.5 Drenkhan et al. (2015): Hydropower in the Vilcanota Basin, Water Resources and Conflicts

Peru's energy demand has been rising at 6.5 % annual average in the last 9 years (2004–2013). Since energy supply strongly relies on hydropower, which accounts for 54 % of the entire national energy production (MINEM 2014), in the Vilcanota basin several hydropower schemes are extended or newly constructed to balance out rising energy demand. They face two main challenges as follows (Drenkhan et al. 2015 and references therein):

First, they require certain minimum discharges to produce at full capacity (e.g., the installation Machu Picchu II, run by the public-private company EGEMSA, produces 101 MW at 31 m^3/s, while the close-by installation Santa Teresa I, run by the private operator Luz de Sur produces 98 MW at 61 m^3/s, OSINERGMIN 2014). Currently, dry season discharge (approx. 45 m^3/s) falls below this threshold for a few of the installations. A planned new installation (Santa Teresa II) would require an even higher discharge of 105 m^3/s, which is not available from natural discharge between May and September. With ongoing glacier shrinkage, particularly upper basin streamflow will significantly reduce during the dry season, while discharge variability increases (Chevallier et al. 2011; Baraer et al. 2012), further aggravating this imbalance.

Second, plant feasibility studies for the construction of Santa Teresa II have provoked broad resistance by local communities which express their distrust towards operator and state agencies but also complain of exclusion from decision-making processes. They fear negative effects for agriculture and the touristic value of close-by hot springs, both related to the planned construction of a diversion tunnel. Local residents are also blocking attempts further upstream, where new dams and water level elevation for high-Andean lakes had been considered in feasibility studies by EGEMSA in order to guarantee sufficient year-round Vilcanota river discharge.

3.2.6 Huggel et al. (2015a): Disaster Databases and Risk Reduction

Disasters related to natural hazards affect many regions in Peru, including Cusco and Apurímac. The understanding, however, both in terms of occurrence and causes, is limited. Therefore, within PACC efforts have been undertaken to analyse the occurrence of disasters in a more systematic way. Huggel et al. (2015a) have compiled data on disasters from two main sources: DesInventar is a multi-national scale database and inventory system for a wide range of disasters, their characteristics and impacts since 1970 (DesInventar 2013). The other data source is a national scale disaster database, the Peruvian National Information System for the Prevention and Response of Disasters (SINPAD) of the National Institute of Civil Defense

(INDECI). SINPAD consists of inventories of events since 2001 at the regional, provincial and district level and reports data such as the number of fatalities and people affected, infrastructure damage, surface area affected, etc. (INDECI 2013). Huggel et al. (2015a) considered different climate related processes and hazards, including cold spells, droughts, precipitation events, floods, and vaious types of landslides from 1970 to 2009.

Results based on the analysis of DesInventar data showed no clear patterns of change in disaster occurrence since 1970 for Cusco even though a higher number of events could be observed for the 1990s and for individual years such as 1994 or 2003. For Apurímac disaster occurrence was clearly higher in the 1990s than in the 1970s and 1980s. The higher number of disasters in the 2000s is mainly related to the first 5 years of this decade, with a peak of events in 2001. The missing, or limited, trend in occurrence of disasters over four decades may be surprising, given that the discourse about increasing disasters is widespread. However, Huggel et al. (2015a) found several limitations of the disaster databases such as unclear criteria for inclusion of events in the disaster databases, differences in reporting, documentation and quality of disaster record. For example, a comparative analysis of available data on the big January–March 2010 floods in Cusco revealed striking differences and inconsistencies in reporting between DesInventar and SINPAD.

While the study could not substantially contribute to designing improved risk reduction efforts in view of climate and socio-economic change, it allows for important conclusions (Huggel et al. 2015a): The differences between disaster metrics and their change through time and space can be enormous, depending on the disaster database analyzed, and the conclusions drawn in terms of risk management could hence be different as well. Both in research and risk management, a careful evaluation of type, method and source of documentation of the analyzed disaster database is essential. Furthermore, efforts to improve the consistency of disaster reporting should be strengthened, including information on the loss and damage due to disasters.

This scientific assessment links to the other phases of the framework. Specifically, the conclusions and recommendations should guide the scoping and framing during the first phase of the framework, raising awareness for the contingency inherent to different databases. On the other hand, the suggested improvement in disaster documentation could be considered as part of a set of adaptation measures, which are foreseen within the third phase of the framework. This study thus also emphasizes the importance of iterations within and between the different phases of the framework, which is basically a dynamic process between different stakeholders.

3.2.7 Flores Moreno and Valdivia Corrales (2012): Perceptions of the Role of Climate (Change) for the Livelihood of Rural Populations

Flores Moreno and Valdivia Corrales (2012) show that for the rural population of the small basin Huacrahuacho (Cusco region), impacts of climate variability and climate change have to be understood in a broader context taking into account risks and dangers in the farmers' everyday life. Their answers to these risks are embedded in the past and the present and go beyond the territorial borders of their group (and the borders of the research area, the small basin of Huacruacho). For example, their livelihood strategies are not restricted to on-site income generation through agriculture and livestock, but include also kinship and (temporal) migration to other places in Peru, such as Andahuaylas and Lima. Simultaneously, their livelihood strategies are shaped by the presence of international and national organizations, financed by different donors, realizing different projects according to political, economic and social processes at national and international levels. Such exposure to the context of modernisation and development programs often leads, through different structural processes and cultural conflicts, to subordination and consequent invisibility of local knowledges.

Climate change is thereby only one out of many different constraints for rural livelihoods, which all need to be considered when conceiving adaptation measures. Understanding the many interconnections of climate variability and climate change with social, cultural, economic and political processes at the local, national and international level is essential during all three phases of the framework.

4 Concluding Remarks

This chapter compiles the findings of several studies from the wider context of the Climate Change Adaptation Program (PACC) in the Central Andes of Peru and evaluates them through the lens of the framework for science contribution to climate adaptation proposed in Huggel et al. (2015b). The framework describes the three phases framing and scoping, scientific assessment, and evaluation/implementation of adaptation measures.

From the framing and scoping of phase one (see first part of Sect. 3), two lines of research necessities emerged: One related to climate and hydrology and related natural disasters, another one about risk perceptions of the stakeholders and the relations between them within their specific situation. Both are equally important for an adequate understanding of the context, its vulnerabilities under present-day and future conditions and conceiving appropriate adaptation measures (the latter corresponding to the third phase of the framework, which is not covered here).

Scientific investigations of the second phase (see second part of Sect. 3) show that water scarcity during the dry season is a frequent concern, already today but even more under projected future conditions. Precipitation from the wet season is stored in glaciers, lakes and soil and thereby available for the dry season, when precipitation is negligible. Temperature rises and climate change have already in the past decades eroded significant portions of the glacier storage and are expected to continue. Substantial decreases of precipitation during the wet season projected for the future decrease the overall quantity of available water. A detailed impact analysis of these tendencies for river runoff indicate decreased river water during future dry seasons.

Changing precipitation seasonality can increase the risk of crop failure, and decreasing river runoff in the dry season makes hydropower generation less efficient. Both of these impacts appear as a consistent consequence of reported and projected climatological and hydrological trends. The scientific understanding of observed tendencies in natural disasters is less complete, as the example of the disaster databases shows. If available data are incomplete or inconsistent, capabilities for support by science are limited, which recommends the consolidation of such databases as an adaptation measure to phase three of the framework.

The scientific studies presented in Sect. 3 were undertaken because the framing and scoping of phase one clearly indicated water scarcity as an important challenge for societies. The studies have contributed to a better conceptual and quantitative understanding of the related physical processes, however, they did not address the implications of their findings for society. Only the study on hydropower, although not providing a thorough anthropologic investigation, mentions explicitly the immediate relevance of hydropower installations for local people, their livelihoods and social cohesion. The many links between expanding mining activities and local livelihoods require a more in-depth investigation as well.

As the social sciences study (see end of Sect. 3) on local people's perceptions of the role of climate change for their livelihood shows, climate change is only one out of many determinants that shape their lives. Temporal migration for extra income generation, a variety of different development programs from national (governments, mining companies etc.) and international (e.g. international cooperation) stakeholders and the general political setting are at least as important for current livelihoods. We observe that findings from natural sciences also need to be contextualized in this sense to become fruitful for adaptation.

Unfortunately, the scientific base for such contextualization in the setting of PACC is still incomplete, as in-depth anthropological analyses of these contexts are rare and scattered in space. The few studies that exist concordantly emphasize the necessity of providing such analyses with denser coverage. This imbalance between studies from natural versus social sciences is a frequent problem of comparable programs. It will be addressed during the remaining time of the program.

Nevertheless, science activities in the context of PACC have produced a new body of insight for the two focus regions, which provides regional decision makers with an improved base and an extended set of tools to evaluate their situation and adaptation options. Thinking of optimizing the scientific support of such programs, a more serious inclusion and engagement of social sciences throughout the entire process, in particular from the very beginning of the framing phase, would render the entire scientific contribution more valuable. If these lessons learnt are applied in other similar projects, the overall impact of PACC will be positive beyond its direct contributions.

Acknowledgements The Climate Change Adaptation Program (Programa de Adaptación al Cambio Climático, PACC) is funded by the Swiss Agency for Development and Cooperation (SDC) through the Global Program Climate Change. The authors acknowledge the work from the practice and science sides of PACC, on which this chapter is based. The practice side is led by the Swiss-Peruvian NGO HELVETAS Swiss Intercooperation together with its Peruvian partners Libélula and PREDES. The science side is supported by a Consortium of Swiss scientific institutions, including the universities of Zurich, Fribourg and Geneva, Meteodat, Agroscope and the Swiss Federal Institute for Forest, Snow and Landscape Research (WSL), in close collaboration with universities in Cusco (UNSAAC) and Apurímac (UNAMBA).

References

Andres, N., Vegas Galdos, F., Lavado Casimiro, W. S., & Zappa, M. (2014). Water resources and climate change impact modelling on a daily time scale in the Peruvian Andes. *Hydrological Sciences Journal, 59*(11), 2043–2059. doi:10.1080/02626667.2013.862336.

Baraer, M., Mark, B. G., McKenzie, J. M., Condom, T., Bury, J. T., Huh, K.-I., Portocarrero, C., Gomez, J., & Rathay, S. (2012). Glacier recession and water resources in Peru's Cordillera Blanca. *Journal of Glaciology, 58*(207), 134–150. doi:10.3189/2012JoG11J186.

Bradley, R. S., Vuille, M., Diaz, H. F., & Vergara, W. (2006). Threats to water supplies in the tropical Andes. *Science, 312*, 1755–1756. doi:10.1126/science.1128087.

Bury, J., Mark, B. G., Carey, M., Young, K. R., McKenzie, J. M., Baraer, M., French, A., & Polk, M. H. (2013). New geographies of water and climate change in Peru: Coupled natural and social transformations in the Santa River watershed. *Annals of the Association of American Geographers, 103*, 363–374. doi:10.1080/00045608.2013.754665.

Casassa, G., López, P., Pouyaud, B., & Escobar, F. (2009). Detection of changes in glacial run-off in alpine basins: Examples from North America, the alps, central Asia and the Andes. *Hydrological Processes, 23*(1), 31–41. doi:10.1002/hyp.7194.

Chevallier, P., Pouyaud, B., Suarez, W., & Condom, T. (2011). Climate change threats to environment in the tropical Andes: Glaciers and water resources. *Regional Environmental Change, 11*(1), 179–187. doi:10.1007/s10113-010-0177-6.

COSUDE. (2013). *Programa de Adaptacion Al Cambio Climatico Peru*. Agencia Suiza para el Desarollo y la Cooperacion COSUDE, Lima. Accessed Nov 2015. http://www.cooperacionsuizaenperu.org.pe/images/documentos/cosude/publicaciones/publicaciones2/brochurepacc.pdf

DDP. (2015). *Reporte de Conflictos Sociales* (Vol. 141). Lima: Defensoria del Pueblo.

DesInventar. (2013). *DesInventar: DesInventar – Inventory system of the effects of disasters*. Corporación OSSA, Cali, Colombia. Accessed Nov 2014. http://desinventar.org

Dilling, L., & Lemos, M. C. (2011). Creating usable science: Opportunities and constraints for climate knowledge use and their implications for science policy. *Global Environmental Change-Human and Policy Dimensions, 21*, 680–689. doi:10.1016/j.gloenvcha.2010.11.006.

Drenkhan, F., Carey, M., Huggel, C., Seidel, J., & Oré, M. T. (2015). The changing water cycle: Climatic and socioeconomic drivers of water-related changes in the Andes of Peru. *Wiley Interdisciplinary Reviews: Water, 2*(6), 715–733. doi:10.1002/wat2.1105.

Flores Moreno, A., & Valdivia Corrales, G. (2012). *Las Percepciones de La Poblacion Rural Campesina de La Microcuenca Huacrahuacho Sobre La Incidencia Del Cambio Climatico En Su Forma de Vida*. Lima: PACC.

Fussel, H. M. (2007). Adaptation planning for climate change: Concepts, assessment approaches, and key lessons. *Sustainability Science, 2*, 265–275. doi:10.1007/s11625-007-0032-y.

Garreaud, R. D., Vuille, M., Compagnucci, R., & Marengo, J. (2009). Present-day South American climate. *Palaeogeography Palaeoclimatology Palaeoecology, 281*, 180–195. doi:10.1016/j.palaeo.2007.10.032.

Haeberli, W., Schaub, Y., & Huggel, C. (in press). Increasing risks related to landslides from degrading permafrost into new lakes in de-glaciating mountain ranges. *Geomorphology*. Available on http://www.sciencedirect.com/science/article/pii/S0169555X16300381

Hegger, D., Lamers, M., Van Zeijl-Rozema, A., & Dieperink, C. (2012). Conceptualising joint knowledge production in regional climate change adaptation projects: Success conditions and levers for action. *Environmental Science & Policy, 18*, 52–65. doi:10.1016/j.envsci.2012.01.002.

Huggel, C., Raissig, A., Rohrer, M., Romero, G., Diaz, A., & Salzmann, N. (2015a). How useful and reliable are disaster databases in the context of climate and global change? A comparative case study analysis in Peru. *Natural Hazards and Earth System Science, 15*(3), 475–485. doi:10.5194/nhess-15-475-2015.

Huggel, C., Scheel, M., Albrecht, F., Andres, N., Calanca, P., Jurt, C., Khabarov, N., et al. (2015b). A framework for the science contribution in climate adaptation: Experiences from science-policy processes in the Andes. *Environmental Science & Policy, 47*, 80–94. doi:10.1016/j.envsci.2014.11.007.

INDECI. (2013). *SINPAD – Sistema de Información Nacional Para La Respuesta Y Rehabilitación*. Insituto Nacional de Defensa Civil, Lima. Accessed Nov 2014. http://sinpad.indeci.gob.pe

INEI. (1995). *III Censo Nacional Agropecuario 1994 – Cuadros Estadísticos*. Lima: Instituto Nacional de Estadística e Informática. http://censos.inei.gob.pe/bcoCuadros/IIIcenagro.htm

INEI. (2008). *Censos Nacionales 2007: XI de Población Y VI de Vivienda. Perfil Sociodemográfico Del Perú*. Lima: Instituto Nacional de Estadística e Informática.

INEI. (2013). *IV Censo Nacional Agropecuario 2012 – Cuadros Estadísticos*. Lima: Instituto Nacional de Estadística e Informática. http://censos.inei.gob.pe/cenagro/tabulados/

INEI. (2014). *Evolucion de La Pobreza Monetaria 2009–2014*. Lima: Instituto Nacional de Estadística e Informática. http://www.inei.gob.pe/media/cifras_de_pobreza/informetecnico_pobreza2014.pdf

Kaser, G., Juen, I., Georges, C., Gómez, J., & Tamayo, W. (2003). The impact of glaciers on the runoff and the reconstruction of mass balance history from hydrological data in the tropical Cordillera Blanca, Perú. *Journal of Hydrology, 282*(1–4), 130–144. doi:10.1016/S0022-1694(03)00259-2.

Kaser, G., Grosshauser, M., & Marzeion, B. (2010). Contribution potential of glaciers to water availability in different climate regimes. *Proceedings of the National Academy of Sciences of the United States of America, 107*, 20223–20227. doi:10.1073/pnas.1008162107.

Lemos, M. C., & Morehouse, B. J. (2005). The Co-production of science and policy in integrated climate assessments. *Global Environmental Change-Human and Policy Dimensions, 15*, 57–68. doi:10.1016/j.gloenvcha.2004.09.004.

Lynch, B. D. (2012). Vulnerabilities, competition and rights in a context of climate change toward equitable water governance in Peru's Rio Santa Valley. *Global Environmental Change, 22*(2), 364–373.

MINEM. (2014). *Anuario Ejecutivo de Electricidad 2013*. Lima: Ministerio de Energía y Minas.

MINEM. (2015). *Mapa de Proyectos Mineros*. Lima: Ministerio de Energía y Minas. http://www.minem.gob.pe/minem/archivos/file/Mineria/PUBLICACIONES/MAPAS/2015/MAPA2015-2.pdf

Minvielle, M., & Garreaud, R. D. (2011). Projecting rainfall changes over the South American Altiplano. *Journal of Climate, 24*(17), 4577–4583. doi:10.1175/JCLI-D-11-00051.1.

Neukom, R., Rohrer, M., Calanca, P., Salzmann, N., Huggel, C., Acuña, D., Christie, D. A., & Morales, M. S. (2015). Facing unprecedented drying of the Central Andes? Precipitation variability over the period AD 1000–2100. *Environmental Research Letters, 10*(8), 084017. doi:10.1088/1748-9326/10/8/084017.

OSINERGMIN. (2014). *Central Hidroeléctrica Santa Teresa*. Lima.

Salzmann, N., Huggel, C., Rohrer, M., Silverio, W., Mark, B. G., Burns, P., & Portocarrero, C. (2013). Glacier changes and climate trends derived from multiple sources in the data scarce Cordillera Vilcanota region, southern Peruvian Andes. *The Cryosphere, 7*, 103–118. doi:10.5194/tc-7-103-2013.

Sanabria, J., & Lhomme, J. P. (2013). Climate change and potato cropping in the Peruvian Altiplano. *Theoretical and Applied Climatology, 112*(3–4), 683–695. doi:10.1007/s00704-012-0764-1.

Sanabria, J., Calanca, P., Alarcón, C., & Canchari, G. (2014). Potential impacts of early twenty-first century changes in temperature and precipitation on rainfed annual crops in the central Andes of Peru. *Regional Environmental Change, 14*(4), 1533–1548. doi:10.1007/s10113-014-0595-y.

Schwarb, M., Acuña, D., Konzelmann, T., Rohrer, M., Salzmann, N., Serpa Lopez, B., & Silvestre, E. (2011). A data portal for regional climatic trend analysis in a Peruvian high Andes region. *Advances in Science and Research, 6*, 219–226. doi:10.5194/asr-6-219-2011.

Sietz, D., Mamani Choque, S. E., & Lüdeke, M. (2012). Typical patterns of smallholder vulnerability to weather extremes with regard to food security in the Peruvian Altiplano. *Regional Environmental Change, 12*(3), 489–505.

Urrutia, R., & Vuille, M. (2009). Climate change projections for the tropical Andes using a regional climate model: Temperature and precipitation simulations for the end of the 21st century. *Journal of Geophysical Research-Atmospheres* 114 (January). doi:10.1029/2008jd011021.

Weichselgartner, J., & Kasperson, R. (2010). Barriers in the science-policy-practice interface: Toward a knowledge-action-system in global environmental change research. *Global Environmental Change-Human and Policy Dimensions, 20*, 266–277. doi:10.1016/j.gloenvcha.2009.11.006.

Wilby, R. L., Troni, J., Biot, Y., Tedd, L., Hewitson, B. C., Smith, D. M., & Sutton, R. T. (2009). A review of climate risk information for adaptation and development planning. *International Journal of Climatology, 29*, 1193–1215. doi:10.1002/Joc.1839.

Managing Glacier Related Risks Disaster in the Chucchún Catchment, Cordillera Blanca, Peru

Randy Muñoz, César Gonzales, Karen Price, Ana Rosario, Christian Huggel, Holger Frey, Javier García, Alejo Cochachín, César Portocarrero, and Luis Mesa

Abstract Glacial lakes hazards have been a constant factor in the population of the Cordillera Blanca due their potential to generate glacial lake outburst floods (GLOFs), which can be increased by the effects of climate change. In past decades, the UGRH (Glaciology and Water Resource Unit) successful implemented security infrastructure, however, events like the GLOF of April 11 in Carhuaz highlighted the need to implement new risk management strategies. In response, the Glaciares Project has been carried out to implement three strategies to reduce risks in the Chucchún catchment through: (1) Knowledge generation, (2) building technical and institutional capacities and, (3) the institutionalization of risk management. Strategies focused on strengthening the Municipality of Carhuaz, the Civil Defense Platform and its members, leading to an improvement of risk management and being based under Peruvian laws. As a result, both the authorities and the population have improved their resilience to respond to the occurrence of GLOF. This chapter will discuss and analyze the strategies and actions implemented under the Glaciares Project to build a model of glacier related risk management and climate change adaptation.

Keywords GLOF • Glacier • Risk management • Cordillera Blanca • Hualcán • Chucchún • Lake 513

R. Muñoz (✉) • C. Gonzales • K. Price • A. Rosario • C. Portocarrero
CARE Perú, Jr. 28 de Julio 467, Huaraz, Perú
e-mail: randym3110@hotmail.com

C. Huggel • H. Frey
Department of Geography, University of Zurich, Zurich, Switzerland

J. García
CREALP, Centre de Recherche sur l'Environnement Alpin, Sion, Switzerland

A. Cochachín
Glaciological Unit UGRH, National Water Authority, Huaraz, Peru

L. Mesa
Civil Defense, Municipality of Carhuaz, Carhuaz, Peru

N. Salzmann et al. (eds.), *Climate Change Adaptation Strategies – An Upstream-downstream Perspective*, DOI 10.1007/978-3-319-40773-9_4

1 Introduction

The Cordillera Blanca, crowning the Santa River, is the source of the water supply to many towns and cities located in the valley of the Callejon de Huaylas. At the same time, it is a source of danger due the occurrence of ice and rock avalanches and GLOFs. Throughout history there have been many disasters that have destroyed towns and villages. Carhuaz is one of the recently affected cities. Located in the catchment of the Chucchún river, a tributary of the Santa river (Fig. 1), it suffered on April 11, 2010 the destruction of public and private infrastructures when an avalanche of rock and ice from Mount Hualcán (6100 m.a.s.l.) impacted on Lake 513 triggering a GLOF (Schneider et al. 2014). Lake 513 is 1 of the 14 lakes declared as posing high hazard in the Cordillera Blanca (ANA 2014) endangering the growing population of the Callejón de Huaylas. This chapter discusses and analyzes the strategies and actions implemented under the Glaciares Project implemented by CARE Peru, the University of Zurich and further partners funded by the Swiss Agency for Development and Cooperation (SDC) to build a model case of glacier related risks management and adaptation to climate change in Chucchún catchment.

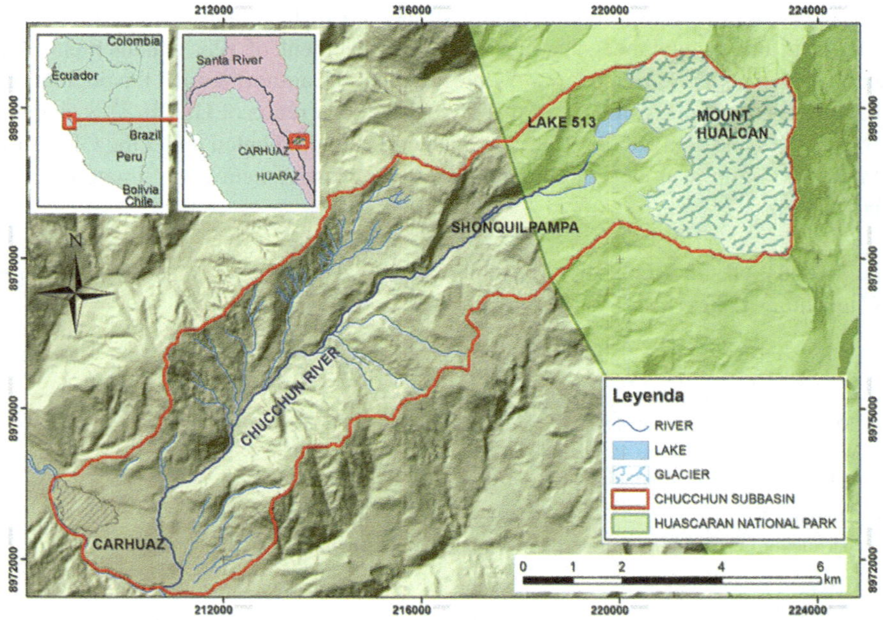

Fig. 1 Location of Chucchún catchment, with the Lake 513 and Mount Hualcán in the uppermost part

2 The Mount Hualcán and the Lake 513 Hazard

Glacier related hazards in the Cordillera Blanca have existed for centuries. However, mainly due on the one hand to the population growth and increasing urbanization, and on the other hand due to the glacier retreat and the formation of glacial lakes (Ames 1998; Zapata 2002), its most disastrous effects have been observed during the second half of the twentieth century. In this current situation, many glacial lakes are particularly dangerous due to the existence of unstable moraine dams and their susceptibility to be impacted by avalanches of rock and ice that might trigger GLOFs (Kershaw et al. 2005).

This is the case of Mount Hualcán and the Lake 513 (Fig. 2). This lake started forming in early 1960 as a result of retreat of Glacier 513A. Rapid growth in 1980 was notable, reaching a depth of 120 m. In 1988, it had a freeboard of only 1 m as a result of increased water level and erosion of the moraine material deposited on the bedrock dam. In addition, rock and ice avalanches from the south-west face of Hualcán regularly impacted the lake. All this aspects catalog the lake as extremely dangerous (Carey et al. 2012).

As result of the imminent hazard of Lake 513, the UGRH conducted in the years 1992/1993 an excavation of a 155 m long tunnel through the rock dam of the lake, creating a freeboard of 20 m (Reynolds et al. 1998) which was considered enough to classify Lake 513 as safe by the authorities, residents of Carhuaz and UGRH specialists at that time (INDECI 2004).

On April 11, 2010, a rock and ice avalanche of 200,000–400,000 m^3 impacted Lake 513 causing an impact wave of 25 m which exceeded the freeboard (Fig. 3). The resulting flow was deposited on Shonquilpampa, a lacustrine plain, reducing the energy of flow. The flow finally became a debris flow as a result of Chucchún river erosion (Schneider et al. 2014). The debris flow reached the outskirts of the city of Carhuaz and although there was no loss of human life, fields and public and private infrastructures (Fig. 4) were destroyed (Carey et al. 2012) causing panic

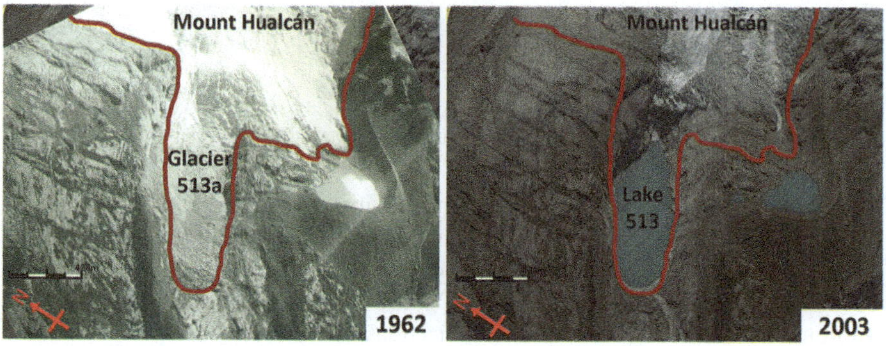

Fig. 2 Evolution of the Lake 513 from 1962 to 2002 due to glacial retreat. Diagrams performed over aerial photographs from the National Aerial Photography Service Peru (*left*) and Google Earth (*right*)

Fig. 3 Detachment zone and trajectory of the rock and ice avalanche from Mount Hualcán (*left*). Lake 513 bedrock dam, the *circles* show the location of drainage tunnel intakes. Avalanche ice floating in the lake is still observed

Fig. 4 Hyperconcentrated flow on the Shonquilpampa during the April 2010 GLOF. In the *circle* the plant of drinking water for the city of Carhuaz that was partially damaged

among the population. Despite what happened, it is considered that the impacts of flood could have been much higher without the existence of the tunnel, which is thus considered a successful risk mitigation measure (Portocarrero 2013b).

In its 5th Assessment Report, the IPCC states with a high level of confidence that warming has caused the reduction of the ice masses with the consequent formation of glacier lakes (IPCC 2014). Warming in the Cordillera Blanca observed in different studies (Mark and Seltzer 2005; Racoviteanu et al. 2008; Vuille et al. 2008) can have negative effects on the stability of glacierized walls (Huggel 2009). Therefore, the implementation of risk management and climate change adaptation becomes an urgent matter.

3 Strategies for Glacier Related Risk Management

Over the past decade, risk management in Peru has made significant progress in legislation, which has been consolidated with the creation of the National System for Disaster Risk Management (SINAGERD) in 2011 (Law No. 29664), the approval of the National Policy on Disaster Risk Management in 2012 and the National Plan for Disaster Risk Management (PLANAGERD) in 2014 (PCM 2014). This legislation defines the risk management mechanisms and responsibilities at all government levels. With all this legislation, the Peruvian state seeks to reduce the risks, facilitating an adequate preparation of people and authorities, and increasing the capacities of response, rehabilitation and reconstruction in disaster situations.

However, translating the legal framework into risk management practice has not yet been achieved. The Peruvian state acknowledges that there is still limited operational capacity of the institutions responsible for the implementation of SINAGERD passing through a slow process of adaptation of standards, limited capacity of authorities and officials, and a slow process of incorporating risk management in planning processes (PCM 2014).

This situation is much more critical when discussing the glacial related risks. Firstly, glacier related hazards are not considered as one of the most important in the PLANAGERD although the Ministry of Environment suggests that the rapid retreat of glaciers by climate change creates new, or increases hazards, with probably more frequent landslides and GLOFs (PCM 2014). Secondly, despite the PLANAGERD promotes the implementation of measures to reduce risks, institutional barriers still exist. In some way the State fights against itself to execute investment projects in the field of risk. It is actually observed that the current application of regulations substantially depends on the will and interest of politics and decision makers.

In response, taking into reference the events of April 11, 2010, and under the Glaciares Project, three key strategies were identified to facilitate risk management: (1) knowledge generation, (2) building technical and institutional capacities in the population and its authorities, and (3) the institutionalization of risk management. These strategies strengthen the three risk management components – prospective, corrective and reactive (according to the SINAGERD Law) – which facilitate the implementation of its seven processes – estimation, prevention, mitigation, preparedness, response, rehabilitation and reconstruction – such that decision making is the more suitable for reducing the risks (Fig. 5).

Glacier risk knowledge generation is extremely important to reduce the risk. Although it was believed the Lake 513 hazard was controlled by safety infrastructure implemented in the years 1992/1993 (INDECI 2004), the 2010 event suggested the immediate need for further studies. First, determine the possibility of occurrence of new events and its magnitude. Secondly, the full GLOF risk analysis, and third to inform and sensitize the population and authorities about the real state of the hazard they face and eventually reduce vulnerability and exposure through technology and social efforts.

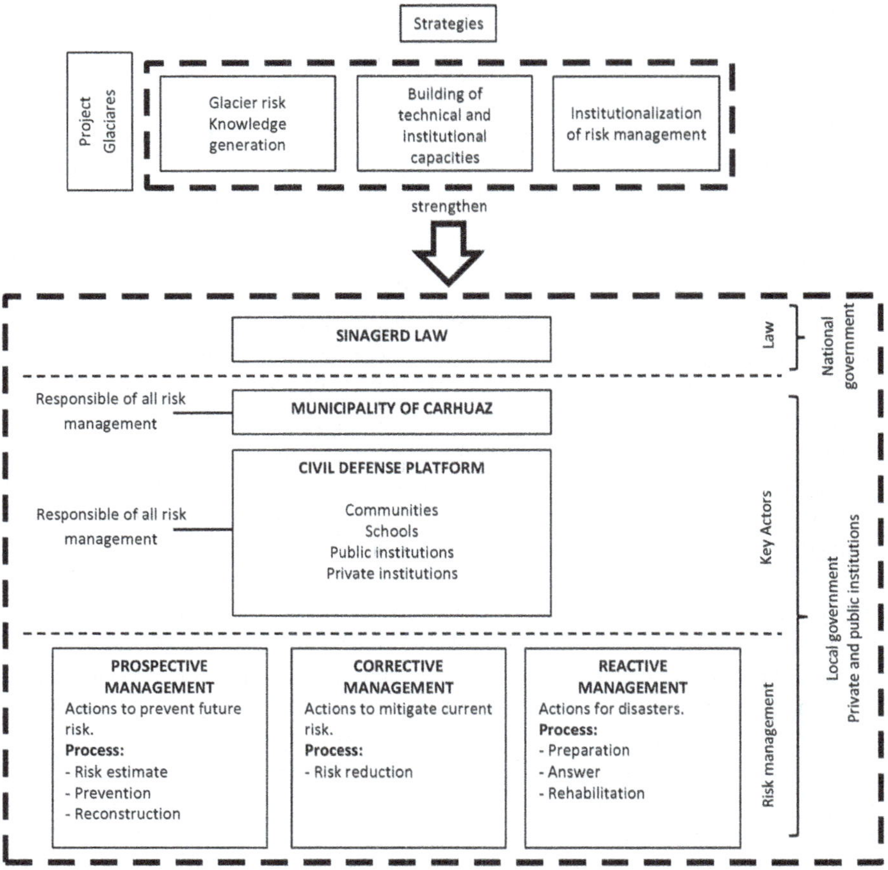

Fig. 5 Strategies for glacial related risk in Chucchún catchment

Enhancing technical and institutional capacities is based on the fact that the Peruvian state recognizes that its capabilities to properly implement risk management are limited (PCM 2014). These limitations are related to the lack of knowledge that each authority has on current regulations, their roles and responsibilities. For instance, during the 2010 event, authorities were unclear about the exact tasks and responsibilities and tried to talk to higher level authorities to ask for help. Technical managers did not know what kind of information to be sent and to whom, despite the existence of tools provided by INDECI since 2004 for risk management. The same applies to public and private institutions and civil organizations that ignore their obligations and mechanisms for participation in risk management. Thus, institutional strengthening begins with training the authorities and population in their role in risk management and from that, creating technical capabilities to prevent, care, rehabilitation and reconstruction of disaster situations.

The strategy of institutionalization has been aspired to disaster risk management in a sustainable manner by the local government, so that they can be achieved with clear and transparent rules and simplified procedures for the implementation of risk management. This reinforcement is given by: (1) Agreements formalized by municipal regulations and resolutions, which involves all public institutions. (2) Implementation of Risks Management under Peruvian law, (3) Capacity building of officials of local governments in risk management and the (4) Implementation of instruments used to direct the risk management, like the risk management plan, the emergency operations plan, contingency plan, among others.

4 Managing Risks at Carhuaz

Obviously such a broad risk management process cannot be resolved overnight or cover all elements in detail. It is important to prioritize what is most urgent. In this context, glacial related risk management in Chucchún catchment prioritized estimation, prevention, mitigation and preparedness processes. The prioritization of these processes is necessary to know the risk that people face and then to establish immediate actions to prevent or reduce the occurrence of disasters. This does not mean that other processes are not important. In fact many of the results in priority processes are inputs for others. For example, modeling for identifying the hazard zones and their use in Urban Management Plan of the City of Carhuaz is a basic input for the reconstruction process after the occurrence of a flood.

Similarly, the prioritization of stakeholders was conducted. The Municipality of Carhuaz is the highest local authority and leader of all risk management processes. Another priority actor is the Civil Defense Platform as the space of coordination and participation of people and institutions. Finally, work with schools was prioritized in order to encourage a culture of risk management in children and adolescents.

One of the critical aspects of risk management is the political leadership from the highest local authority. They lead all risk management processes. Their participation, involvement and leadership are essential to achieve adequate risk management. That is why the Glaciares Project emphasized the awareness of the Mayor, so that he becomes the leader of all processes, his presence and support at every stage, action and meeting of risk management was essential for success (Fig. 6). When the Mayor is not involved, people, public and private institutions must demand his participation through the Civil Defense Platform.

The new Peruvian legislation states that those responsible for risk management are the local governments led by the Mayor. To assume that role or not depends on the interest that shows the authority, so the strategy is often to explain what can happen if the Mayor or any officer do not do their job. This has been understood so the State has begun issuing amendments to existing standards that explicitly sanction non-compliance. This does not yet ensure adequate risk management, since the authority may only comply with the law by convening Civil Defense Platform meetings and simply delegate to an officer its address without the necessary commitment. Internalization is important, and this is achieved through a continuous

Fig. 6 On the *left*, the Mayor of Carhuaz (seated) leads one of the meetings of the Civil Defense Platform. On the *right*, the Mayor leads an earthquake and GLOF emergency exercise. Glaciares Project photographic archives

awareness that strategically should start with the Mayor with political decisions influencing the local government and population.

4.1 Securing Technical and Social Knowledge

Carey et al. (2012) emphasized that when the population does not participate in the processes of risk management, the result is seen as an imposition, causing it to resist management actions even though these help them save their lives. Therefore, the whole process of risk management integrated scientific and local knowledge. The Glaciares Project has worked to identify which studies are the bases for implementing appropriate risk management.

It started collecting the perceptions of the population and authorities through the Climate Vulnerability and Capacity Analysis[1] of the hazards to which they are exposed. GLOF hazards were clearly prioritized by all communities. The analysis also identified what livelihoods are affected by the floods, highlighting infrastructure, farmland and natural resources (Gonzales et al. 2012).

The first technical studies (Schneider et al. 2013, 2014) focused on assessing the stability conditions of the Mount Hualcán and the occurrence of potential avalanches that impacts lakes in Chucchún catchment. The results showed that potential avalanches are still possible and could even result in flood volumes up to 3 million m^3. From this, models identifying potential flooding areas, flow heights and arrival times were performed and hazard maps generated (Schneider et al. 2014) (Fig. 7).

Once finished the glacial related hazard analysis, the vulnerability and risk analysis were performed (Fig. 8). For the vulnerability analysis, it was found that the socioeconomic conditions of the area were relatively homogeneous. The location of the assets (e.g. residential houses, schools) relative to the river bed was also consid-

[1] Climate Vulnerability and Capacity Analysis is an internationally validated tool of CARE

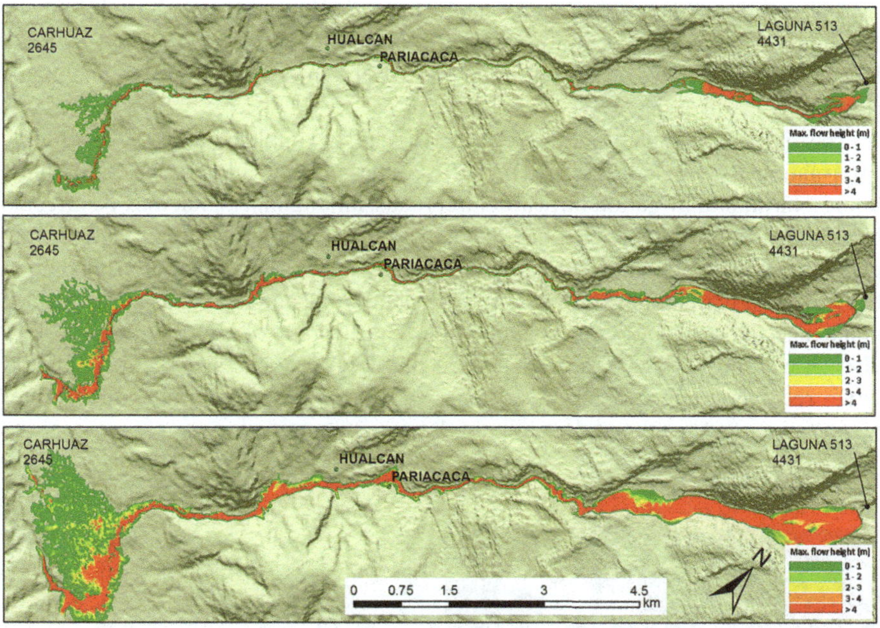

Fig. 7 Models developed for Chucchún catchment, including small, medium and large scenarios where the small scenario is similar to the 2010 event, and the large scenario includes a GLOF volume of up to 3 million m^3 (Source: Schneider et al. 2014)

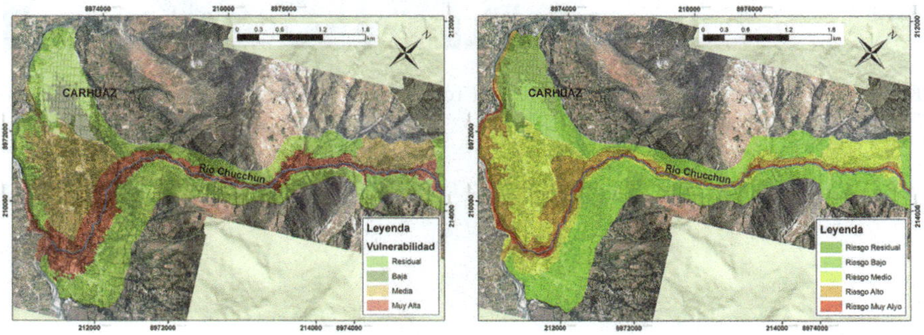

Fig. 8 Glacial related vulnerability (*left*) and risk maps (*right*) taken as reference the maximum risk scenario of 3 million m^3 (Portocarrero 2013a)

ered. The analysis of risk from floods resulted in a map with a bigger extent than the GLOF process due to secondary effects such as destruction of bridges, roads or irrigation infrastructure among others (Portocarrero 2013a).

In addition to the hazard, vulnerability and risks studies, several other studies on biophysical, socioeconomic and hydro-meteorological conditions of the region were developed. Together, they allow for a detailed knowledge not only about the socio-environmental conditions of the Chucchún catchment and adjacent areas.

The results of the studies were mapped socialized and discussed with all the communities in the Chucchún catchment, through meetings of the Civil Defense Platform. This step was essential to internalize the information on the inhabitants (Fig. 9).

The operability of all studies was performed by implementing a Geographic Information System GIS (Fig. 10). Its usefulness lies in identifying beforehand homes, farmland, irrigation infrastructure, roads, among others that may be affected by GLOFs. Similarly, the GIS facilitates the development of zoning regulations to prevent the construction of infrastructure in hazard areas. The continuous update of the current situation in the GIS will help to identify the amount of resources needed to manage a disaster and support population. The diversity of information in the GIS also allows the local government to take the necessary decisions and to address other recurring climatic hazards such as frost and facilitate integrated water resources management and the optimum land use, among others.

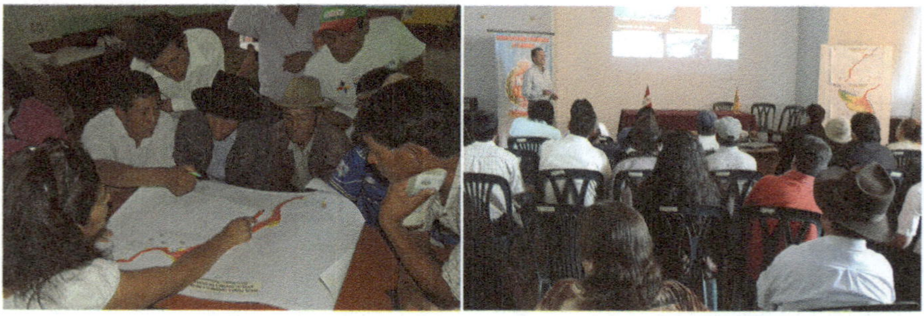

Fig. 9 Socialization and evaluation of glacial related vulnerability and risk maps workshops. Glaciares Project Photographic Archives

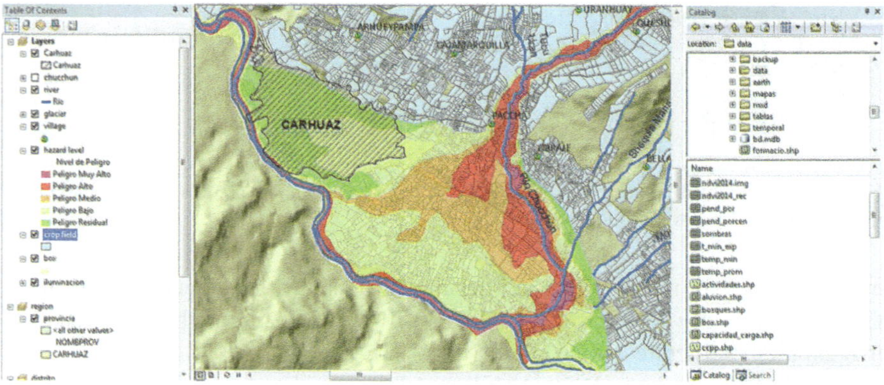

Fig. 10 Chucchún geographic information system. Rural properties (*sky-blue blocks*) superimposed on the GLOF model

4.2 *Early Warning System of Carhuaz*

The studies and results of the risk assessment confirmed the need to implement immediate measures to prevent a GLOF related disaster. Under the risk management Preparation Process, the SINAGERD proposes the implementation of Early Warning Systems (EWS) as a set of skills, tools and procedures articulated to monitor hazards and alert authorities and the population and prepare an eventual evacuation. In response, CARE Peru's experience implementing community early warning systems, and the University of Zurich with a large knowledge in glaciers, facilitated the design of an EWS in face of GLOFs as an innovative and unique design in Latin America. (EWS for other types of mass movements such as debris flows and landslides have been developed in other parts of the Andes and Latin America, see e.g. Bulmer and Farquhar 2010; Huggel et al. 2010).

In this framework, the EWS implemented in Chucchún catchment consists of four components according to the SINAGERD EWS model: Knowledge of possible GLOF risks incurred in Lake 513, their monitoring and alert, broadcasting and communication and, response.

Knowledge of risks has been achieved through various studies that analyze the glacial related hazards, vulnerabilities and risks. From these studies and the respective maps, population can identify evacuation routes and safe areas (Fig. 11). Evacuation has been tested during the execution of simulation organized and

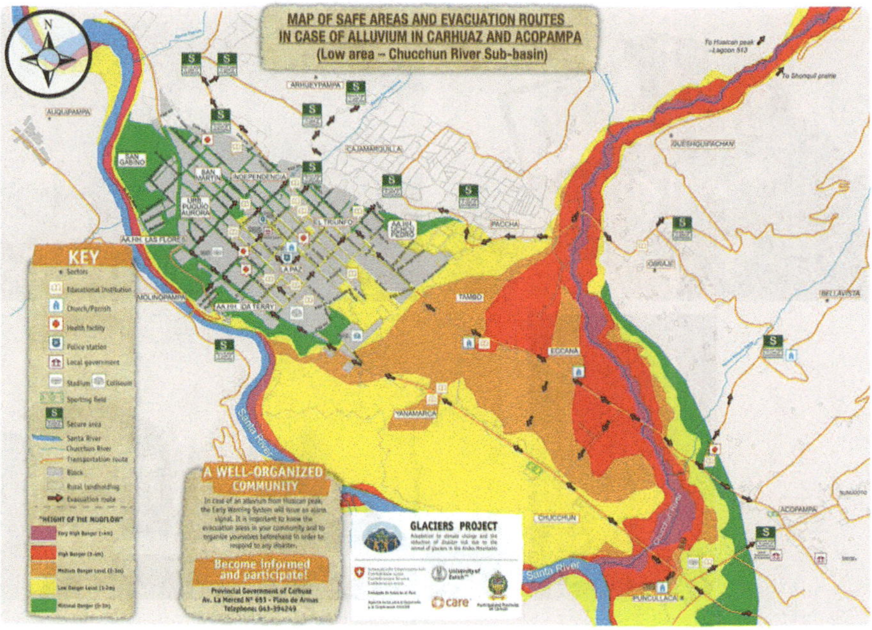

Fig. 11 Map of safe areas and evacuation routes to GLOF for Carhuaz and Acopampa. Glaciares Project Archive

evaluated by the Civil Defense Platform. Such simulations allow improving evacuation routes, avoiding agglomeration of population and identifying critical risk areas. All this information also allows identifying public investment projects for risk management.

The monitoring and alert system has been achieved through the installation of a network of five stations (Fig. 12) that monitor in real time Mount Hualcán and Lake 513 in order to detect possible GLOF occurrence and subsequent alert of the public (Frey et al. 2014). The first is the glacier monitoring station located at Lake 513, having four geophones that capture the noise of avalanches and their impact on Lake 513. Additionally it has two cameras oriented to Mount Hualcán and Lake

Fig. 12 Location map of the EWS monitoring network (*above*). *Below*, *left* to *right*: Data Center (1), Lake 513 station (3), Shonquilpampa Station (4), Pariacaca Station (5). Glaciares Project Photographic Archives

513, allowing assessing their real-time status. The second is a hydro-meteorological monitoring station. Located in Shonquilpampa, it includes a complete weather station and a water pressure sensor installed in the Chucchún River. The third station is the Data Center located in the Municipality of Carhuaz. It receives in real time all information collected by the monitoring stations. From this information, the Mayor can take the decision to evacuate the population. It is complemented by a fourth repeater station that receives and sends information from distant stations. The fifth is a warning station located in the community of Pariacaca. This station consists of monitoring equipment that allows the community to view real-time status of the Lake 513. This station has two sirens that are activated from the Data Center station when a GLOF is confirmed. The information collected in the Data Center is processed by the UGRH who determines thresholds for the geophones. The system warns, through text message, the Secretary of Civil Defense when a threshold is exceeded, so he can analyze and compare it with the photos from the cameras (taken during daylight), and the pressure sensor (day and night). The Secretary of Civil Defense has the responsibility to analyze the information any time when the system warns, for that he can access to the system through a web site wherever he is.

Training of technical personnel such as from the UGRH and the Secretary of Civil defense on the EWS was essential.

The broadcasting and communication component consists of the implementation of protocols that allow the Municipality of Carhuaz alerting the population and communicate with first responding institutions such as police, health centers and schools so that they are prepared to execute the evacuation plans. This component is facilitated by the Data Center that is designed to send text messages to the first responding institutions. Additionally, the Data Center is connected to Regional Government and the Center for National Emergency Operation at Lima, which is the main national risk manager in Peru to access to an immediate response in case of disaster.

Finally, the response component is to get authorities, institutions and population able to respond to GLOF hazards through the development of evacuation plans and organization and execution of related evacuation and emergency exercises.

4.3 Key Actors of Response Action

As result of strategies for capacity building and institutional incorporation of risk management, both authorities and population improved their capacity to respond to GLOFs.

Leadership begins with the highest local authority, in this case the Mayor of Carhuaz, sensitive and committed, leads all processes of risk management. He is leading the Civil Defense Platform meetings, the Emergency Operations Center (EOC), simulations of earthquakes producing GLOFs, promotes the development of projects to reduce risk and especially encourages its employees to fulfill their responsibilities. This active leadership translates into a more active and ongoing participation of public and private institutions for political commitment (Fig. 13).

Another key player is the Secretary of Civil Defense of the Municipality. Its function is to advise the authorities for decision-making. He also has to be a leader to convene the population and assume responsibilities. Technical decisions are the basis for the Mayor's decisions. The technical secretary accompanies the EOC, its members being officers of the Municipality and responsible to manage the emergency.

This institutional set up is largely a consequence of political decision. To ensure continuity each of the roles and responsibilities has been codified by local regulations which are legally binding for any future local governments and their officials.

Strengthened authorities facilitate capacity building in the Civil Defense Platform. In this case the building is led by its qualified authorities, so that the people and institutions see the process of risk management as a policy of local government. To ensure compliance of their duties, regulations and operational plans were developed, being one of its main actions the organization and evaluation of simulations.

An important element to adequately respond to the glacier risks is education. Children are among the most vulnerable groups during disasters, and a learning process on these topics helps students play an important role when it comes to saving lives and protecting members of the community (Fig. 14). Therefore, the teacher's role is to develop a set of capabilities of students to establish a harmonious relationship with the natural and cultural environment. Based on the National Environmental Education Policy of the Ministry of Education, the Glaciares Project decided to improve the curriculum development of basic education with adaptation to climate change and disaster risk management content.

In Chucchún catchment, there are 26 schools with more than 4500 students. 40 % of the students are placed in 7 schools located at high glacier related risk. This is why a plan was developed in order to improve the participation of the educational community in climate change adaptation and disaster risk management. The next

Fig. 13 Emergency Operation Center (EOC) meetings led by the Mayor (*left*). EOC staff evaluates the actions during the execution of a night simulation (*right*). Glaciares Project Photographic Archives

Fig. 14 Participation of school in simulation (*left*). Students explain the GLOF process in a model contest (*right*). Glaciares Project Photographic Archives

actions taken were: Review and improvement of management tools such as education project, institutional curriculum, risk management and contingency plan, and training of teachers. It was important that teachers properly understand the concepts of risk and climate change, especially contextualized in their location. With this they can determine the degree of risk of every school to prioritize public investment. As result, students understand the risk they face and how to respond, thus becoming examples for other schools in the province.

4.4 Multipurpose Infrastructure Project

All of the above actions will certainly help to reduce vulnerabilities. The EWS, a Civil Defense Platform organized and strengthened, a skilled and sensitive population will be more responsive, thus reducing the risk of loss of lives. Municipal management should be able to implement zoning regulations and reduce the loss of public and private investment. However, these activities do not prevent the occurrence of new GLOFs. As many studies demonstrate (Huggel 2009; ANA 2014; Mark and Seltzer 2005; Portocarrero 2013b) GLOFs will continue occurring as new lakes are formed as a result of glacier retreat. It is important, therefore, the implementation of security infrastructure.

For decades, the UGRH has implemented safety infrastructure, most of them under the same approach: reduce the volume and control the drainage of lakes. These works were successful because they reduced, and in some cases, eliminated the hazard through draining the lake. However it has a side effect to reduce the water storage capacity. Today and in the future, glacier shrinkage reduces water storage and availability, especially during the dry season. As result new conflicts over water use emerge (Drenkhan et al. 2015). In 1966 the first multipurpose project idea was born with the execution of the safety infrastructure at Lake Parón. It would reduce the glacier related risks and facilitate the water management for many uses. Despite the project reduced the glacier related risks, the water management was not

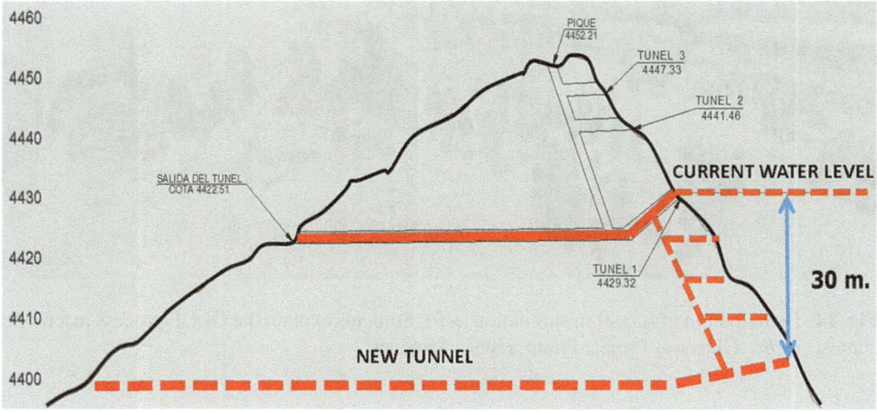

Fig. 15 Profile view of dam of Lake 513 with the current tunnel built in 1992/1993 (*full line*) and the future tunnel for multipurpose project (*dashed line*)

successful due the non-existing management rules of the lake and disagreements of every water user (Portocarrero 2013b; French et al. 2015).

In the Chucchún catchment, the water balance study demonstrated the water deficit during dry season (Tarazona et al. 2013). Consequently the traditional approach to reduce the risk by reducing the volume of water is not a viable alternative. A multipurpose project was therefore developed to reduce the water level by 30 m below its present level through a tunnel drilled in the rocky dam (Fig. 15). At the end of the tunnel a gate to facilitate the regulation of the lake would be installed. During the rainy months, the gate would remain closed to a level 10 m below the current one, thus the risk is reduced because the freeboard increases by 10 m so that events such as occurred in 2010 should be contained. During the dry season the stored water should be sufficient to meet the demands.

In this context, the questions need to be tackled how social conflicts over water such as those emerged at Lake Parón and other locations can be avoided in the case of Lake 513. Governance and coordination among all water users is a key element: population and agriculture demands through a permanent leadership of the Municipality. This coordination space has been called the Integrated Water Resources Management Committee, and aims to achieve a harmonious management of the resource avoiding conflicts and maximizing the efficiency of water use. The leadership of the Municipality is again important because it has the resources to manage infrastructure such as reservoirs or canals to reduce water deficit. This does not imply that the Municipality is the main actor, constant communication and equality among all water users is the key to avoid conflicts. There are still many regulations, ideological, social and policy barriers that must be overcome to build these kinds of committees. However the experience gained in Chucchún shows the importance of linking glacier related risk to water management.

5 Summary and Conclusions

During many decades, the UGRH has led the glacier related hazard reduction by lowering the probability of occurrence and magnitude of GLOF through reducing the volume of water of the lakes. These infrastructures are considered successful and one example of this is the tunnel built in 1992/1993 for Lake 513. The tunnel reduced the effects of the April 2010 GLOF from Lake 513 in the Chucchún catchment substantially. Despite this achievement many private and public infrastructure were damaged.

When the 2010 event occurred, local authorities did not know how to adequately respond, although tools for risk management were developed by INDECI in 2004 as the national authority for risk management. It was clear that both authorities and public and private institutions did not know their responsibilities.

In this context the Glaciares Project was developed. The aim was to implement strategies and actions for risk management and adaptation to climate change. These strategies were knowledge generation, building technical and institutional capacities in the population and its authorities, and institutionalization of risk management.

The knowledge generation as a first step aims to analyze the risks and identify immediately actions to reduce and manage it. Two studies conducted within the project were key: the Mount Hualcán slope stability and the GLOF modeling studies, indicating the importance of the implementation of an EWS. The EWS was designed and implemented based on the SINAGERD model for EWS which implies not only the technology but also and most important the organization of people and the lead of municipality's Mayor. This EWS for glacier related risks is a pilot experience and unique so far in Latin America.

Thanks to a number of additional studies, interaction with and participation by the communities of the catchment it became apparent that water issues were a major concern of the local population. It thus was recognized that efforts should not only concentrate on GLOF risks but also include the management of water resources. One of the approaches in this direction was the design of a multipurpose project to reduce the risk and at same time facilitate the water management of the glacier fed catchment, in particular in view of vanishing glacier ice. Technically, the project was sufficiently advanced to be implemented but important hurdles were met on the legal side (constraints for engineering works in protected National Park areas), and partly on the social side. Efforts have therefore been initiated to more systematically address the legal and financial aspects of such multipurpose projects, and deeper social and cultural aspects that influence any such effort.

For building technical and institutional capacities, one of the key elements was the sensitization of the Mayor of Carhuaz, to become him the leader of risk management. Part of this sensitization was to teach him his responsibilities in the Peruvian risk management context and indicating consequences of failure to comply with them. Another important actor was the Secretary of Civil Defense, as the technical expert for risk management. He also leads the Civil Defense Platform, as the umbrella organization endorsing public and private institutions with a role in risk management. This Platform organizes and evaluates the simulation of GLOFs using the information from EWS.

The institutionalization of risk management was to anchor every action, function and responsibility related to risk management in local law, regulations and management plans both in the Municipality and schools.

As a result, more than 8000 people found tranquility to be found in areas of low risk, while others know the risks they face and how to respond. The studies conducted are now part of land management of the Municipality of Carhuaz. The success of these results was only possible by combining comprehensively scientific and local knowledge and the involvement of people.

Finally, the results of the experiences generated in Chucchún catchment can serve as a model to reduce the glacier related risks in other high-mountain regions. Even though the local context and conditions are important and distinct, some of the experiences made at Lake 513 and in the Glaciares Project are of more general relevance. The importance of achieving political leadership belongs to these experiences. Knowledge can be generated, capabilities can be developed and enhanced, but missing commitment by the local government commitment can jeopardize all these efforts. The institutional policy aspects are therefore among the key elements. Overall, the experiences made at Lake 513 in the Chucchún catchment clearly show that technical, institutional and social aspects of risk management and early warning need to be comprehensively approached in an integrated and concerted way to achieve full effect. Furthermore, in view of ongoing climate change and effects on glaciers and water resources and local communities a stronger emphasis should be put on approaches that combine management of risks and water resources to achieve multiple benefits.

Acknowledgements Studies, activities and experiences described in this paper are part of the Glaciares Project, funded by the Swiss Agency for Development and Cooperation (SDC). Thanks to the Municipality of Carhuaz due their collaboration, to the Glaciological Unit (UGHR) for their technical and scientific support, and CARE Perú as the responsible of implementation of the Glaciares Project. We are furthermore thankful for the thorough and constructive review by Jan Klimeš.

References

Ames, A. (1998). A documentation of glacier tongue variations and lake development in the Cordillera Blanca, Peru. *Zeitschrift für Gletscherkunde und Glazialgeologie, 34*(1), 1–36.

ANA. (2014). *Inventario Nacional de Glaciares y Lagunas*. Huaraz: Autoridad Nacional del Agua.

Bulmer, M. H., & Farquhar, T. (2010). Design and installation of a prototype geohazard monitoring system near Machu Picchu, Peru. *Natural Hazards and Earth System Sciences, 10*, 2031–2038.

Carey, M., Huggel, C., Bury, J., Portocarrero, C., & Haeberli, W. (2012). An integrated socio-environmental framework for glacier hazard management and climate change adaptation: Lessons from Lake 513, Cordillera Blanca, Peru. *Climatic Change, 112*, 733–767.

Drenkhan, F., Carey, M., Huggel, C., Seidel, J., & Oré, M. T. (2015). The changing water cycle: Climatic and socioeconomic drivers of water-related changes in the Andes of Peru. *Wiley Interdisciplinary Reviews Water, 2*, 715–733.

French, A., Barandiarán, J., & Rampini, C. (2015). Contextualizing conflict. Vital water and competing values in glaciated environments. In C. Huggel, M. Carey, J. J. Clague, & A. Kääb (Eds.), *The high-mountain cryosphere* (pp. 315–336). Cambridge: Cambridge University Press.

Frey, H., García-Hernández, J., Huggel, C., Schneider, D., Rohrer, M., Gonzales Alfaro, C., Muñoz Asmat, R., Price Rios, K., Meza Roman, L., Cochachin, A., & Masias, P. (2014). An early warning system for lake outburst floods of the Laguna 513, Cordillera Blanca, Peru. In *Proceedings of the international conference on the analysis and management of changing risks for natural hazards*. Padua, Italy, 2014.

Gonzales, C., Ocaña, D., Marlene, A., Valverde, W., & Huamán, Z. (2012). *Análisis de capacidad y vulnerabilidad climática en la subcuenca Chucchún*. Carhuaz: CARE Perú, Proyecto Glaciares 513.

Huggel, C. (2009). Recent extreme slope failures in glacial environments: Effects of thermal perturbation. *Quaternary Science Reviews, 28*, 1119–1130.

Huggel, C., Khabarov, N., Obersteiner, M., & Ramírez, J. M. (2010). Implementation and integrated numerical modeling of a landslide early warning system: A pilot study in Colombia. *Natural Hazards, 52*, 501–518.

INDECI. (2004). *Mapa de peligro, plan de uso del suelo y medidas de mitigación ante desastres Ciudad de Carhuaz*. Instituto Nacional de Defensa Civil.

IPCC. (2014). Climate change 2014: Impacts, adaptation, and vulnerability. Part A: Global and sectoral aspects. Contribution of Working Group II to the Fifth Assessment Report of the Intergovernmental Panel on Climate Change. Cambridge/New York: Cambridge University Press.

Kershaw, J., Clague, J., & Evans, S. (2005). Geomorphic and sedimentological signature of a two-phase outburst flood from moraine-dammed Queen Bess Lake, British Columbia, Canada. *Earth Surface Processes and Landforms, 30*(1), 1–25.

Mark, B., & Seltzer, G. (2005). Evaluation of recent glacier recession in the Cordillera Blanca, Peru (AD 1962–1999): Spatial distribution of mass loss and climatic forcing. *Quaternary Science Reviews, 24*, 2265–2280.

PCM. (2014). *Plan Nacional de Gestión del Riesgo de Desastres PLANAGERD 2014–2021*. Lima: Presidencia del Consejo de Ministros – Perú.

Portocarrero, C. (2013a). *Mapas de vulnerabilidad y riesgo de la subcuenca Chucchún ante la posible ocurrencia de un proceso aluviónico procedente de la Laguna 513*. Huaraz: CARE Perú, Proyecto Glaciares.

Portocarrero, C. (2013b). *Technical report: The glacial lake handbook, reducing risk from dangerous glacial lakes in the Cordillera Blanca, Peru*. Washington, DC: USAID.

Racoviteanu, A., Arnaud, Y., Williams, M., & Ordoñez, J. (2008). Decadal changes in glacier parameters in the Cordillera Blanca, Peru, derived from remote sensing. *Journal of Glaciology, 54*(186), 499–510.

Reynolds, J., Dolecki, A., & Portocarrero, C. (1998). The construction of a drainage tunnel as part of glacial lake hazard mitigation at Hualcán, Cordillera Blanca, Peru. In M. Eddleston & J. G. Maund (Eds.), *Geohazards in engineering geology* (pp. 41–48). London: The Geological Society.

Schneider, D., Frey, H., García, J., Giráldez, C., Guillén, S., Haeberli, W., Huggel, C., Rohrer, M., Salzmann, N., & Schleiss, A. (2013). *Línea base de la cuenca del río Chucchún (Ancash) Mapeo y modelamiento de amenazas*. Huaraz: Proyecto Glaciares.

Schneider, D., Huggel, C., Cochachin, A., Guillén, S., & García, J. (2014). Mapping hazards from glacier lake outburst floods based on modelling of process cascades at Lake 513, Carhuaz, Peru. *Advances in Geosciences, 35*, 145–155.

Tarazona, E., Ocaña, D., Gonzales, C., Gonzales, R., & Altamiza, K. (2013). *Balance hidrológico de los usos agrícolas y poblacionales de la subcuenca del río Chucchún*. Huaraz: CARE Perú, Proyecto Glaciares 513.

Vuille, M., Francou, B., Wagnon, P., Juen, I., Kaser, G., Mark, B., & Bradley, R. (2008). Climate change and tropical Andean glaciers: Past, present and future. *Earth-Science Reviews, 89*, 79–96.

Zapata, L. (2002). La dinámica glaciar en las lagunas de la Cordillera Blanca. *Acta Montana (Czech Republic), 19*(123), 37–60.

Climate Change Adaptation in the Carpathian Mountain Region

Saskia Elisabeth Werners, Sándor Szalai, Henk Zingstra, Éva Kőpataki, Andreas Beckmann, Ernst Bos, Kristijan Civic, Tomas Hlásny, Orieta Hulea, Matthias Jurek, Hagen Koch, Attila Csaba Kondor, Aleksandra Kovbasko, M. Lakatos, Stijn Lambert, Richard Peters, Jiří Trombik, Ilse van de Velde, and István Zsuffa

Abstract The Carpathian mountain region is one of the most significant natural refuges on the European continent. It is home to Europe's most extensive tracts of montane forest, the largest remaining virgin forest and natural mountain beech-fir forest ecosystems. Adding to the biodiversity are semi-natural habitats such as hay meadows, which are the result of centuries of traditional land management. Like other mountain regions areas, the Carpathian mountain region provides important ecosystem goods and services such as water provision, food products, forest products and tourism. But these ecosystem services are feared to be under threat from climate change.

This chapter reports on climate trends, impacts and adaptation options. Analysis of climate trends show an increase in annual mean temperature of 1.1–2.0 °C over the last 50 years (1961–2010), further increasing by 3.5–4.0 °C towards the end of the century. Precipitation changes are dispersed with an increase of 300–400 mm in the north and decrease of 100–150 mm in the south regions. Summer precipitation is projected to reduce by 20 %, whereas winter precipitation is projected to increase in most areas by 5–20 % by the year 2100. Both future scenarios and observations

S.E. Werners (✉)
Wageningen University and Research Centre, Wageningen, The Netherlands
e-mail: saskia.werners@wur.nl

S. Szalai
Szent Istvan University, Gödöllő, Hungary

H. Zingstra
Zingstra Water and Nature, Wageningen, The Netherlands

É. Kőpataki • A.C. Kondor
Aquaprofit Ltd, Budapest, Hungary

A. Beckmann • O. Hulea • A. Kovbasko
WWF DCP, Bucharest, Romania

E. Bos
LEI-Wageningen-UR, Den Haag, The Netherlands

N. Salzmann et al. (eds.), *Climate Change Adaptation Strategies – An Upstream-downstream Perspective*, DOI 10.1007/978-3-319-40773-9_5

show high spatial variability and uncertainty. The same holds for the impacts on the investigated sectors water resources, forests, wetlands, grasslands, agriculture and tourism.

The review of climate trends and adaptation options, inspired a strategic agenda on adaptation to be implemented under the regional Carpathian Convention. Planning for climate change adaptation benefits from transnational cooperation because many impacts relate to seasonal and geographical shifts across borders. This is true for the natural system (e.g. shifts in species distribution and snow cover) as well as for socio-economic activities like agriculture, forestry and tourism (e.g. shifting opportunities for growing crops and changes in the tourist season). Examples of adaptation exist, yet need to be communicated for wider adoption. Essential components of adaptation will be capacity building and information sharing, climate-proofing of infrastructure and investments, promotion of eco-system based adaptation measures and making biodiversity management more dynamic.

Keywords Carpathian region • Climate change • Water • Forest • Grassland • Wetland • Agriculture • Tourism • Impact • Adaptation • Vulnerability

---

K. Civic
ECNC, Tilburg, The Netherlands

T. Hlásny
National Forest Centre – Forest Research Institute, Zvolen, Slovakia

M. Jurek
UNEP Vienna Office, Vienna, Austria

H. Koch
Potsdam Institute for Climate Impact Research, Potsdam, Germany

M. Lakatos
Hungarian Meteorological Service, Budapest, Hungary

S. Lambert
Arcadis, Brussels, Belgium

R. Peters
Arcadis, Brussels, Belgium

Swiss Federal Research Institute for Forest, Snow and Landscape Research (WSL), Zürcherstrasse 111, CH-8903 Birmensdorf, Switzerland

J. Trombik
Czech University of Life Sciences, Prague, Czech Republic

I. van de Velde
Ecorys, Rotterdam, The Netherlands

I. Zsuffa
VITUKI Hungary Ltd., Budapest, Hungary

1 Introduction

The Carpathian mountain region covers an area of about 210,000 km^2. It is one of the three most extensive mountain system in Europe besides the Alps and the Scandinavian mountains. The Carpathian region is shared by seven Central and Eastern European countries,[1] five of which are member of the European Union (Fig. 1). The Carpathian region support Europe's most extensive tracts of montane forest (between the heights of 950 and 1350 m above sea level), the largest remaining natural mountain beech and beech/fir forest ecosystems and the largest area of virgin forest left in Europe. No less than one-third of all European vascular plant species can be found in this region – 3988 plant species, 481 of which are found only in the Carpathians. A bridge between Europe's northern and south-western forests, the range serves as a corridor for the dispersal of plants and animals throughout Europe. The mountains are home to Europe's largest populations of brown bears, wolves, lynx, European bison and rare bird species including the globally threatened imperial eagle. Next to the richness of natural diversity, the semi-natural habitats such as montane pastures and hay meadows, which are the result of centuries of traditional management of the land, are of great ecological and cultural importance. WWF included the Carpathian region in the "Global 200" worldwide list of eco-regions noted for exceptional levels of biodiversity.

The Carpathian region is also a major source of freshwater, delivering water to the Danube, Dniester and Vistula river. In addition, the wider Carpathian region,

Fig. 1 The Carpathian region covers territory in seven countries (Source: GRID-Arendal)

[1] Czech Republic, Hungary, Republic of Poland, Romania, Republic of Serbia, Slovak Republic and Ukraine.

including forelands, is home to millions of people, ranging from small communities to middle size urban centres (Csagoly 2007).

This unique living environment is threatened by natural and human impacts, such as land abandonment, habitat conversion and fragmentation, deforestation, pollution and -more recent- climate change. Worries about the impacts of climate change have initiated a series of projects. This chapter synthesises findings and recommendations about climate change impacts and adaptation from three linked research projects CARPATCLIM, CARPIVIA and CarpathCC, called for by the European Parliament and funded by the European Commission. CARPATCLIM harmonized climate data from 1961 to 2010 and made available a gridded database (for data and data harmonisation see www.carpatclim-eu.org). CarpathCC provided a series of vulnerability studies and evaluated adaptation measures. CARPIVIA assessed the vulnerability of the Carpathian region's main ecosystems and ecosystem-based production systems and proposed adaptation options.

This chapter is based on project reports, in particular the final report of the CARPIVIA project (Werners et al. 2014b) and the summary report for policy makers (Werners et al. 2014a). Section 2 discusses past climate trends and future projections. Sections 3 and 4 present impacts and proposed adaptation measures respectively for the sectors Water resources, Forests and Forestry, Wetlands, Grasslands, Agriculture and Tourism. Conclusions are offered in Section 5.

2 Climate Change in the Carpathian Region

2.1 Historic Climate Trends

Temperature and precipitation are changing throughout the Carpathian region (Szalai 2012). The average annual temperature has increased, particularly in the summer. Over the last 50 years, the strongest increase was observed in the west and east, and in the lower regions, where the warming was between 1.1 and 2.0 °C. Higher elevations saw less change. Figure 2 shows the seasonal temperature changes between 1961 and 2010. Whereas most warming is seen in summer (1.0–2.4 °C) warming was less pronounced during winter, when temperature increase was less than 0.4 °C everywhere. Some areas even show a slight cooling during the investigation period.

Compared to temperature, changes in precipitation have a more mosaic pattern with large spatial variation. The main trend is decreasing precipitation in the western and south-eastern parts of the region and an increase in the north, especially in the north-east. Seasonally there are also large spatial differences (Fig. 3). Wetter and dryer areas are observed in each season, but overall increasing precipitation is found in winter and summer, while a decrease occurs in spring.

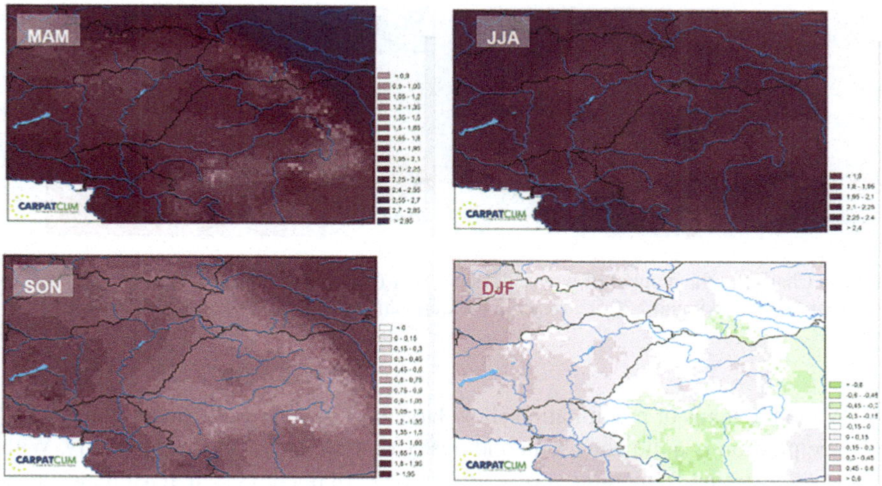

Fig. 2 Seasonal temperature changes, 1961–2010 (spring *upper left*, summer *upper right*, autumn *lower left*, winter *lower right*)

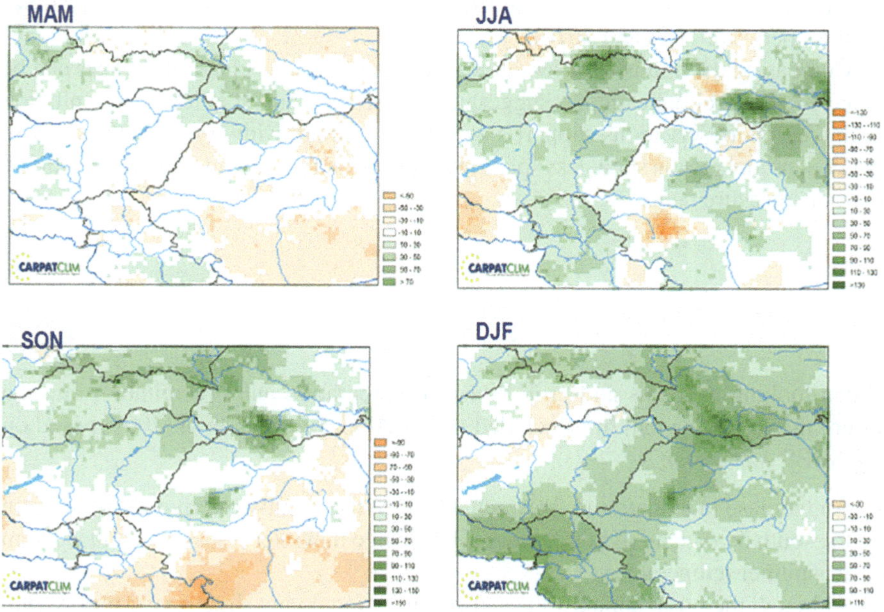

Fig. 3 Change in seasonal precipitation 1961–2010 (spring *upper left*, summer *upper right*, autumn *lower left*, winter *lower right*)

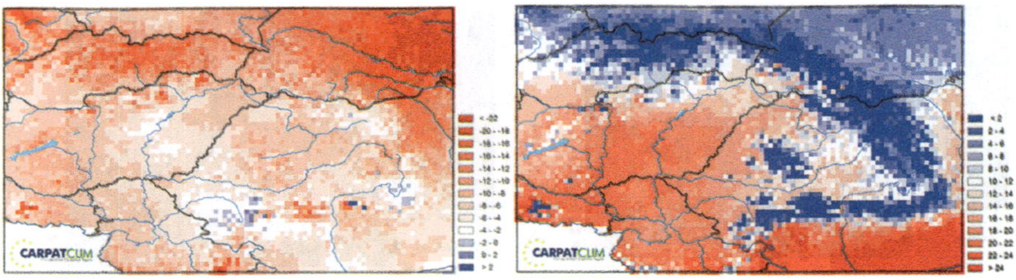

Fig. 4 Change in the number of winter days per year (daily maximum <0 °C, *left*) and hot days per year (daily maximum ≥30 °C, *right*) in the Carpathian region in the period of 1961–2010

While the maps of precipitation show large spatial variability, the intensification of precipitation is quite consistent and heavy rainfall occurs more frequently. With respect to extreme temperature events, the number of hot days is increasing, whereas extreme cold days are decreasing. Figure 4 shows that the number of winter days decreased everywhere in the Carpathian region with very few exceptions. The greatest decline can be seen in the north-west Carpathians (a reduction of 18–20 days between 1961 and 2010). The change in the number of hot days correlates strongly with topography – fewer hot days are seen at higher levels in the mountains than at lower altitudes. The largest increase in hot days is found in the south and east (over 24 days between 1961 and 2010).

Comparing the trends in observations and model results, clear differences can be detected. While climate models suggest a north-south gradient of lower precipitation in the region, observations support a west-east gradient, with drying in the west. In general, model results show less spatial variability and present more unified patterns than the measurements.

2.2 Future Trends

When looking at trends and projections of temperature and precipitation into the future the following can be concluded (Láng 2006; Bartholy et al. 2007, based on the SRES scenarios):

- An increase in temperature between 3.0 °C in the north-western part to 4.5 °C in the south by the end of this century. The change in winter is less pronounced than in summer temperature.
- Although the mean annual values of precipitation do not show a very clear trend, reductions in summer precipitation are projected of over 20 % and increases in winter precipitation in most areas of between 5 % and 20 % by the year 2100. A shortened period of snowfall is projected.
- An increase in maximum daily precipitation and in the number of days with over 20 mm precipitation (Fig. 5).

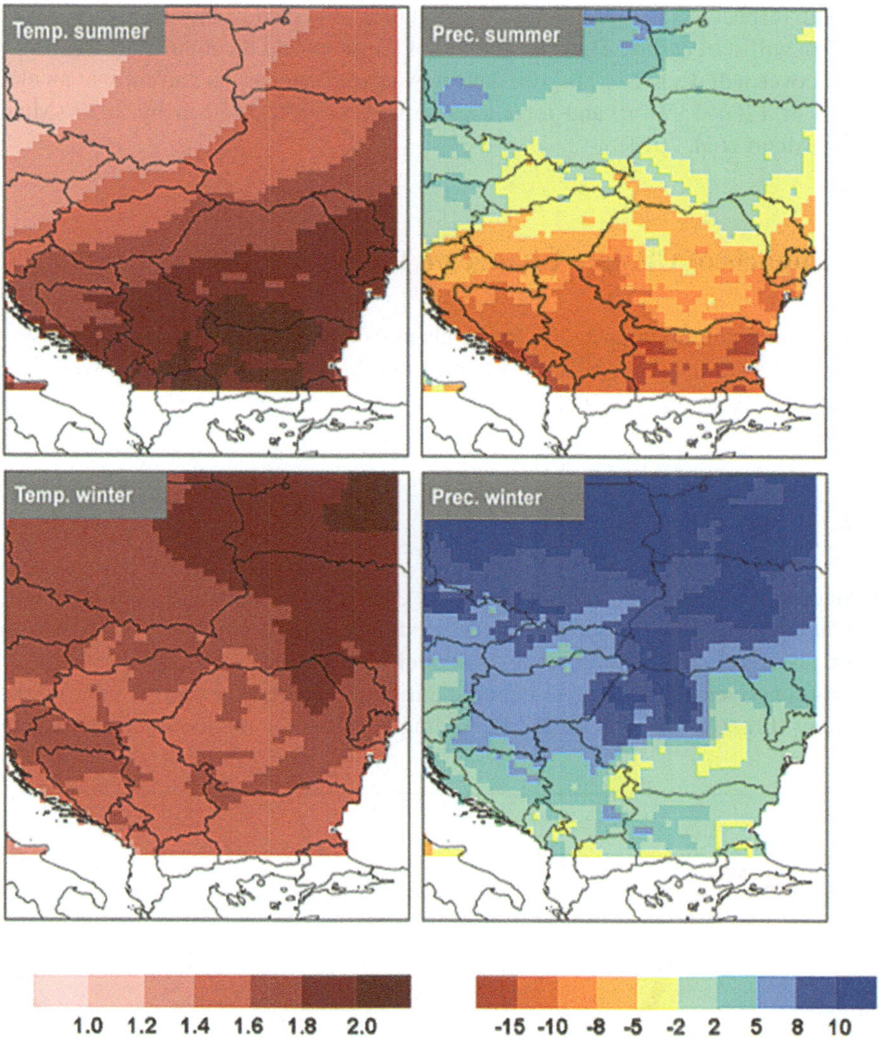

Fig. 5 Changes in daily mean air temperature (°C) (*left*) and precipitation (%) (*right*) in the greater Carpathian region in winter (DJF) and summer (JJA) as the multi-model mean for the years 2021–2050 relative to 1971–2000 (absolute differences in mm), for the A1B greenhouse gas emissions scenario with 14 different GCM-RCM combinations from the ENSEMBLES project (Albert et al. 2013)

These changes will have a substantial impact on the occurrence of severely dry years. The lowest impact on drought is projected for the higher parts of the north-western and north-eastern Carpathians, while the lower elevated inner skirt of the mountain ranges, the basin area and the Southern ranges of the Carpathian region are most likely to suffer more from drought events (Albert et al. 2013).

With respect to snow, the period of snowfall is projected to shorten and the snow melting will start earlier. Thus the amount of the snow and the number of days with snow cover will decrease. The 100 day snow cover boundary is currently at an elevation of 1250–1350 m, and forecasted to rise to 1350–1450 m by 2050 (Micu 2009; Albert et al. 2013).

3 Impacts of Climate Change on the Key Sectors

This chapter present impacts of climate change on the key ecosystems and ecosystem based socioeconomic sectors: Water resources, Forests and Forestry, Wetlands, Grasslands, Agriculture and Tourism.

3.1 Impacts on Water Resources

Reduced snow cover, heavy rains caused by climate variability, and changes in precipitation patterns are projected to alter flood regimes and increase the risk of both river and flash floods. According to model-based projections, the discharge of Carpathian rivers is expected to increase during the winter and decrease during the summer (Fig. 6) (Albert et al. 2013). Adding to this is the reduced water buffering

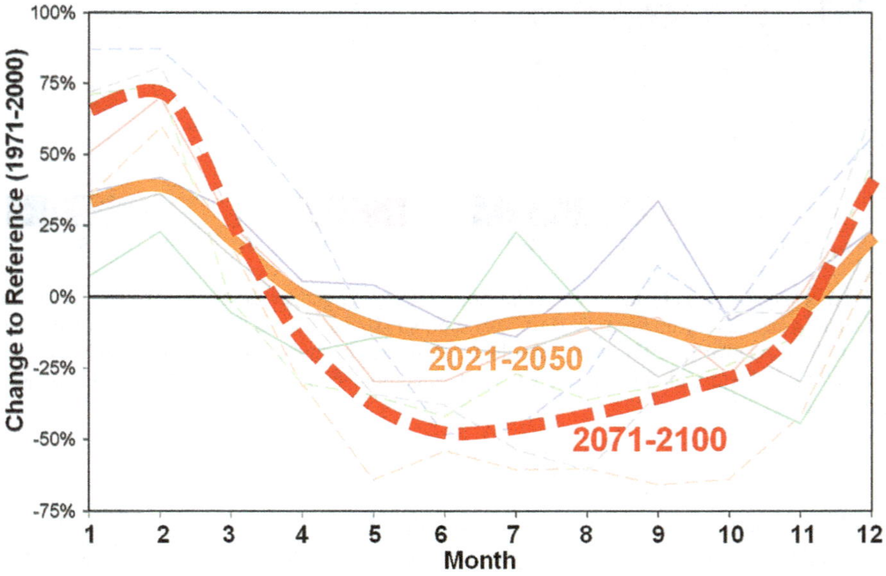

Fig. 6 Expected changes in monthly mean discharges of a large Carpathian river (Mures) – average of climatic scenario-based hydrological model projections (using alternative climate scenarios from the ENSEMBLES Project (thin lines)) (Zsuffa et al. 2013)

effect of snow (Kostka and Holko 2004; Peters et al. 2013). Overall, a decline in annual river discharge is predicted for the southern and eastern parts of the Danube basin, while western and northern parts might experience increases. In particular, southern parts of Hungary and Romania, and the Republic of Serbia are expected to face severe droughts and water shortages. Extreme high precipitation over a short period of time will add to erosion and the risk of land slides. There may be impacts on the quantity and quality of drinking water for communities dependant on mountain streams.

Annual water temperature is projected to change by 1.7 °C between the periods 1971–2000 and 2021–2050. Projections of average water temperature change in the summer months reach 4 °C and above. In addition, the number of days with extreme water temperature (>28 °C) increases. This would have definite impacts on aquatic ecosystems. There is clear spatial variation in the identified climate-induced trends in thermal and runoff conditions, and consequently in the impacts on aquatic ecosystems. While the northern part of the Carpathian region is insignificantly or moderately affected, the southern part is expected to be highly affected. As opportunities for adaptation are relatively well shared yet depend on financial resources, the southern (sub-)basins are expected to be the most vulnerable as well (Ignjatovic et al. 2013).

3.2 Impacts on Forestry

Recently, forest damage in the Carpathian region has increased. Wind damage followed by insect pest outbreaks, outbreaks of defoliating insects as well as the effects of drought have been observed to compromise the stability of Carpathian forest ecosystems and the sustainability of forest ecosystem services. Climate change is expected to make this situation worse although interactions between climate, forest disturbances and forest management are not yet thoroughly understood (Barcza et al. 2013).

The actual impact of the predicted climate changes depends on a variety of factors including tree species composition, forest structure, forest management and elevation. In general the lower elevation forests, mainly in south Slovakia, Hungary, Romania and Serbia are especially prone to drought and temperature rise. Once the health situation of the forest is impacted by droughts and windstorms the possible outbreak of bark beetles and defoliating insects are increasing along with the occurrence of new pest species (e.g. Northern spruce bark beetle throughout Romania) (Korpeĺ 1995; Szewczyk et al. 2011; Barcza et al. 2013).

Another impact of climate change and especially of increasing temperatures will be an upward moving treeline and changes in the tree composition and productivity (Hlásny et al. 2011). In addition, intensive forest management with clear cutting in combination with an increase in peak precipitation events may increase the risk of landslides (Barcza et al. 2013).

3.3 Impacts on Wetlands

Little research exists on the effects of climate change on Carpathian wetlands. Most reported are the effect of increasing temperatures and precipitation changes (Barcza et al. 2013). Increased temperatures can lead to drying out of wetlands, compounded by higher incidence of drought. If precipitation declines and groundwater is extracted for human needs, shallow and temporary areas that often harbour rare species, can be lost entirely. In addition, climate change will affect the carbon cycle and the emission and uptake of greenhouse gasses by wetlands.

The most vulnerable wetland habitats are peat lands because they have limited resilience to climate variability and are sensitive to human activities and changes in land use. Less vulnerable are halophytic habitats (where plants are adapted to saline soils), steppes and marches. These habitats can adapt to climate fluctuations, yet are highly sensitive to human activities and changes in land use. The lowest vulnerability is found in habitats already subjected to regular flooding, subterranean wetlands and some riverbank and water habitats. They are most likely to be able to cope with even more extreme fluctuations in climate. However, human intervention can represent important threats also in this case (Barcza et al. 2013).

3.4 Impacts on Grasslands

Typical for the Carpathian region are traditionally managed multifunctional grasslands, 'green-veined' by hedges, woodland, forests and watercourses. Such landscapes are rich in biodiversity and provide numerous ecosystem services, including food (milk, cheese, meat) and wool, clean water, mitigation of climate change by absorbing greenhouse gases, pollination, biodiversity conservation, recreation and aesthetic and cultural values (Vandewalle et al. 2010).

Carpathian grasslands host 40 species listed in the EU Habitat Directive, 11 of which are endemic. In total, 19 grassland habitat types of European importance occur in the Carpathian region. However, these landscapes and grasslands are under serious pressure. Most significant are land abandonment, forestry and intensified agriculture but also climate change (Peters et al. 2013).

The impacts of climate change on grasslands are both direct and indirect. Direct because of changes in CO_2 concentration, in mean temperature, precipitation and extreme weather conditions. Indirect, through the climbing tree line, which adds to encroachment together with the abandonment of the grasslands by farmers (UNDP 2005). Changes in species composition occur because of the appearance of 'new' species rather than the intolerance of 'original' grassland species to climate change. As the changes proceed, species diversity may increase in the first years (when the 'old' and 'new' species are present), but thereafter decreases as new species take over habitats. Grasslands on calcarous substrate – the most species rich habitats – are found to be more sensitive, thus more threatened, than grasslands on other substrates. Nardus grasslands for example, are less sensitive, than Festuco-Brometelia grasslands. Yet, it is very difficult to make accurate projections as drivers interact with each other and can reinforce or counteract specific impacts. Next to abandonment, overgrazing remains a threat to the protection of grassland biodiversity (Peters et al. 2013).

3.5 Impacts on Agriculture

Agricultural land covers over one third of the Carpathian region and the proportion of the rural population working in agriculture can be as high as 50 %. The structure of the agricultural sector is rapidly changing; overall crop and livestock production is declining and abandoned cropland lies fallow. This trend is expected to continue (Kuemmerle et al. 2007).

Next to the changes in the agricultural sector caused by socio-economic processes, agriculture will be impacted by climate change through changing precipitation patters, changes in temperature, extreme events and changes in the length of seasons. As a result agriculture may become feasible at higher altitudes. In some parts of the Carpathians maize and wheat yields will decline, while elsewhere sunflower and soya yields might increase due to higher temperatures and the possibilities to grow these crops more northward. Likewise, winter wheat is expected to increase. In general a shift towards winter crops will be possible (Halenka 2010).

Unfortunately, vulnerability to pests is predicted to rise, and productivity losses are expected due to soil erosion, groundwater depletion, and extreme weather events. Deeper analysis of socio-economic trends is necessary to identify the most vulnerable areas in the Carpathian region but preliminary results show that small-scale farmers in remote villages in Romania and Serbia could be among the most vulnerable. Pastures region are especially vulnerable through the combined impacts of climate change and socio-economic dynamics (see also under grasslands) (Werners et al. 2014b).

3.6 Impacts on Tourism

Changes in the tourism industry depend more on the general state of the region's economy than on climate change, at least in the short term. Climate change has both positive and negative impacts. Ecotourism, summer tourism, health tourism and vocational tourism can be positively influenced by climate change. Rising temperatures can bring more tourists to the mountains. On the other hand, the possibilities of winter sport will become more limited because the amount of snow and the number of days with snow cover will decrease in the area. However, snow cover and snow depth changes will not impact tourism as much as formerly supposed. This is because the Carpathian region has a wide range of tourism activities and winter sports are not as important as, for example, in the Alps. Due to the lower altitude, ski resorts appear better prepared for changing snow conditions. All in all it is estimated that climate change can bring 60–75,000 additional tourists per year with 9.6–12 million EUR additional revenue for the region (approx. 1 % of the total revenue from tourism), with substantial differences between regions and activities (Peters et al. 2013).

4 Adaptation Measures

Adaptation measures have been reviewed for the ecosystems and sectors discussed in the previous section, i.e. water resources, forestry, wetlands, grasslands, agriculture and tourism. Possible measures were identified from literature and policy review. Measures were selected in consultation with stakeholder groups, taking into account costs and benefits and long-term sustainability (Arany et al. 2013; Werners et al. 2014b). The most promising measures are reported in this section.

4.1 Adaptation Measures for Water Resources

One of the most efficient adaptation measures against the combined threat of droughts and floods in the Carpathian mountains is local water storage. Where it adds to recharge aquifers, it also delays discharge onto rivers compared to discharge through (surface) drainage systems. In addition, water storage measures can reduce surface erosion and counteract the desiccation of forests.

Storage capacity can be increase by blocking (old) drainage canals which were dug in the past but which often do not serve a purpose anymore. Eliminating road networks can also encourage storage, especially in the Eastern Carpathians. Intensively used dirt roads act as drains accelerating runoff and causing local erosion problems. However, eliminating roads necessitates adjusting land use. Activities requiring frequent transportation (e.g. hay production) have to be replaced by transportation-free uses, such as grazing or nature conservation (Zsuffa et al. 2013). (Re)creating wetlands and ponds increases storage capacity and allows to harvest rainwater (Kravcík et al. 2007) (See Fig. 7). Structural measures such as constructing dams, water tanks and subsurface reservoirs also help. However, dam construction has to be carefully planned so as not to damage river ecosystems.

Sub-surface water storage can be enhanced by protecting and restoring natural grasslands so more rainwater can infiltrate into the deeper soil layers. This land use measure is especially recommended for the karstic systems in the Carpathian region, where grasslands are the primary sources of water supply for the sub-surface water resources (Zsuffa et al. 2013).

A clear legal framework is crucial. The Water Framework Directive (2000/60/EC) together with the Flood Risk Directive (2007/60/EC) are important vehicles to

Fig. 7 Example of small-scale on-farm water storage (Location: Nyzhnye Selyshche, Transcarpathia, Ukraine, Photo: Saskia Werners)

streamline climate change adaptation activities for water resources (Ignjatovic et al. 2013). Adaptation measures can be an integral part of river basin management plans. Such adaptation measures could include non-technical actions such as floodplain restoration, afforestation of selected catchment areas, adjustment of permits for water abstraction and use and pollution discharge. Other actions include setting up warning systems and awareness programmes. Other mountainous areas like the Alps provide lessons in increased efficiency of water use, infiltration and water saving.

4.2 Adaptation Measures for Forests and Forestry

The Carpathian countries have limited capacity to take measures to help forests and forestry adapt to climate change. None of them has yet directly addressed climate change in its forestry legislation (although the issue is usually included in national strategies). Adaptive capacity related to socio-economic development is substantially lower in the Romanian and Serbian part of the Carpathian region compared to the Western Carpathians.

Adaptation should be geared to practical forest management and legislation, and ensuring that risk assessment in forest management planning. This is becoming increasingly important and there is a need to change the traditional timber production-oriented management towards an adaptive risk-responsive management. Adaptive forest management uses concepts such as continuous-cover-forestry and

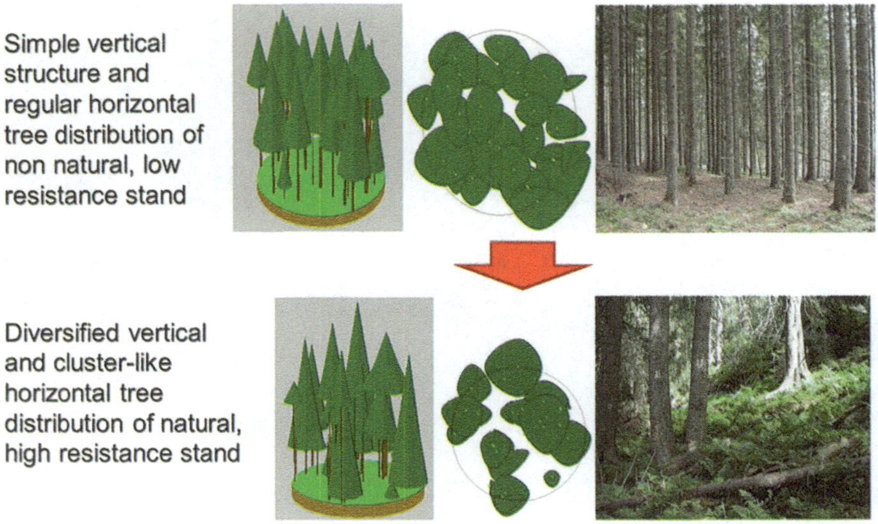

Fig. 8 An extended cutting regime leads to adaptation of forest structure (Source: CarpathCC Project presentation)

close-to-nature forestry to increase adaptive capacity of forests and lower anticipated risks (see Fig. 8). It supports drought tolerant species, for example oaks, and reduces the number of vulnerable water demanding conifers and beech at lower elevations (Lindner et al. 2010; Hlásny et al. 2014).

Another important line of action is to consolidate and harmonize forest monitoring systems, in order that they can provide timely information to support adaptive forest management. This includes the trans-national monitoring of pests and diseases.

At a landscape scale awareness is to be created of the indispensable role of forests in integrated watershed management, particularly for biodiversity, water regulation and erosion control. Landscape level policies are required to avoid forest fragmentation and maintain the connectivity of larger forest areas to support species' natural migration and gene flows.

4.3 Adaptation Measures for Wetlands

Adaptation strategies for wetlands are closely linked to measures for making hydrological systems more resilient. Maintenance and restoration of wetlands in higher altitudes play an essential role in increasing retention capacity and reducing impacts of droughts and excess precipitation. Floodplain restoration, including the re-creation of wetlands, will restore important functions such as water purification, nutrient retention and will be a buffer for droughts and floods. Thus protection of wetlands is a "no-regrets" strategy even in times of high uncertainty about the extent and location of specific impacts of climate change. Wetland protection needs to be integrated with flood control practices and support programmes aiming for wetland and peat land restoration, floodplain rehabilitation and creation of new wetlands and lakes.

In places where wetland restoration is difficult, it is highly recommended to reduce external non-climate pressures such as land-use change and pollution. Improving connectivity between wetlands and water bodies can help species to move, as well as preserve habitat heterogeneity and biodiversity, which can provide genetic diversity for successful adaptation. As information on wetlands is scarce a priority action is also to monitor the state of waters and aquatic ecosystems.

Box 1: Adaptation Example: Maintenance of Wetlands' Riparian Forests
Forest is the natural vegetation along streams and wetlands. Near-natural riparian forests have virtually disappeared from Central Europe as many riparian forests have been cleared and transformed into pasture. Riparian forests have high recreational value, store water, offer flood protection and improve groundwater quality. Maintenance of the riparian forest is discussed for the wetland in Divici Pojejena, Iron Gates national park, Romania. Covering an area of about 55 ha, total project cost are estimated 56,000 euro. Benefits include: reducing nutrients and pollutants, erosion control, species diversity and offering a wintering and nesting habitat for birds such as the pygmy cormorant and ferruginous duck (protected species).

Source: Arany et al. (2013)

4.4 Adaptation Measures for Grasslands

An important measure for grasslands is to support small scale farming. Agro-environmental programmes can be a tool for the development of local green economies by paying farmers for the ecosystem services they produce. Agro-environmental programmes aim to harmonize relations between the production of food and environmental conservation. A parallel goal is to strengthen village communities. The general approach is to issue compensation payments for environmental friendly management practices, which maintain and increase the biodiversity of agricultural areas, and enhance the protection of soil against degradation, especially erosion. This requires among others location-specific solutions for unresolved property rights, especially for grasslands used for common grazing.

After abandonment most grasslands gradually turn back into forest. Removing trees (for instance Robinia) and shrubs should be the first priority in restoration but this is costly and can only been done with financial support from for instance agri-environment measures or Life projects when the area is designated as a N-2000 site.

Fig. 9 Traditional grassland management (Location: Huklyvyi, Transcarpathia, Ukraine, Photo: Saskia Werners)

Additionally targeted management such as grazing and mowing will further enhance the development towards a divers grassland habitat type (Fig. 9).

Box 2: Adaptation Example: Restoration of Degraded Grasslands

The Bükk region in Hungary (part of the Vár-Hegy-Nagy-Eged Natura 2000 Habitat Directive Site) is restored by the KEOP project. Before the start of the restoration project in 2012 the area was essentially an abandoned grassland overgrown by shrubs. The KEOP project restores this site to mowed grassland with fruit trees. The interventions started in 2012 by manually removing bushes and shrubs (costs 1400 euro/ha). This was followed by the mechanized crushing of stalk left in the soil (costs 340 euro/ha). Remaining grasslands are preserved by mowing on a regular basis (grazing will not be allowed because of drinking water wells downstream in the valley) (initial costs during 3 years period 500 euro/ha). In addition, fruit trees are planted. Traditional, autochthonous (endemic) fruit trees are selected, which are more resistant against climate stresses than new breeds. The National Park purchased a total area of 4.2 ha within the frame of the KEOP project for 1700 euro/ha. The costs of planting fruit trees (apple: 159, quince: 494, cherry: 2, pear: 28, plum: 112): 3745 euro. The estimated annual yields of these fruit trees is 38,000 euro. Other benefits include water infiltration and provisioning, improved pollination capacity, species diversity and touristic value.

Source: Arany et al. (2013)

4.5 Adaptation Measures for Agriculture

Farmers are used to adapt to changing conditions. For small-scale farmers potential adaptation options can include changes in sowing dates and crop varieties, improved water-management and irrigation systems, adapted plant nutrition, protection and tillage practices. To achieve the broader goal of sustainable agriculture and rural development in a changing climate, policies should support farmers who are looking to adapt. The current economic market model leaves small-scale traditional farms, typical for the Carpathian region, at a disadvantage. Farming activities such as grazing on high altitude grasslands, are economically no longer feasible. Farmers need technical and financial support for example through agri-environment measures to maintain their activities and to avoid grasslands becoming forest.

Given developments in the Carpathian region (including land abandonment, overgrazing, aging population, and limited budgets for government action), promotion is needed of the countryside as an attractive place to live, work and do business. Activities are underway in which non-governmental organizations help local communities to connect to markets. Here, the role of volunteers should not be underestimated because they can aid with reporting, data collection, and with elaborating on and implementing ideas.

Box 3: Adaptation Example: High Nature Value Farming

High nature value farming (HNV) supports farming activity and farmland that has high biodiversity or contains species and habitats of conservation concern. The Romanian Government has implemented a High Nature Value Grassland agri-environment scheme as part of its National Rural Development Plan (NRDP) in an attempt to limit both agricultural abandonment and intensification. Farmers can voluntarily enter into a 5-year agreement and receive payments, currently set at €124 per ha, in return for adhering to a specified set of management requirements, for example no chemical fertilizers. Farmers can also apply for the Traditional Farming option whereby additional payments can be obtained in return for not using any mechanization.

An agri-environment scheme was piloted in the Tarnava Mare region in Romania, an area consisting mostly of small-scale semi-subsistence farms, producing meat, dairy products and honey. The pilot found that, along agri-environment payments, actions were needed that support local markets. These included improvement of milk collection points for small-scale milk producers in the Tarnave region and the creation of a brand for local products. Opportunities were also created for agro-tourism. The adaptation measure thus creates a "multiplier effect" whereby tourism becomes an extra form of "payment" to local people for landscape conservation.

Source: Arany et al. (2013)

4.6 Adaptation Measures for Tourism

In many places the potential for the development of the tourist sector is underutilized and there is a lack of resilience to cope with change or capitalise on opportunities that exist. In the light of adaptation to climate change, it is advised to base the development of tourism on the specific natural beauty and culture of the Carpathians, whilst limiting the development of mass tourism. This implies that tourism development should be integrated into wider planning to continue to diversify resorts and markets and promote sustainable development. Specific actions are to promote year-round, resilient destinations with good accommodations (e. g. wellness and conference hotels), support climate-friendly winter sport projects (e.g. alternative design of ski slopes), and to develop ecotourism, health tourism, active tourism (such as cycling and hiking). In addition, measures are proposed to support the development of tourism information networks in the region involving accommodations, suppliers and tourism organizations. These networks would provide up-to-date information and warnings about conditions relevant for tourism (weather, snow depth, hazards, road and traffic conditions, etc.).

4.7 Strategic Adaptation Planning in the Carpathian Region

Adaptation plans will need to be developed and implemented across national borders and have to be integrated into wider planning for sustainable development of the Carpathian region. Therefore, in close cooperation with the research described above, the Framework Convention on the Protection and Sustainable Development of the Carpathians (*Carpathian Convention*) decided to adopt a "Strategic Agenda on Adaptation to Climate Change in the Carpathian Region".[2] The aim of the agenda is to assist Member States of the Carpathian Convention, local and regional authorities and other stakeholders in formulating responses to climate change as a way to secure sustainable development in the region. The Agenda was endorsed at the Fourth Meeting of the Conference of the Parties to the Carpathian Convention (COP4) 23–26 September 2014.

The Strategic Agenda includes recommendations for policy development, institutional change and ecosystem-based adaptation measures. In summary, it infers that linking different policies of nature conservation, river basin management, and sustainable farming could substantially strengthen the Carpathian region and its resilience to climate change. The added value of increased transnational cooperation and joint activities is especially important when planning for climate change adaptation, since many of the predicted impacts of climate change, such as seasonal changes in temperature and precipitation, will occur over vast geographical areas,

[2] For further information and the full text of the Strategic Agenda, please visit www.carpathianconvention.org

affecting several countries at once. Many of the possible measures are best planned scaled to the eco-region rather than the nation-state. Further, many of the tools and capacities required for climate change adaptation are currently lacking, such as the ability to designate and map future refuge habitats for wetlands and grasslands. This may need to be developed at the transnational level, with the support of externally funded joint initiatives that could fill the gaps and build cooperative capacity. To succeed, the working group on adaptation, working towards implementation of the Agenda, looks for new partnerships between governments, civil society, research and education institutions, the private sector and international organisations.

5 Conclusions and Recommendations

This section offers the main conclusions and recommendations. Water resources are mostly impacted by increasing temperature and winter precipitation, and decreasing summer precipitation. Periods when ecological water demands will not be met will increase, leading to damage to aquatic and riparian ecosystems. At the same time, increasing wintertime flows will likely exacerbate existing flood issues. The impact on water resources shows a clear south-north gradient with higher values in south, modified by topographic, climate and economic factors.

The main impact on forest ecosystems stems from drought events, pests (partly because of the reduced fitness of the trees), low adaptive capacity, and the changing composition of forests. The Carpathian wetlands are very sensitive to natural as well as anthropogenic pressures. The most vulnerable wetland habitats are peat lands because they have limited resilience to climate variability and are sensitive to human activities and changes in land use. Grasslands in the region have strong cultural associations, provide a wide range of ecosystem services and associated economic benefits, and are rich in wildlife and biodiversity. Grasslands in (sub)mountain areas, where management possibilities exist, have lower vulnerability than (sub) alpine grasslands, especially those on calcareous substrate. Climate change has both positive and negative impacts on tourism. Summer tourism will benefit, where changes in snow cover can demote winter tourism.

A strategic agenda on adaptation to climate change was elaborate for implementation under the Carpathian Convention. This agenda underlines linking different policies of nature conservation, river basin management and sustainable farming at regional level. Regional cooperation platforms, like the Carpathian Convention, can be a critical vehicle to mainstream this in member countries as well as for cooperation and information sharing with other mountainous region. The added value of transnational cooperation and joint activities is especially strong when planning for climate change adaptation, as much of the impacts of climate change relate to seasonal and geographical shifts. This is true for species and communities (forests, tree-lines, northern limits) as well as for socio-economic aspects (tourist arrivals, tourism seasons).

Financial resources are limited. A key action is to create flexible and equitable financial instruments that facilitate benefit – and burden-sharing, and that support a diverse set of potentially better-adapted new activities rather than compensate for climate impacts on existing activities. To succeed, new partnerships between government, civil society, the research and education institutions, the private sector and international organisations will be key. Essential components of such partnerships will be capacity building and information sharing, climate-proofing of infrastructure and investments, climate-cross compliance, design of eco-system based adaptation measures, and to make biodiversity management more dynamic.

References

Albert, K., Bálint, G., Baracza, Z., Bakacsi, Z., Chendes, V., Deák, G., Dobor, L., Dobos, E., Farkas, C., Gelybó, G., Hegedűs, A., Hlásny, T., HlavČová, K., Horel, Á., Horvát, O., Ivan, P., Koch, H., Kohnová, S., Lipták, G., Macurová, Z., Matreata, M., Szalai, S., Szolgay, J., Tóth, E., Vágó, J., Zempléni, A., & Zsuffa, I. (2013). In-depth study on the key climate change threats and impacts on water resources (Final report – Module SR1). CarpathCC, report to European Commission – DG Environment http://carpathcc.eu/node/35. REC – The Regional Environmental Center For Central Eastern Europe, AQUAPROFIT, INCDPM, CAR HAS, ARTELIA, Budapest, HU.

Arany, I., Kondor, A., Orosz, B., Kőpataki, É., Adriansens, V., Lambert, S., & Szalai, S. (2013). Insights in costs and benefits of forest, wetlands and non-agricultural grassland (FINAL Report for SR 4 Task 3). CarpathCC, report to European Commission – DG Environment http://carpathcc.eu/node/35. REC – The Regional Environmental Center For Central Eastern Europe, ARCADIS, INCDPM, INHGA, ARTELIA, Budapest, HU.

Barcza, Z., Bodea, L., Csoka, G., Dodor, L., Galvánek, D., György, D., Hidy, D., & Hlásny, T. et al. (2013). In-depth study on the impacts of climate change threats on ecosystems (Final report – Module SR2). CarpathCC, report to European Commission – DG Environment http://carpathcc.eu/node/35. REC – The Regional Environmental Center For Central Eastern Europe, AQUAPROFIT, INCDPM, CAR HAS, ARTELIA, Budapest, HU.

Bartholy, J., Pongrácz, R., & Gelybó, G. (2007). Regional climate change expected in Hungary for 2071–2100. *Applied Ecology and Environmental Research, 5*(1), 1–17.

Csagoly, P. (Ed.) (2007). Carpathian Environment Outlook, UNEP/DEWA – Europe, Geneva, Switzerland.

Halenka, T. (2010). *CECILIA project: Central and eastern Europe climate change impact and vulnerability assessment*. Prague: CECILIA Project, Charles University.

Hlásny, T., Barcza, Z., Fabrika, M., Balázs, B., Churkina, G., Pajtík, J., Sedmák, R., & Turčáni, M. (2011). Climate change impacts on growth and carbon balance of forests in Central Europe. *Climate Research, 47*, 219–236.

Hlásny, T., Mátyás, C., Seidl, R., Kulla, L., Merganičová, K., & Trombik, J. (2014). Climate change increases the drought risk in Central European forests: What are the options for adaptation? *Forestry Journa, 60*, 5–18.

Ignjatovic, J., Vassilev, V., & Koszta, I. (2013). Implementation of the Water Framework Directive and Flood Directive under climate change. Contribution to SR1 – In-depth study on the key climate change threats and impacts on water resources. CarpathCC, report to European Commission – DG Environment http://carpathcc.eu/node/35. REC – The Regional Environmental Center For Central Eastern Europe, Budapest, HU.

Korpeĺ, Š. (1995). *Die Urwälder der Westkarpaten*. Stuttgart: Gustav Fischer Verlag.

Kostka, Z., & Holko, L. (2004). Expected impact of climate change on Snow Cover in a Small Mountain Catchment. In TTL Conference on 'Snow', Vienna University of Technology, Austria.

Kravcík, M., Pokorný, J., Kohutiar, J., Kovác, M., & Tóth, E. (2007). Water for the recovery of the climate – A new water paradigm. People and Water NGO, the Association of Towns and Municipalities of Slovakia, the community help society ENKI and the Foundation for the Support of Civic Activities, Slovakia.

Kuemmerle, T., Hostert, P., Radeloff, V. C., Perzanowski, K., & Kruhlov, I. (2007). Post-socialist forest disturbance in the Carpathian border region of Poland, Slovakia, and Ukraine. *Ecological Applications, 17*(5), 1279–1295.

Láng, I. (2006). The project "VAHAVA", Executive summary. Ministry for the Environment and Water Management (KvVM) and the Hungarian Academy of Sciences (MTA), Budapest.

Lindner, M., Garcia-Gonzalo, J., Kolström, M., Green, T., Reguera, R., Maroschek, M., Seidl, R., Lexer, M. J., Netherer, S., Schopf, A., Kremer, A., Delzon, S., Barbati, A., Marchetti, M., & Corona, P. (2008). *Impacts of climate change on European forests and options for adaptation*. Joensuu: European Forest Institute.

Lindner, M., Maroschek, M., Netherer, S., Kremer, A., Barbati, A., Garcia-Gonzalo, J., et al. (2010). Climate change impacts, adaptive capacity, and vulnerability of European forest ecosystems. *Forest Ecology and Management, 259*, 698–709.

Micu, D. (2009). Snow pack in the Romanian Carpathians under changing climatic conditions. *Meteorology and Atmospheric Physics, 105*(1), 1–16.

Peters, R., Lambert, S., Kőpataki, É., Varga, R., Arany, I., Iosif, A., Moldoveanu, M., Laslo, L., Interwies, E., Görlitz, S., Kondor, A., Orosz, B. (2013). In-depth study on the impact of climate change on ecosystem-based production systems (Final report – Module SR3). CarpathCC, report to European Commission – DG Environment http://carpathcc.eu/node/35. REC – The Regional Environmental Center For Central Eastern Europe, ARCADIS Belgium N.V., Budapest, HU.

Szalai, S. (2012). The CARPATCLIM project: Creation of a gridded climate atlas of the Carpathian regions for 1961–2010 and its use in the European Drought Observatory of JRC. In ECAC2012, Lodz.

Szewczyk, J., Szwagrzyk, J., & Muter, E. (2011). Tree growth and disturbance dynamics in old-growth subalpine spruce forests of the Western Carpathians. *Canadian Journal of Forest Research, 41*(5), 938–944.

UNDP. (2005). Conservation of biological diversity of Carpathian Mountain grasslands in the Czech Republic through targeted application of new EU funding mechanisms. United Nations Development Programme (UNDP).

Vandewalle, M., Sykes, M. T., Harrison, P. A., Luck, G. W., Berry, P., Bugter, R., Dawson, T. P., Feld, C. K., Harrington, R., Haslett, J. R., Hering, D., Jones, K. B., Jongman, R., Lavorel, S., Martins da Silva, P., Moora, M., Paterson, J., Rounsevell, M. D. A., Sandin, L., Settele, J., Sousa, J. P., & Zobel, M. (2010). Review of concepts of dynamic ecosystems and their services. The RUBICODE project—Rationalising Biodiversity Conservation in Dynamic Ecosystems.

Werners, S., Szalai, S., Kőpataki, É., Kondor, A. C., Musco, E., Koch, H., Zsuffa, I., Trombik, J., Kuras, K., Koeck, M., Lakatos, M., Peters, R., Lambert, S., Hlásny, T., Adriaenssens, V., Jurek, M., Crump, J., & Maréchal, J. (2014a). *Future imperfect. Climate change and adaptation in the Carpathians*. Arendal: GRID-Arendal.

Werners, S., Zingstra, H., Bos, E., Hulea, O., Jones-Walters, L., Velde, I. V. D., Vliegenthart, F., Beckmann, A., Delbaere, B., Civic, K., Kovbasko, A., Moors, E., & Schipper, P. (2014b). Integrated assessment of vulnerability of environmental resources and ecosystem-based adaptation measures (Final report). CARPIVIA, report to European Commission – DG Environment – www.carpivia.eu. Alterra – Wageningen UR, ECORYS, ECNC, Grontmij, WWF-DCP, Wageningen, NL.

Zsuffa, I., Laslo, L., Musat, C., Dorobanțu, G., Deák, G., Matei, M., Koch, H., Hattermann, F. F., Stagl, J., Werner, P. C., Liersch, S., Lange, S., Lobanova, A., Kundzewicz, Z. W., Coiron, B., Dobos, E. Vágó, J., Hegedűs, A., Kondor, A. C., Orosz, B. (2013). Final report for SR6: Integral vulnerability assessments in focal areas: Tatra Mountains, Rodna-Maramureş, Târnava Mare, Iron Gate Parks and Bükk Mountains. CarpathCC, report to European Commission – DG Environment http://carpathcc.eu/node/35. REC – The Regional Environmental Center For Central Eastern Europe, INCDPM, PIK, ARTELIA, Aquaprofit, Budapest, HU.

Community Forest Management as Climate Change Adaptation Measure in Nepal's Himalaya

Rabin Raj Niraula and Bharat K. Pokharel

Abstract This paper analyses the community forest management practice and its outcome, the increased forest cover and quality, as climate change adaptation measure in Nepal's Himalaya. Change in forest cover is measured by applying Geographic Information System (GIS) and Remote Sensing comparing 1990s and 2010, and 1992 and 2014 satellite imageries based on supervised land cover classification of four sites covering all three types of forest management regimes – community, government and privately managed forests. The paper demonstrates that community forest management practice has resulted in a change in forest cover in a relatively better way, mainly in the creation of new forest, improvement in forest condition, reduction in the rate of deforestation and degradation. This change has contributed directly to the conservation of biological diversity, ecosystem management, integrated water resource management, forest-agriculture interface and ultimately to overall livelihoods of people dependent on forest resources. The paper concludes that community forest management practice has not only contributed to increase the formation of natural capital, forest cover, but also other livelihood capitals such as human and social capital, strong grassroots level organizations and capacity of local communities to manage forests, which eventually contribute to increase adaptive capacity of local communities to the increased threats of climate change.

Keywords Community forest management • Forest cover change • Climate change adaptation • Deforestation • Degradation

---

R.R. Niraula (✉)
Department of Environmental Science and Engineering,
Kathmandu University, Dhulikhel, Nepal

Helvetas Swiss Intercooperation Nepal, Kathmandu, Nepal
e-mail: rabin.niraula@helvetas.org.np

B.K. Pokharel
Helvetas Swiss Intercooperation Nepal, Kathmandu, Nepal
e-mail: bharat.pokharel@helvetas.org.np

N. Salzmann et al. (eds.), *Climate Change Adaptation Strategies – An Upstream-downstream Perspective*, DOI 10.1007/978-3-319-40773-9_6

1 Literature Review

Current practices dealing with climatically driven variability can be referred as adaptation (Le Goff et al. 2005). Adaptation to climate change refers to adjustments in ecological, social, and economic systems in response to the effects of changes in climate (Adger et al. 2005; Smit et al. 2000; Spittlehouse and Stewart 2003). Adaptation to climate change in forest management encompasses forest protection and regeneration, silvicultural management, forest operations, management of non-timber resources and conservation of biodiversity and management of park and wilderness area (Spittlehouse and Stewart 2003). Forest itself is vulnerable to climatic change and adaptation is needed for forest to maintain their functional status while it plays a role in adaptation of communities and the society at large (Locatelli et al. 2011). The global share of forestry in terms of carbon emission through deforestation and forest degradation is about 17 % (IPCC 2007). On the other hand, forestry has high potential for reducing the emissions as well. According to IPCC Fourth Assessment Report (2007), forestry sector can reduce up to 30 % of the total emission globally by 2020 (4.2 G tonnes in 14 G tonnes).

While global research has marked climate change significantly, Nepal is likely to be impacted as temperature rise and precipitation pattern alters (Duncan et al. 2013; Gentle et al. 2014; Karki and Gurung 2012; Shrestha et al. 2012). The impacts are mainly floods, landslides and drought that could affect the 30 million population dependent on agriculture (Bartlett et al. 2010; Dahal and Hasegawa 2008; Karki and Gurung 2012). Nepal faced severe deforestation in the 1990s (Niraula et al. 2013), which led to myriads of landslides and flood related problems (Shroder and Bishop 1998). The mountainous landscape is fragile and geological processes are highly active in the region (Ives and Messerli 1989). Moreover, deforestation aggravated the situation by intensifying such disasters (Marston et al. 1998). Such problems will increase, particularly when climate change related problems are adding up (Gurung and Bhandari 2009). Increased temperature and change in precipitation, mainly increasing extreme events of high rainfall in short durations and long drought periods will impact the landscapes as well as the communities (Dahal and Hasegawa 2008; Gentle et al. 2014; Shrestha et al. 2012). Ives and Messerli (1989) attempted to challenge The Theory of Himalayan Degradation. According to them, local people are not the problems rather they are part of the solution of Himalaya degradation. The anthropogenic causes of erosion were relatively minor compared with natural causes. There is now a growing recognition of the adaptation and mitigation activities by farmers who are seen to contribute positively to environmental conservation rather than being the cause of the problems (Oven 2009).

Such issues exist, but communities in these fragile landscapes are not just waiting for increasing disasters, but have shown an adaptation tendency adopting different survival strategies like farmer managed irrigation system, contour farming in private farmland, agro forestry and community forest management (Gurung and Bhandari 2009). Nepal has 39.6 % forests, 21 % cultivated agricultural land and 3 % land covered with agro-forest areas of its 147,181 km^2 landmass (MoFSC 2009). Of the total forest area about 30 % forest land is being protected and managed under community forest management regime. Forests serve rural people to supply more than 80 % fodder and 60 % of their energy requirement and nourishes farmland. Forests also play an important role in enhancing local community's capacity to adapt to changing climate (Pokharel and Byrne 2009) because forests provide diverse ecosystem services that contribute to human well-being and reduce social vulnerability (Locatelli et al. 2011). In fact, forest management, particularly community forest management recently has been an example of climate change adaptation in the mountainous areas (Khatri et al. 2013; Pokharel and Byrne 2009; West 2012).

Community forest management practice has been officially recognized following Nepal's forest policy shift that took place in 1988, and subsequent legal frameworks and national guidelines paved the way to recognize the local communities's role in managing forests together with the changing role of the forest service from policing to participation (DoF 2008; HMGN 1990; HMGN/USAID 1993, 1995). As per the policy and legal framework mentioned above, local communities, private tree owners, leasehold groups and religious bodies were recognized as potential partners of government and real managers of forest resources (MoFSC 2013; NFA 2007; Pokharel and Nurse 2004). Various literature indicates that due to such policy reform, there are positive changes taking place in forest cover, especially in community forests (Gautam et al. 2003, 2004; HELVETAS and RRI 2011; Niraula and Maharjan 2011; Niraula et al. 2013) and the impacts on local people's livelihoods (Campbell 2012; LFP 2009; MoFSC 2013; Pokharel et al. 2007).

Measurement of the performance of community forest management is crucial in climate change adaptation as it enables or disables the outcomes of biodiversity protection, ecological management, habitat conservation, climatic regulation, ecosystem services (Chhatre and Agrawal 2009; Cudlín et al. 2013; Schlaepfer et al. 2002). Creation of new forests and improved forest quality are considered to the positive outcomes of forest management; whereas deforestation and forest degradation are the negative forest management outcomes which are global concerns aggravating impacts of climate change. Deforestation contributes to carbon emission, loss of biodiversity and degradation of ecosystem services and agricultural landscapes. This further deteriorates the resilience and adaptation capacity at all scales. Positive outcomes of forest management contribute directly to build ecosystem function and adaptation capacity of biological as well as socioeconomic systems. In this context, community forest management in Nepal has been found to be an effective tool for environmental and social benefits in terms of the improvement in forest condition and reduction in the trend of deforestation and degradation (Campbell 2012; Gautam et al. 2003, 2004; HELVETAS and RRI 2011; Pokharel et al. 2007; Pokharel and Byrne 2009).

Following the United Nations Conference on Environment and Development (UNCED) – *The Earth summit*, 1992 sustainable management of forest with the nonbinding authoritative statement of forest principles (McDonald and Lane 2004; Siry et al. 2005) was emphasized. The assumption was that appropriate forest management regime significantly changes the landscape as well as the ecosystem and contributes to ecological, economic and social development, including ecosystem services of climate and water regulation, flood and erosion control, production of timber and non-timber products (Cudlín et al. 2013; Thomas and Andreas 2013; Miyamoto and Sano 2008; Schlaepfer et al. 2002; Shifley et al. 2008; Spittlehouse and Stewart 2003; Dave and White 2013) and so on. However, review of sustainable development goals has identified that sustainable management of forest resources remained as fundamental challenge (Schlaepfer et al. 2002). Mountain forests and watersheds are fragile ecosystems of the planet and are topographically vulnerable system where role of forest management is high. In such watersheds, soil erosion, landslide and flood are major risks that intensify with climate change (Bartlett et al. 2010; Dahal and Hasegawa 2008; Karki and Gurung 2012). Climate change poses further challenges making the lives of the forest dependent communities more difficult. In Nepal, nevertheless, community forest management with the active participation of local communities, has contributed to the positive change in forest cover, conservation of biological diversity, ecosystem management, integrated water resource management, people's livelihoods and more. All these contribute to increase capacity of local communities to the increased threats of climate change.

Despite the positive impact of Nepal's community forest management, literature shows that forest dependent local people are heterogeneous in terms of the scale of marginalisation on the basis of class, caste, ethnicity, gender and geographical proximity to the service centre and power to have access to livelihood resources of various types, and community forest management also has a disproportionate impact (Khadka 2010; MoFSC 2013; Pokharel et al. 2007; Pokharel and Nurse 2004). This calls upon the need for equity and social justice in development in general and community forestry in particular so that the majority of these forest dependent communities, which have low income, suffer from mulit facet poverty, geographically marginalized and are vulnerable to stresses and shocks (Becken et al. 2013) can benefit. The increased role of community forest management to have better access to natural capital such as forest resources and the exercise of local democracy in rural areas have helped to reduce vulnerability (Gentle and Maraseni 2012) and increase adaptive capacity to climate change. This can only be ensured effectively through local control of forests such as community forest management practice, a paradigm shift in theoretical and practical forestry (Schoene and Bernier 2012). The outcomes of which can be measured both in terms of the scale of change in environmental and social capital formation over the years. Changes in forest have been studied globally. Findings from various studies have already identified major hotspots of improvement as well as loss of forest. Other studies nevertheless support that adequately designed and implemented community forest management regimes can avoid deforestation and restore forest cover and forest density (Klooster and Masera 2000) contributing directly to adaptation to climate change. Moreover, community forests are found to have lower and less variable annual deforestation rates than protected forests (Porter-Bolland et al. 2011). Similarly, a study with original data on 80 forest commons in ten countries across Asia, Africa, and Latin America, shows that larger forest size and greater rule-making autonomy at the local level are associated with high carbon storage and livelihood benefits (Chhatre and Agrawal 2009) which reflects the strength of community forest management over any other forest management regimes. A study by Hansen et al. (2013) has also confirmed major gain and loss in the global forest cover. For which, the use of remote sensing has emerged as a nearly mandatory tool for land use land cover studies. In this paper, we have attempted to show the results of the measurement of the impact of Nepal's community forest management by using such remote sensing tools. The limitation of this study, however, is that analysis lacks clear environmental linkages between upstream and downstream study areas.

2 Objective and Methodology

The main objective of the study is to understand the outcome of community forest management, a viable means of climate change adaptation, across Nepal's upstream and downstream, and measure the scale of change of forest cover through a quantitative method of GIS application.

The study covers four different sites from Nepal which are located in Dolakha, Sindhupalchok, Sindhuli and Udayapur districts. These districts are selected to represent upstream and downstream of Koshi basin, one of the largest river basins in the central Himalayas. Dolakha and Sindhupalchok are located in the upstream of Koshi basin also have the oldest community forests registered in the country. Whereas Sindhuli and Udayapur districts located in the downstream of the same Koshi basin have recently handed over community forests. This study assumes 1990s as the baseline as community forest management was initiated in this decade after widespread deforestation in the 1980s and comparison is made with the most recent time (i.e. 1990–2010 or 1992–2014). The study includes synthesis of four case study sites of forest cover change analysis carried out by the authors in three phases during 2010–2014. Study of Dolakha and Sindhupalchowk, middle mountainous districts in the upstream, for example, were carried out in 2010 and 2011 respectively, whereas the study of Sindhuli and Udayapur, lower hill districts representing downstream was carried out in 2014. The study applies the Geographic Information System (GIS) and Remote Sensing data comparing forest covers during the 1990 and 2010, and 1992 and 2014 satellite imageries of four sites accompanied by selected ground truthing exercise with local communities and the Federation of Community Forest Users Nepal (FECOFUN) through Focus Group Discussion. The latter was conducted at three levels, mainly (i) at a central FECOFUN meeting where people were asked for their perception of changes in forest and its consequences (ii) at the district level where existing information and maps were discussed and (iii) at community and settlement levels where the impact of policies and perception of changes were discussed.

Located in the middle mountainous region of the country in an elevation range of 1000–2500 m, the study sites, Dolakha and Sindhupalchok are characterized by steep slopes, deep valleys and terraced farming. Similarly, areas in downstream

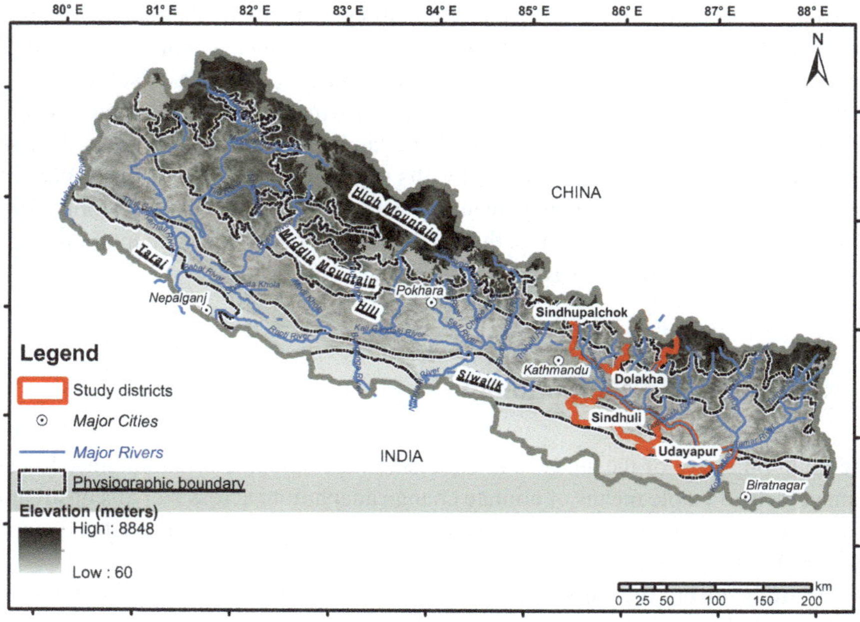

Fig. 1 Map of Nepal showing study areas which covers Sindhupalchok, Dolakha, Sindhuli and Udayapur districts in Koshi Basin

districts Sindhuli and Udayapur covers physiographic zone of Churia also known as the Siwalik with the altitude range from 200 to 1500 m (See Fig. 1).

Landsat TM images from 1990, 1992, 2010 and 2014, acquired from the United States Geological Survey (USGS) Environmental Resource Observation and Science Centre (EROS) archive served as the main source of data. The cloud free images taken during the same season were used in the study in order to minimize the effect of climatic conditions. Also, field based forest boundary mapping by using Global Positioning system (GPS) provided the boundary of the community forests as they were clearly demarked. Landsat 8 imagery has been recently available and was used for the year 2014. Available images were prepared at 30 m resolution and the image subsets were independently classified using supervised maximum likelihood classification. The dense forest, sparse forest and bushes/grass in forest categories, and cultivated land, barren land, river bed and water bodies (Niraula et al. 2013; Niraula and Maharjan 2011). Forest cover change is measured in five different change processes of change, namely ***New forest area***, ***Improved forest area and Unchanged forest area that represent the positive outcome***; ***and Deforested area and Degraded forest area as negative outcomes***.

Similarly, assessment of the change in forest cover as forest management outcome is based on the classification matrix of various categories of land. Based on crown cover, imageries for the month of April were classified into seven categories of land, namely ***Dense forest*** with >40 % crown cover; ***Sparse forest*** with 10–40 %

crown cover, ***Bushes/Grass land*** with <10 % crown cover, ***Barren land, Cultivated Land, Riverbed and Water bodies***.

Measurement of the outcomes of forest management regimes in this study is the spatial analysis of land cover and forest cover compared between 1990/1992 and 2010/2014. The rate of change in forest cover provides trend of change calculated for the study period. In this study, annual rate of change is calculated using the formula provided by the (FAO 2006).

$$\text{Rate of change (q)} = ((\frac{A2}{A1})^{\frac{1}{t2-t1}} - 1) \times 100$$

Where,

A1 = Forest cover at time t1 (1990 or 1992 in this study) and
A2 = Forest cover at time t2 (2010 or 2014 in this study).

The following diagram indicates that the outcomes, ***positive or negative*** forest management are shaped by the nature of collective action among members of local communities, and their interface with forest resources, and appropriateness of the forest policies and legal frameworks together with the increasing trend of vulnerability context such as climate change. The conceptual diagram indicates that the positive outcomes of forest management contribute to improve livelihood conditions and increase adaptive capacity of local people; whereas the negative outcomes of forest management further increase vulnerability and deteriorates the livelihood condition.

The Master Plan for the Forestry Sector (MPFS) 1989 (HMGN 1990), the Forest Act 1993, Forest Regulations 1995 and community forestry national guidelines 2008 are the forest policy and legal framework which has provisioned the hand over about 61 % of Nepal's forests to local management as community forests and created an environment to emerge grassroots level institutions such as Community Forest User Groups (CFUGs), which became a vehicle of community forest management. The latter shapes the outcomes of forest cover change and the livelihood situation of the local people.

Positive outcomes of community forest management practice constitute the formation of various livelihood capitals as highlighted below.

Social Capital Formation Local communities residing in and around forest resources form a legal, autonomous, self sustained local institution called Community Forest User Group (CFUG) which secures full tenure rights to access, use, manage forest resources, identifies the boundary of its members and elect its executive committee; makes rules, award and punish its members as per local rules; amends rules and ensure compliance with these rules; promotes gender equality, social inclusion and voices against economic and social discrimination based on income inequality, gender, caste, ethnicity, regional remoteness and physical and climate vulnerability.

Natural Capital Formation Protects, manages, utilizes forest resources as per the provisions being mutually agreed with the government state authority and makes optimal use of forests, and contributes to create new forest, improve existing forests, adopts measures to reverse the rate of deforestation and degradation.

Financial Capital Formation Generates, reserves and utilizes funds that are generated internally and externally; distributes forest products to its members and market surplus forest produces to generate revenue; establishes enterprises, own business and makes profits, redistribute profits to its members.

Physical Capital Formation Reinvests on local community infrastructure development, notably such as local school building, road/trail, temple, drinking water, community building, community utensils, teacher salary, school furniture, health facilities, local wooden bridge, small irrigation channel, community pond, supplies wooden poles to electrification, toilet construction, sanitation, hygiene and stretcher devices, water mills, telephone lines, improved cooking stove and so on.

Human Capital Formation Reinvests to build capacity of its members on leadership skill, economic and social empowerment process to uplift its members which are economically poor and socially marginalized; promotes good governance; voices for constitutional and legal rights of local communities and clarity of tenure rights over forest resources; demonstrates good governance practice by itself in terms of transparency, accountability and responsive governance system, empowers local community members to voice for equity, justice and better and coordinated service delivery of the state, NGOs and private sector service providers

3 Results

3.1 Forest Management in the Upstream

Analysis of forest cover by using Geographic Information System (GIS) and Remote Sensing comparing 1990 and 2010 shows that there is a positive outcome of forest management. Increased in forest cover and its quality is relatively better in community forests in upstream than the downstream districts. This change of the formation of such natural capital is multiplying positive effects of forest, agriculture and watershed.

The result indicates that the area of agricultural land has slightly declined, whereas forest area and its quality has increased. Figures 2, 3, 4, 5, 6, 7, 8, and 9 and Tables 1 and 2 represent the change in forest and land cover in the study areas. Agriculture land of Dolakha and Sindhupalchok districts, for example, have decreased by 0.04 % and 0.7 % per year, respectively, whereas, in the same study area, dense forest has been found to increase by 1.6 % and 3.1 % per year respectively.

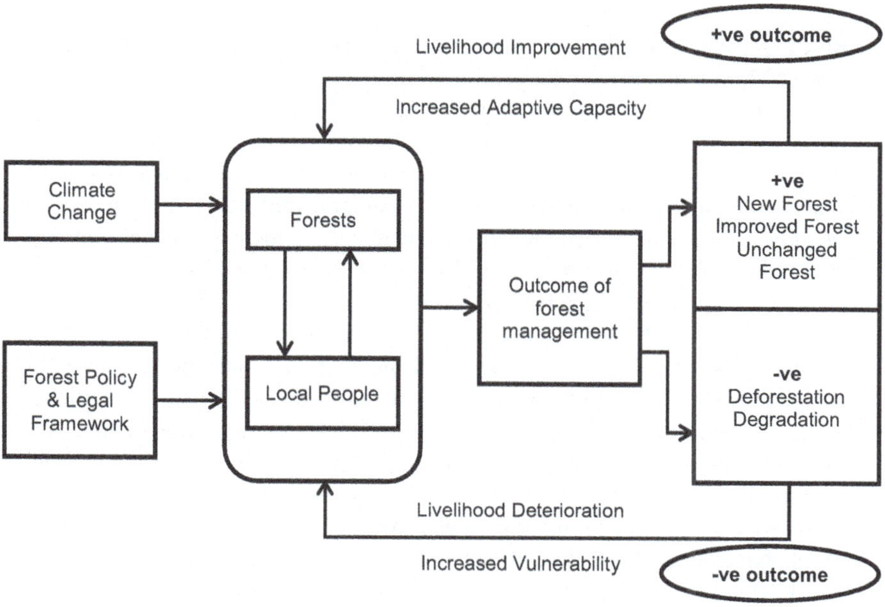

Fig. 2 The conceptual diagram of the study

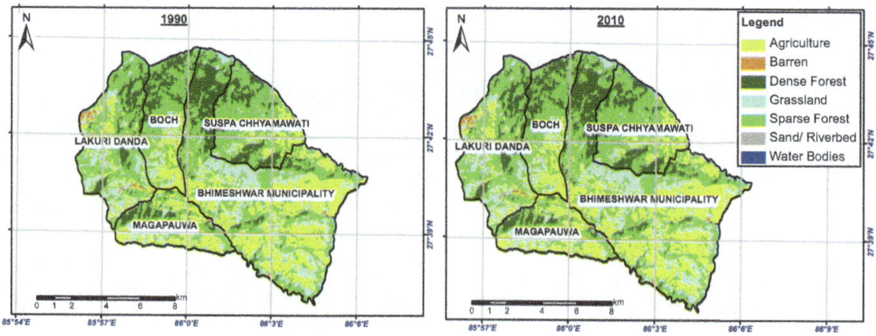

Fig. 3 Land cover change in study area of Dolakha District (Niraula et al. 2013)

It is observed that part of the sparse forest that existed in 1990 must have graduated to a dense forest in 2010. Decrease in the grassland and barren land coverage could indicate that some part of grassland and barren land have been developed in forest areas mainly through community forest management practice described earlier.

Comparison of the patterns of forest cover change among community forests and others show that coverage of community forest managed under community forest management practice mobilizing Community Forest User Groups (CFUGs) have been found relatively better than other management regimes in terms of the increased rate of afforestation, and decreased rate of deforestation (see Fig. 5).

This finding is supported by the perception of local communities (see Box 1 below).

Box 1: Perception of Local Communities on the Outcome of Community Forest Management

Formerly denuded hills around villages are covered with forests and greenery – much better than before; forest condition has improved in terms of regeneration, number of stems per unit area, wildlife, species diversity, off-take of litter and organic manure on their farmland; cultivation of more varieties of cash crops are seen in the farmland than before, increased availability of grass and fodder from community forests have encouraged local people to practice stall feeding which have reduced grazing pressure and to some extent saved the time of children to herd cattle to the forest, enrollment of girls children at local school have also increased; number of water springs and discharge of water volumes have increased; soil nutrition and moisture conditions in their agricultural land during the dry season have improved; incidences of forest fire have reduced; encroachment of forest land adjacent to the private farm is stopped; illegal felling of trees and stealing of forest products have decreased; number of complaints and forests offences have reduced; local people have become self-disciplined and in some cases have been able to fine forest offenders by local rules; community members have come together for forest tending and harvesting operations leading to efficient use of forest resources. There is a decline in slash and burn practice. Regular plantation in watersheds and increased awareness for the protection of trees in private land and terraces of farmland have been observed as a phenomenon.

Source: Findings of group discussion organized by Federation of Community Forest Users Nepal (FECOFUN), September 2014

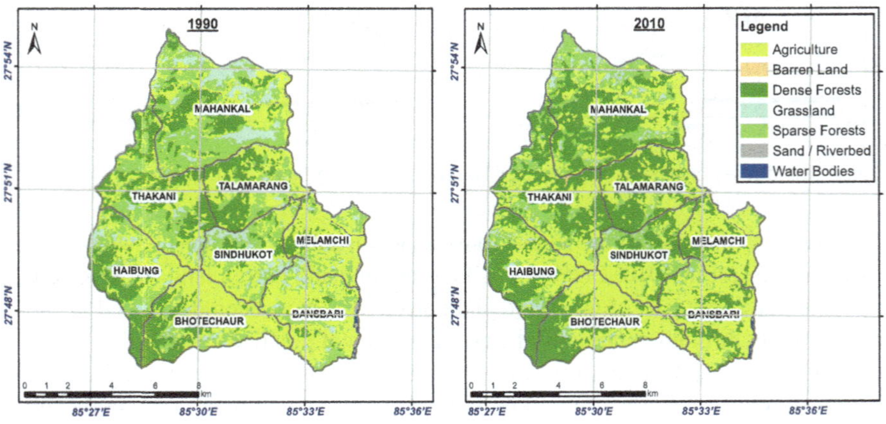

Fig. 4 Land cover change in study area of Sindhupalchok district

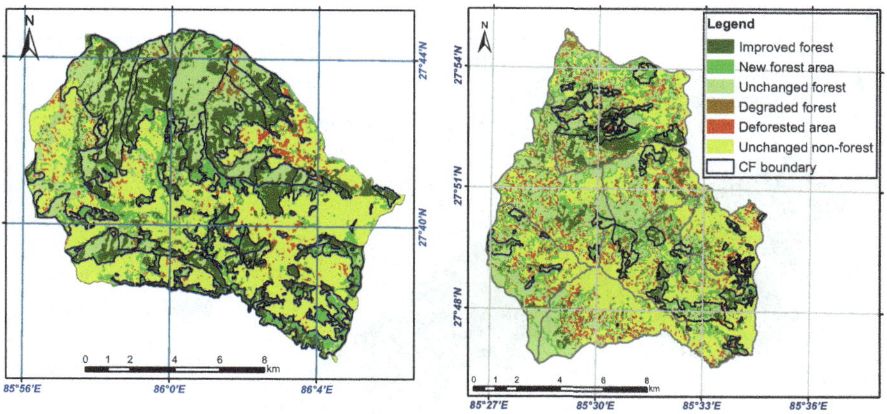

Fig. 5 Forest cover change in the study area of Dolakha (*left*) and Sindhupalchok (*right*)

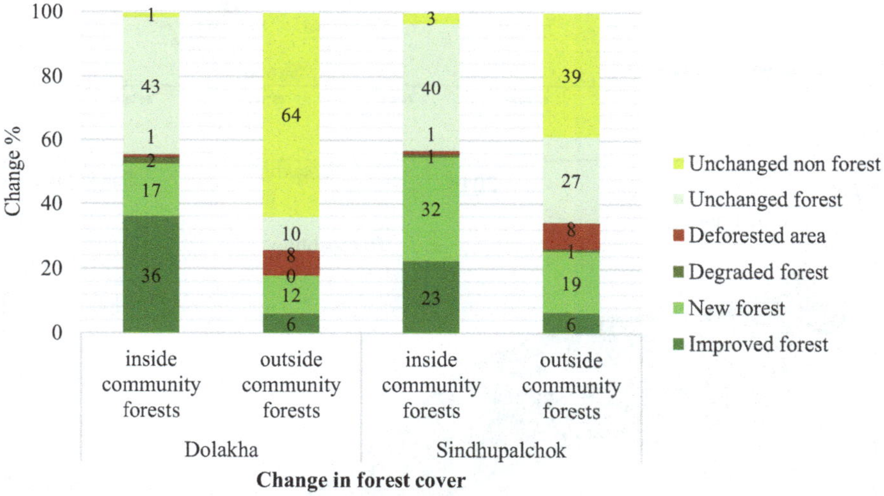

Fig. 6 Forest cover change comparison in the upstream areas

3.2 *Forest Management in the Downstream*

Sindhuli and Udayapur district are located in the downstream side of Mahabharat and Siwalik range where agricultural lands and settlements are mostly concentrated in the valleys. The Churia has sub-tropical deciduous forest with comparatively less agricultural land, except in the inner Terai valley. The region continuously extends along the south of the Mahabharat from the east to west. Landslides are a major source of forest disturbance in this area (DFRS 2014). This range has a unique geography, scarcity of water and economically vulnerable people, mostly migrants

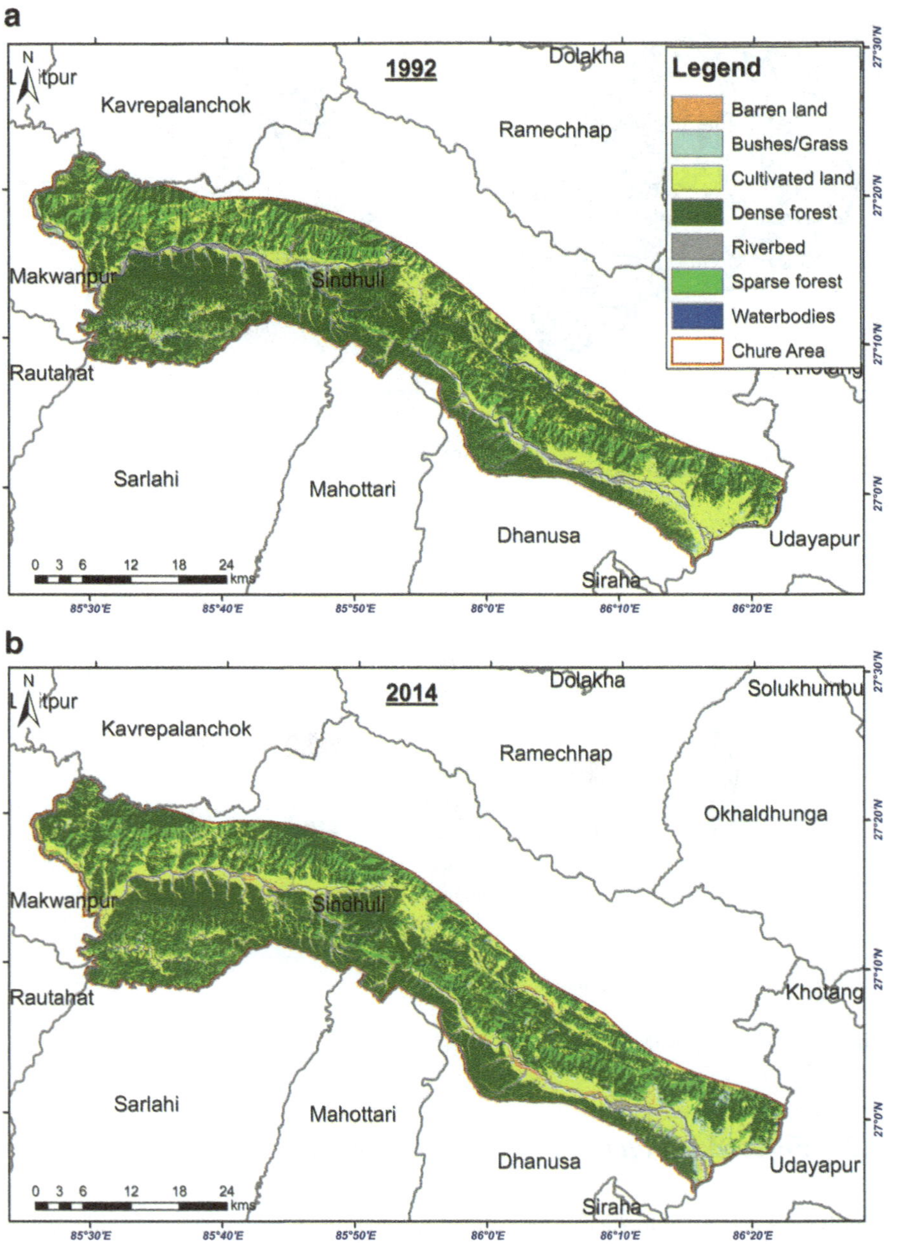

Fig. 7 Land cover change in Churia area of Sindhuli district

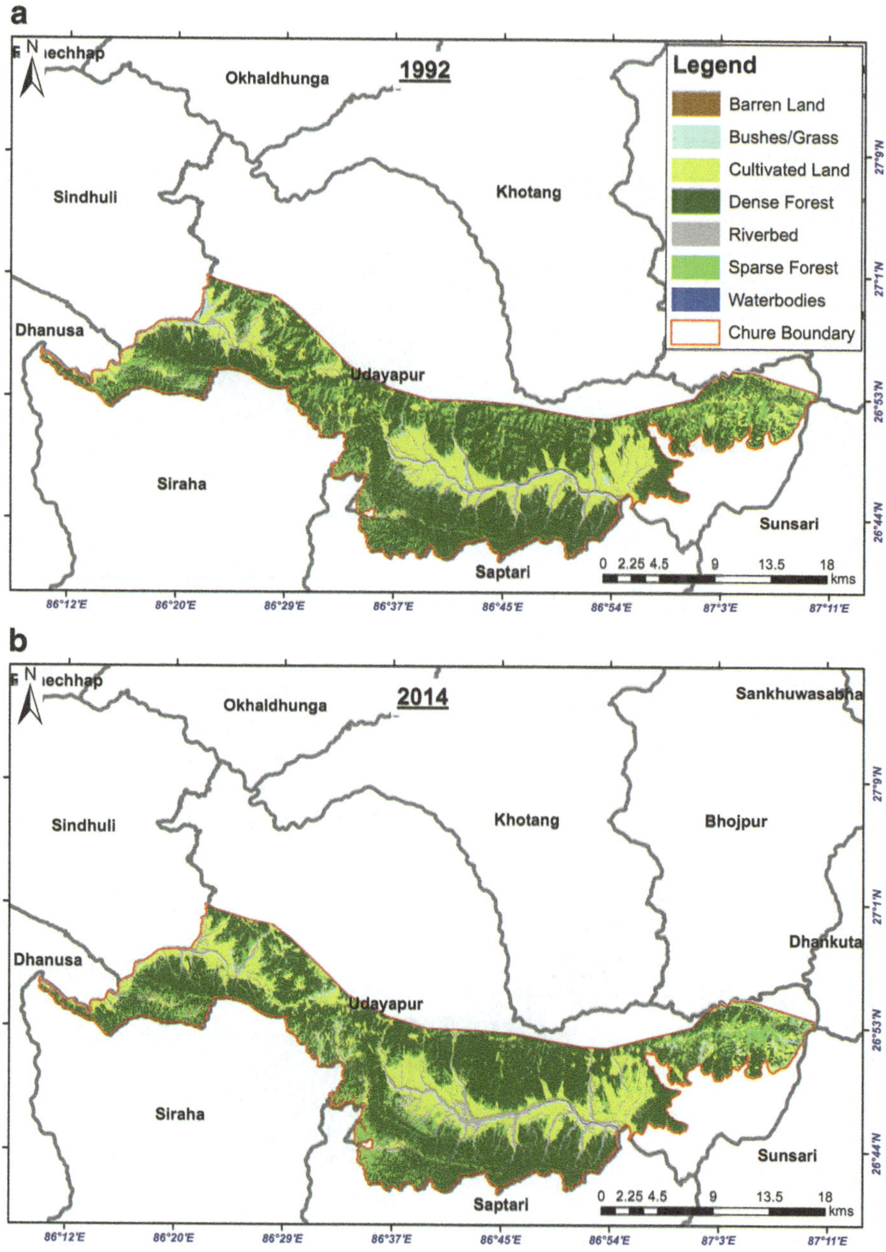

Fig. 8 Land cover change in Churia area of Udayapur district

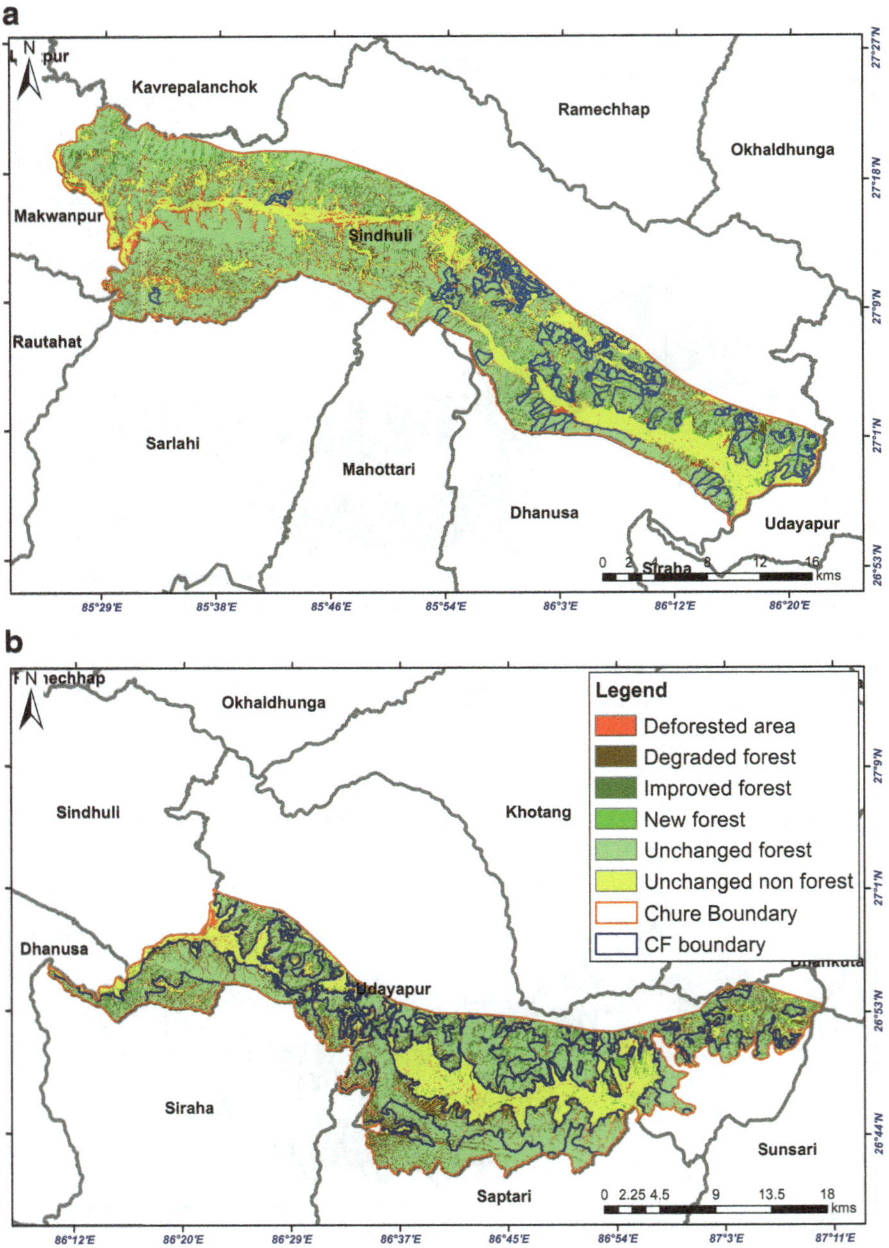

Fig. 9 Forest cover change in downstream areas – Sindhuli (*top*) and Udayapur (*bottom*)

Table 1 Land cover change in the upstream areas

Landcover (area in hectare)	Dolakha			Sindhupalchok		
	1990	2010	Rate of change (%)	1990	2010	Rate of change (%)
Dense forest	2829.9	3870.8	1.6	2569.2	4718.9	3.1
Sparse forest	5738.9	4833.2	−0.9	3191.1	2749.0	−0.7
Grassland	496.5	481.3	−0.2	1314.6	395.9	−5.8
Agriculture	5749.6	5698.8	0.0	5580.8	4812.9	−0.7
Bare soil	200.0	113.8	−2.8	10.4	34.6	6.2
Sand	0.3	0.3	0.0	33.9	6.5	−7.9
Water bodies	17.5	34.7	3.5	90.2	72.5	−1.1
Total	15,032.6	15,032.9		14,780.3	14,800.3	

Table 2 Land cover change in the downstream areas

Land cover (area in hectare)	Sindhuli			Udayapur		
	1992	2014	Rate of change (%)	1992	2014	Rate of change (%)
Dense forest	77,542.1	76,489.8	−0.1	62,971.6	61,232.0	−0.1
Sparse forest	20,132.1	22,755.0	0.6	20,596.7	21,270.2	0.1
Bushes/grass	5092.4	3746.1	−1.4	3342.3	2394.5	−1.5
Cultivated land	30,985.7	3,2120.2	0.2	24,889.5	24,603.7	−0.1
Barren land	763.4	1457.0	3.0	403.4	866.0	3.5
Riverbed	5549.9	5425.7	−0.1	4868.8	6979.9	1.7
Waterbodies	2388.7	1240.6	−2.9	555.8	195.2	−4.6
	142,454.3	143,234.4		117,628.1	117,541.5	

live in this area. Major of settlements in the Churia hills is believed to be started from the middle of the nineteenth century and gradually increased with the eradication of malaria and the government's policy of resettling people. Now about five million population are directly/indirectly connected with Churia area. In many parts of the Churia hills, disaster victims and poverty enforced migrants are settled without land ownership and their livelihood primarily depend on nearby forest (HELVETAS and RRI 2015). Thus, dynamic nature of the Churia region has been pushing the already marginalized community towards more sever conditions.

Analysis of forest covers by using Geographic Information System (GIS) and Remote Sensing comparing 1992 and 2014 shows that unlike in the upstream districts dense forest area has decreased, so is cultivated land (see Figs. 7 and 8). Percentage of newly created and improved quality forest is less than the degraded and deforested forest area. Large part of the forest has been seen unchanged means simply a protection, but missed opportunity for further increment, growth and income, therefore more room for improved forest management (see Fig. 8). The main drivers of such a scenario in this downstream area could be less engagement of local communities for community forest management practice. Unlike in the hills and mountains, there are less number of Community Forest User Groups (CFUGs) in this region because forest policy and institutional arrangement for this geographical area is crafted in such a way that government controlled forest protection system is encouraged rather than community forest management model described earlier.

In contrary to the government's policy assumption that community forest management approach does not work well on the downstream, this study shows that performance of community forest management practice is relatively better than the other practice of government and privately managed forests. Despite the trend of deforestation to a certain extent existed in all regimes, in the community forest area the deforestation is three times lower than outside (see Fig. 9). See example of the Sindhuli district in Fig. 7. This trend however is not consistent to all downstream districts. Community forest management in Udaypur district, the other study district, for example, is not found to be a good performing district in comparison to other regimes due to tenure insecurity that local communities perceive (Ganesh Karki, Chair of FECOFUN during Focus Group Discussion in September 2014) (Fig. 10).

4 Conclusion

The study provides a quantitative measurement of change in forest cover in community forest and other forest management regimes and supports the findings of earlier studies of community forest management. Improvement in forest condition

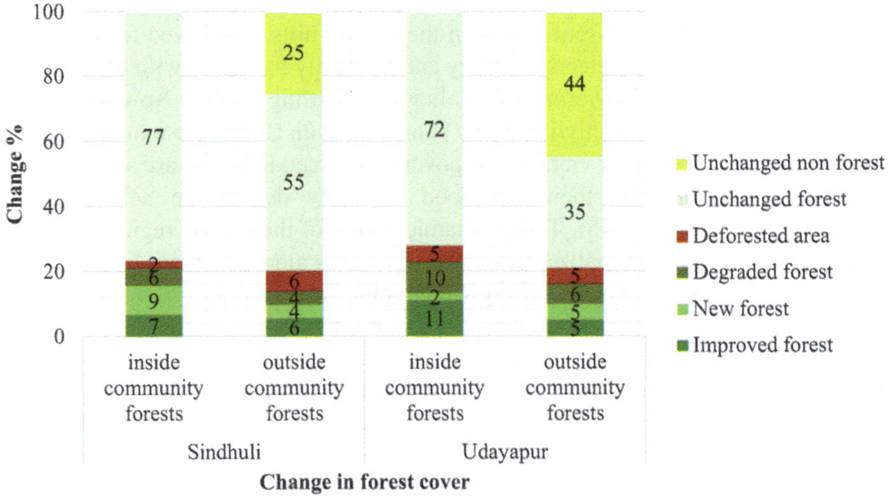

Fig. 10 Forest cover change comparison in the downstream areas

and the creation of new forest dominate the changing pattern. The latter contributes to increase adaptive capacity of local people as well as the resillience of forests to changing climate. The study brings results from upstream and downstream of Koshi basin where forest management was applied in different ways during various timescales but under similar policies and the context of climate change. The interaction between people and forest, the effect of policy and legal framework and the context of changing climate comes up with mixed outcomes, positive and negative.

Major findings from the study provides an indication as follows:

- Nepal's middle mountains in the upstream demonstrates better forest management with positive results of improvement in forest quality as well as the increase in coverage forest area compared to those in downstream lower areas.
- Community forest management regimes have done relatively well in terms of both in creating new forest and in the improvement of forest quality.
- Substantial areas of forests in upstream and downstream show unchange in forest condition. It clearly indicates that forest is protected, but active management is required to allow forest to put further growth for more production.
- In all forest regimes, there are some areas which are exposed to both 'deforestation' and 'degradation' due to both natural and physical infrastructure development processes such as landslides, road construction and urbanization. However, the extent of the area of deforestation and degradation is relatively much lower than the forest area with 'improvement' and 'unchanged'condition.
- The most feared and exaggerated deforestation in Nepal by 1.7 % per year reported by the FAO as the national average of deforestation for the period of 1978–1994 (MoFSC 2009) is now a history. The overall forest cover change in upstream and downstream areas of Nepal is now positive.

Acknowledgements HELVETAS Swiss Intercooperation Nepal, Right and Resource Initiatives (RRI) and International Development Research Centre (IDRC) deserve special thanks for providing authors opportunity to invest some of their time to do this study.

References

Adger, W. N., Arnell, N. W., & Tompkins, E. L. (2005). Successful adaptation to climate change across scales. *Global Environmental Change, 15*(2), 77–86. doi:10.1016/j.gloenvcha.2004.12.005.

Bartlett, R., Bharati, L., Pant, D., Hosterman, H., & McCornick, P. (2010). *Climate change impacts and adaptation in Nepal* (p. 35) (IWMI Working Paper 139). Colombo: International Water Management Institute. doi:10.5337/2010.227.

Becken, S., Lama, A. K., & Espiner, S. (2013). The cultural context of climate change impacts: Perceptions among community members in the Annapurna Conservation Area, Nepal. *Environmental Development, 8*, 22–37. doi:10.1016/j.envdev.2013.05.007.

Campbell, J. G. (2012). *Development assistance in action : Lessons from Swiss and UK funded forestry programmes in Nepal*. Kathmandu: Multi Stakeholder Forestry Programme (MSFP). Retrieved from http://www.msfp.org.np/uploads/publications/file/DevelopmentAssistanceinAction_20121121023451.pdf.

Chhatre, A., & Agrawal, A. (2009). Trade-offs and synergies between carbon storage and livelihood benefits from forest commons. *Proceedings of the National Academy of Sciences of the United States of America, 106*(42), 17667–17670. doi:10.1073/pnas.0905308106.

Cudlín, P., Seják, J., Pokorný, J., Albrechtová, J., Bastian, O., & Marek, M. (2013). Forest ecosystem services under climate change and air pollution. *Developments in Environmental Science, 13*, 521–546. doi:10.1016/B978-0-08-098349-3.00024-4.

Dahal, R. K., & Hasegawa, S. (2008). Representative rainfall thresholds for landslides in the Nepal Himalaya. *Geomorphology, 100*(3–4), 429–443. doi:10.1016/j.geomorph.2008.01.014.

Dave, R., & White, P. C. L. (2013). Chapter One: Ecosystems and their services in a changing world: An ecological perspective. In W. Guy & E. J. O'Gorman (Eds.), *Advances in Ecological Research* (Vol. 48, pp. 1–70). Academic Press. http://dx.doi.org/10.1016/B978-0-12-417199-2.00001-X, ISSN 0065–2504, ISBN 9780124171992.

DFRS. (2014). *Churia forests of Nepal*. Forest Resource Assessment Nepal Project/Department of Forest Research and Survey. Babarmahal, Kathmandu, Nepal. Retrieved from http://www.franepal.org/wp-content/uploads/downloads/publications/Chure_report.pdf.

DoF. (2008). *National community forestry operational guidelines*, Kathmandu.

Duncan, J. M. A., Biggs, E. M., Dash, J., & Atkinson, P. M. (2013). Spatio-temporal trends in precipitation and their implications for water resources management in climate-sensitive Nepal. *Applied Geography, 43*, 138–146. doi:http://dx.doi.org/10.1016/j.apgeog.2013.06.011.

FAO. (2006). Global forest resources assessment 2005. *FAO Forestry Paper*. Retrieved from http://www.fao.org/documents/show_cdr.asp?url_file=/DOCREP/004/Y1997E/y1997e0d.htm\n, www.fao.org/forestry/site/fra2005/en\n, http://www.fao.org/forestry/index.jsp

Gautam, A. P., Shivakoti, G. P., & Webb, E. L. (2004). Forest cover change, physiography, local economy, and institutions in a mountain watershed in Nepal. *Environmental Management, 33*(1), 48–61. doi:10.1007/s00267-003-0031-4.

Gautam, A. P., Webb, E. L., Shivakoti, G. P., & Zoebisch, M. A. (2003). Land use dynamics and landscape change pattern in a mountain watershed in Nepal. *Agriculture, Ecosystems & Environment, 99*(1–3), 83–96. doi:10.1016/S0167-8809(03)00148-8.

Gentle, P., & Maraseni, T. N. (2012). Climate change, poverty and livelihoods: Adaptation practices by rural mountain communities in Nepal. *Environmental Science & Policy, 21*, 24–34. doi:10.1016/j.envsci.2012.03.007.

Gentle, P., Thwaites, R., Race, D., & Alexander, K. (2014). Differential impacts of climate change on communities in the middle hills region of Nepal. *Natural Hazards, 74*(2), 815–836. doi:10.1007/s11069-014-1218-0.

Gurung, G. B., & Bhandari, D. (2009). Integrated approach to climate change adaptation. *Journal of Forest and Livelihood, 8*(1) (February 2009). Retrieved from www.forestaction.org

Hansen, M. C., Potapov, P. V., Moore, R., Hancher, M., Turubanova, S. A., Tyukavina, A., Thau, D., Stehman, S. V., Goetz, S. J., Loveland, T. R., Kommareddy, A., Egorov, A., Chini, L., Justice, C. O., & Townshend, J. R. G. (2013). High-resolution global maps of 21st-century forest cover change. *Science, 342*(6160), 850–853. doi:10.1126/science.1244693.

HELVETAS, & RRI. (2011). *Does tenure matter ? Assessment of change in forest cover in Nepal. Kathmandu*. Retrieved from http://assets.helvetas.org/downloads/2011_2_does_tenure_matter.pdf.

HELVETAS, & RRI. (2015). In B. K. Pokharel, R. R. Niraula, N. Timalsina, & R. Neupane (Eds.), *Changing face of the Churia range of Nepal : Land and forest cover in 1992 and 2014*. Kathmandu: Helvetas Swiss Intercooperation Nepal & Rights and Resources Initiative.

HMGN. (1990). *Master plan for the forest sector, Nepal, forestry sector policy revised edition*. Ministry of Forests and Environment/Finnida/Asian Development Bank. Retrieved from http://www.forestrynepal.org/images/publications/MPFS_Summary1.pdf.

HMGN/USAID. (1993). *Forest act, 2049 (1993)*. Kathmandu: MFSC and Forest Development Project, HMGN/USAID. Retrieved from http://www.forestrynepal.org/images/Forest_Act_of_Nepal_1993.pdf.

HMGN/USAID. (1995). *Forest regulation, 2051 (1995)*. Kathmandu: MFSC and Forest Development Project, HMGN/USAID. Retrieved from http://www.forestaction.org/app/webroot/js/tinymce/editor/plugins/filemanager/files/Forest_Regulation_1995_2_.pdf.

IPCC. (2007). *IPCC Fourth Assessment Report* (AR4) (Vol. 1). Retrieved from http://www.ipcc.ch/publications_and_data/publications_ipcc_fourth_assessment_report_wg2_report_impacts_adaptation_and_vulnerability.htm.

Ives, J. D., & Messerli, B. (1989). The himalayan dilemma: Reconciling development and conservation. *Land Use Policy, 7*(1), 93–94. doi:10.1016/0264-8377(90)90063-5.

Karki, R., & Gurung, A. (2012). An overview of climate change and its impact on agriculture: A review from least developing country, Nepal. *International Journal of Ecosystem, 2*(2), 19–24. doi:10.5923/j.ije.20120202.03.

Khadka, M. (2010). *Why does exclusion continue? Aid, knowledge and power in in Nepal's community forestry policy process*. Den Haag: International Institute of Social Studies.

Khatri, D. B., Bista, R., & Gurung, N. (2013). Climate change adaptation and local institutions: How to connect community groups with local government for adaptation planning. *Journal of Forest and Livelihood, 11*(1), 14–28.

Klooster, D., & Masera, O. (2000). Community forest management in Mexico: Carbon mitigation and biodiversity conservation through rural development. *Global Environmental Change, 10*, 259–272. doi:10.1016/S0959-3780(00)00033-9.

Le Goff, H., Leduc, A., Bergeron, Y., & Flannigan, M. (2005). The adaptive capacity of forest management to changing fire regimes in the boreal forest of Quebec. *Forestry Chronicle, 81*, 582–592. doi:10.5558/tfc81582-4.

LFP. (2009). *Community forestry for poverty alleviation: How UK aid has increased household incomes in Nepal's middle hills*. Household economic impact study 2003–2008. Kathmandu: Livelihood and Forestry Programme. Retrieved from http://www.msfp.org.np/uploads/publications/file/CF for Poverty.

Locatelli, B., Evans, V., Wardell, A., Andrade, A., & Vignola, R. (2011). Forests and climate change in latin america: Linking adaptation and mitigation. *Forests, 2*(4), 431–450. doi:10.3390/f2010431.

Marston, R. a., Miller, M. M., & Devkota, L. P. (1998). Geoecology and mass movement in the Manaslu-Ganesh and Langtang-Jugal Himals, Nepal. *Geomorphology, 26*(1–3), 139–150. doi:10.1016/S0169-555X(98)00055-5.

McDonald, G. T., & Lane, M. B. (2004). Converging global indicators for sustainable forest management. *Forest Policy and Economics, 6*(1), 63–70. doi:10.1016/S1389-9341(02)00101-6.

Miyamoto, A., & Sano, M. (2008). The influence of forest management on landscape structure in the cool-temperate forest region of central Japan. *Landscape and Urban Planning, 86*(3–4), 248–256. doi:10.1016/j.landurbplan.2008.03.002.

MoFSC. (2009). Nepal forestry outlook study (No. APFSOS II/WP/2009/05). Bangkok.

MoFSC. (2013). *Persistence and change: Review of 30 years of community forestry in Nepal*. Kathmandu: Multi Stakeholder Forestry Programme (MSFP).

NFA. (2007). Democratization, governance and sustainable development of the forestry sector of Nepal, Kathmandu.

Niraula, R. R., Gilani, H., Pokharel, B. K., & Qamer, F. M. (2013). Measuring impacts of community forestry program through repeat photography and satellite remote sensing in the Dolakha district of Nepal. *Journal of Environmental Management, 126*, 20–29. doi:10.1016/j.jenvman.2013.04.006.

Niraula, R. R., & Maharjan, S. (2011). *Forest cover change analysis in Dolakha District (1990–2010). application of GIS and remote sensing*. Kathmandu: Nepal Swis Community Forestry Project.

Oven, K. (2009). *Landscape, livelihoods and risk: Community vulnerability to landslides in Nepal*. Durham University. Retrieved from http://etheses.dur.ac.uk/183/.

Pokharel, B. K., Branney, P., Nurse, M., & Malla, Y. B. (2007). Community forestry: Conserving forests, sustaining livelihoods and strengthening democracy. *Journal of Forest and Livelihood, 6*(2), 8–19.

Pokharel, B. K., & Byrne, S. (2009). *Climate change mitigation and adaptation strategies in Nepal's forest sector: How can rural communities benefit?* Kathmandu: Nepal Swiss Community Forestry Project.

Pokharel, B. K., & Nurse, M. (2004). Forests and people's livelihood: Benefiting the poor from community forestry. *Journal of Forest and Livelihood, 4*(1), 19–29.

Porter-Bolland, L., Ellis, E. A., Guariguata, M. R., Ruiz-Mallén, I., Negrete-Yankelevich, S., & Reyes-García, V. (2011). Community managed forests and forest protected areas: An assessment of their conservation effectiveness across the tropics. *Forest Ecology and Management*. doi:10.1016/j.foreco.2011.05.034.

Schlaepfer, R., Iorgulescu, I., & Glenz, C. (2002). Management of forested landscapes in mountain areas: An ecosystem-based approach. *Forest Policy and Economics, 4*(2), 89–99. doi:10.1016/S1389-9341(02)00009-6.

Schoene, D. H. F., & Bernier, P. Y. (2012). Adapting forestry and forests to climate change: A challenge to change the paradigm. *Forest Policy and Economics, 24*, 12–19. doi:10.1016/j.forpol.2011.04.007.

Shifley, S. R., Thompson, F. R., Dijak, W. D., & Fan, Z. (2008). Forecasting landscape-scale, cumulative effects of forest management on vegetation and wildlife habitat: A case study of issues, limitations, and opportunities. *Forest Ecology and Management, 254*(3), 474–483. doi:10.1016/j.foreco.2007.08.030.

Shrestha, U. B., Gautam, S., & Bawa, K. S. (2012). Widespread climate change in the Himalayas and associated changes in local ecosystems. *PloS One, 7*(5), e36741. doi:10.1371/journal.pone.0036741.

Shroder, J. J. F., & Bishop, M. P. (1998). Mass movement in the Himalaya: New insights and research directions. *Geomorphology, 26*, 13–35. doi:10.1016/S0169-555X(98)00049-X.

Siry, J. P., Cubbage, F. W., & Ahmed, M. R. (2005). Sustainable forest management: Global trends and opportunities. *Forest Policy and Economics, 7*(4), 551–561. doi:10.1016/j.forpol.2003.09.003.

Smit, B., Burton, I., Klein, R. J. T., & Wandel, J. (2000). An anatomy of adaptation to climate change and variability. *Climate Change, 45*, 223–251. doi:10.1023/A:1005661622966.

Spittlehouse, D. L., & Stewart, R. B. (2003). Adaptation to climate change in forest management. *British Columbia Journal of Ecosystems and Management, 4*(1), 1–11. Retrieved from http://www.forrex.org/jem/2003/vol4/no1/art1.pdf.

Thomas, K., & Andreas, H. (2013). Chapter 26: Global change and the role of forests in future land-use systems. In R. Matyssek, N. Clarke, P. Cudlin, T. N. Mikkelsen, J.-P. Tuovinen, G. Wieser, & E. Paoletti (Eds.), *Developments in environmental science* (Vol. 13, pp. 569–588). Elsevier. http://dx.doi.org/10.1016/B978-0-08-098349-3.00026-8. ISSN 1474–8177, ISBN 9780080983493.

United Nations. (1992). Agenda 21. In *United Nations Conference on Environment and Development (UNCED)*. Retrieved from http://www.un.org/esa/dsd/agenda21/.

West, S. (2012). *REDD plus and adaptation in Nepal. REDD-net*. London: Overseas Development Institute.

Ecosystem-based Adaptation (EbA) of African Mountain Ecosystems: Experiences from Mount Elgon, Uganda

Musonda Mumba, Sophie Kutegeka, Barbara Nakangu, Richard Munang, and Charles Sebukeera

Abstract In many developing regions of the world, economies and local communities depend largely on the services provided by ecosystems for their sustenance. Recent evidence has shown that the degradation and possible loss of these vital ecosystem services results in imbalance of both societies and ecosystems resulting in vulnerabilities. Hence resilient ecosystems have been seen as an important foundation to human well-being and also necessary for better adaptive capacity for the communities that depend on them. It's on this premise that the concept of Ecosystem-based adaptation (EbA) is particularly relevant. Evidence has shown that mountain ecosystems are particularly vulnerable to climate change and as such the ecosystem services that they provide for communities and species within and without the proximity of these areas are also threatened. This paper discusses the EbA approach that has been applied to mountain ecosystems of Nepal, Peru and Uganda. Vulnerability Impact Assessments (VIAs) were undertaken to understand community vulnerability, mapping the important ecosystems services provided and options offered for reducing this vulnerability for resilient ecosystems. The EbA implementation in all three countries also demonstrates a move from EbA conceptualization to realization on the ground and at the policy level. Finally this paper will also examine the policy implications of this approach nationally and ability for the work to be up-scaled to other mountain ecosystems and other ecosystems as well.

Keywords Ecosystems services • Climate change adaptation • Elgon • Uganda

M. Mumba (✉) • R. Munang • C. Sebukeera
United Nations Environment Programme (UNEP), P.O. Box 30552-001001, Nairobi, Kenya
e-mail: Musonda.Mumba@unep.org

S. Kutegeka
International Union for Conservation of Nature (IUCN), P.O. Box 10950, Kampala, Uganda

B. Nakangu
University of Makerere, P.O. Box 11064, Kampala, Uganda

N. Salzmann et al. (eds.), *Climate Change Adaptation Strategies – An Upstream-downstream Perspective*, DOI 10.1007/978-3-319-40773-9_7

1 Introduction

Climate change is no longer seen as an abstract issue and the impacts accompanied with it have manifested harshly in many parts of the world and more so in Africa. It's affecting millions of local communities across Africa with devastating effects that have left them in a cycle of poverty (IPCC 2007). Additionally natural ecosystems underpin the existence and sustenance of economies and local communities across the continent and the ecosystem services they provide are vital for human well-being.

This direct dependence on these life-sustaining services provided by ecosystems, is what makes communities and economies particularly vulnerable to the risks of climate change and the inability and lack of capacity by poor communities to adapt to the changes further complicates this vulnerability.

2 The Evolving Nature of Adaptation

The global discourse in recent years on climate change through the United Nations Framework Convention on Climate Change (UNFCCC) has further highlighted that the need for adaptation efforts (UNFCCC 2007) has never been so urgent (Munang et al. 2013). Furthermore parties to this convention have recognized that resilient ecosystems are certainly vital for human well-being and critical for helping communities adapt to climate change impacts. In June 2012 at the Rio+20 United Nations Conference on Sustainable Development, world leaders there explicitly recognised the role of ecosystems as the core element in addressing climate change and hence paving the way for better sustainable development, further noting that sustainable development has its roots in ecosystem maintenance (UNFCCC 2010, United Nations 2012).

There is evidence that ecosystems have substantively changed (Millennium Ecosystem Assessment 2005), affected by various drivers of change with climate change as the additional stressor. As such their capacity to deliver provisioning, buffering and regulatory services have greatly reduced. Its on this premise that the term resilience comes into play and is widely used to describe the ability of a social or ecological system to maintain basic structural and functional characteristics over time despite external pressures (Levine et al. 2012). However, ecosystems differ in the way they respond to change and also their reaction to disturbance – making some more resilient than others.

It's against this backdrop that this chapter addresses the issue of ecosystem resilience, which is a particularly relevant concept in the context of ecosystem-based adaption (EbA). Ecosystem-based adaptation has been described as the use of biodiversity and ecosystem services as part of an integral part of an overall adaptation strategy to help people adapt to the negative effects of climate change at local, national, regional and global levels. Its prudent to highlight that EbA spans many activities, which include forest restoration to storm surges; catchment management to protect communities against effects of droughts and floods; sustainable fisheries to alleviate the threat of food security and much more (e.g. Dudley et al. 2010).

Regardless of the disjointed relationship between the two Rio Conventions – the Convention on Biodiversity (CBD) and UNFCCC – both have in more recent years acknowledged and recognised the relevance and importance of EbA (CBD 2009; UNFCCC 2010). Encouragingly, the governing bodies of both conventions recently sought to encourage parties to implement EbA (Chong 2014).

3 African Mountain Ecosystems and Landscapes in the Face of Changing Climate

African mountain landscapes (Fig. 1) across the African continent are some of the most spectacular ecosystems. These ecosystems, though staggered across the continent in discrete locations, are particularly vital for the 1.1 billion people of Africa because of the ecosystem services they provide. While they are important and vital

Fig. 1 Elevation map of Africa showing mountain areas (UNEP/DEWA)

for these services they provide, recent evidence has shown that they are also vulnerable and sensitive to environmental change (e.g. Knapen et al. 2006). Evidenced by the recognition of ecosystems in sustainable development at Rio+20 (UNDP 2013) mountain ecosystems also received substantive attention with renewed global political attention to these beautiful yet fragile ecosystems.

Furthermore in Africa, a step towards cooperative action and the creation of collaborative platforms for sustainable mountain development was taken at a regional meeting for African mountain regions held in November 2011 in Mbale, Uganda.

4 Understanding the Vulnerabilities of Mount Elgon

Uganda hosts two of Africa's mountain ecosystems – Rwenzori Mountain and Mount Elgon – and has been a champion (an ally in advocating for the issues around climate change in Mount Elgon) in the discourse on sustainable mountain development across Africa and on the global platform. It is on this premise that the German Government in liaison with the Government of Uganda recognised the vulnerability of the Mount Elgon ecosystem to climate change (e.g. Government of Uganda 2009a). The International Climate Risk Report (CIGI 2007) identified Uganda as one of the least prepared and most vulnerable countries in the world. Many parts of Uganda are already experiencing the impacts of climate change (Hepworth 2010) such as frequent droughts, famine, floods and landslides, and their knock on consequences on natural resources, agriculture, food security and livelihoods. Uganda being a signatory to the UNFCCC developed its National Adaptation Programme of Action (NAPA) in 2007. According to this report, drought was identified as the most prominent effect of climate change in Uganda. The NAPA suggested that the frequency of drought is on the increase with seven serious droughts experienced between 1991 and 2000 (Government of Uganda 2007). Floods were also experienced in many parts of the country. For example in 2007, the eastern region of Uganda experienced the worst floods in 35 years (Government of Uganda 2009b). In March 2010, several districts within the Mount Elgon area experienced unusual high rainfall, which resulted in landslides in Bududa District, for example. Hundreds of lives were lost, with the landslides burying three whole villages (e.g. Government of Uganda 2009b). In addition several households were displaced, schools and health facilities destroyed in the wake of this event. In close proximity in the next district, Kapchorwa, about 300 ha of wheat were destroyed. Nationally, coffee exports dropped by 60 % between October and November 1999, partly due to disrupted transport system. Facilities for treated water supply to climate change affected districts were destroyed following flooding of some pumping stations.

The Ugandan government identified the Mount Elgon ecosystem as a particularly vulnerable area because of its high incidences of floods and landslides (Knapen et al. 2006) and also climate change. These events have become more frequent and are emerging issues in this highly productive region with high population densities. Land shortages within the landscape have resulted in inappropriate land-use such as cultivation on steep slopes, lack of contour ploughing and terracing (NEMA 2008) that has also resulted in severe deforestation within the area (Petursson et al. 2013; Sassen et al. 2015). This ecosystem is also increasingly vulnerable to variable rainfall patterns (FAO 2010; Banana et al. 2014). Based on the frequent emergency operations that have become common for this area, it's evident from reports that adaptation is imperative and must be part of community response to climate change. In 2011 deadly landslides as a result of heavy rains as a result of changes in rainfall patterns left several hundreds of people dead. Further investigations showed that critical areas within the landscape that were responsible for providing important ecosystem services (e.g. provisioning services such as water supply) were severely degraded and hence vulnerable. As a result of this, communities in these areas had become very vulnerable.

Mount Elgon ecosystem occupies 772,300 ha, an ecosystem shared between Kenya and Uganda in almost equal parts (Moyini 2007). It is a mixed ecosystem composed of closed canopy tropical hardwood and bamboo forests and a variety of grassland and riverine ecosystems (Scott 1998). The area is composed of two main ethno groups the Sabiny who live on the northern slopes of the mountain and the Bagisu of Bantu origin who are located in the lower slopes of the Mountain districts. The Sabiny were pastoralists and the Bagisu agriculturalists. However, in the present time, they both share similar socio economic characteristics. Most are engaged in intensive cultivation and animal husbandry such that the land outside the national park has all been cultivated (Kazoora 2001). Kazoora (2001) notes that even the formerly protected areas like steep slopes have been farmed. The area of Mount Elgon that makes up the protected area (e.g. Nakakaawa et al. 2015) is approximately 221,401 ha. This protected area has also been highlighted as an important buffer for its dependent population that live adjacent to it. However, following the gazetting of the National Park, local communities lost all access rights in 1993 (e.g. Sassen et al. 2013, 2015).

It is also expected that the temperature rise will mean that the tree cover will move upwards to follow temperature zones. This implies that the prevalent socio economic and population pressures and movements stand to significantly increase and threaten all populations. A potential conflict between communities on the lower slopes is projected as climate and other anthropological factors continue to change the vegetation cover. Therefore without a clear solution to manage the threat, rather than buffer community, the vulnerability of the communities is expected to increase. This is a rather complex situation: the population is undermining the same ecosystem that is viewed as the best buffer and support to its adaptive capacity.

The other critical challenge for Mount Elgon, is that it is denoted by varied micro ecosystems. This has meant that while some communities enjoy favorable climate and exploit it fully especially such as those located on the higher altitude, on the other hand, communities in the lower reaches are exposed and suffer from the impacts of poor land use. Such is the case presented below of Sanzara Parish, which is located in a rain shadow and on the lower slopes of Mount Elgon. The challenge is such that it faces frequent long dry spells, it also faces significant floods and land slides from the upper catchment that is highly inhabited and highly exploited due to its favourable conditions. Hence for such a community, it has been projected that increased long dry spell and floods are expected from the intensifying rainfall on intensely farmed land. This is particularly likely to be enhanced in areas with very poor drainage (e.g. Government of Uganda 2009a, b).

Thus, as this paper will demonstrate, a holistic ecosystem approach is needed that tries to consider all factors that increase risk and all opportunities in the ecosystems mapped and used to determine the best EbA approach for a particular context.

In 2011, Germany (BMUB) supported Uganda to implement the ecosystem-based adaptation approach in the Mount Elgon area over a period of 4 years. The EbA Mount Elgon project is implemented jointly by United Nations Environment Programme (UNEP), United Nations Development Programme (UNDP) and The International Union for Conservation of Nature (IUCN). It was envisaged that this project would complement the efforts by the government of Uganda in implementing the different action plans in line with climate change adaptation and mitigation. However most of these action plans have been hampered by the lack of tools and methodologies that suit the local conditions of Uganda and more particularly Mount Elgon (Fig. 2).

5 Taking Action, the Bottom: Up Approach

While EbA is now widely accepted, evidence from the ground has shown that there is widespread uncertainty as regards the interpretation and measurement of ecosystem resilience (e.g. Levine et al. 2012). However, there is acceptance generally that as part of adaptation planning, some steps have to be followed at the community level in order to address the issue of ecosystem resilience. There is also general consensus that there is no "one-size fits all" approach nor is there an agreed single approach to ecosystem and hence climate resilience.

Detailed review to understand the challenges faced by the communities showed that ecosystem changes have occurred over time within this mountain ecosystem over a long period of time (e.g. IUCN 2014a). While most of these changes can be alluded to direct drivers such as changes in land-use, population growth, degradation of the forest ecosystems as a result of over-harvesting (e.g. Mutekanga et al. 2013), they have been further compounded by variability in climate within the area. Furthermore actions that include transition to other livelihoods, intensification of agricultural (e.g. cultivation on very steep slopes), are examples of factors that accentuate the impacts of climate change (IUCN 2014b).

The factors articulated above made part of the rationale for selecting Mount Elgon as the project site by the Uganda government. Disasters, particularly landslides that have been compounded by extreme rainfall events, further demonstrated the vulnerability of the communities in the area. As such there has been acknowledgement that adaptation to these changes is at best the only way forward and a

Fig. 2 View of Mount Elgon (© UNEP)

necessity (e.g. Levina and Tirpak 2006). The project partners – UNEP, UNDP and IUCN – agreed from the outset that adaptation approaches are certainly wide ranging infrastructure that reduce the risk of extreme events to actions and roles of local communities, individuals to name a few. As such there was consensus that in order to address the challenges faced by the communities within the Mount Elgon area, ecosystem-based approaches that took into consideration climate risk were favourable, hence ecosystem-based adaptation (EbA). This approach as a strategy took into account actions such as integrated management of land, water, natural resources, while promoting conservation and sustainable use in a fair and equitable manner (IUCN 2014a).

Therefore overall aim in implementing EbA in Mount Elgon has been to strengthen capacities of local communities, which are particularly vulnerable to climate change, by working to strengthen ecosystem resilience through the management of the ecosystem services and hence reducing vulnerability.

In order to address the challenges and hence work towards building community and ecosystem resilience, the project took a bottom-up approach and implemented EbA through a layered approach. This layered approach is shown in Fig. 3. This layered approach is iterative and lessons are drawn at every stage.

Firstly, in identifying the problem areas, three micro-catchment areas and four districts (Fig. 4) were identified and prioritised by the communities and their leaders through a participatory approach, for implementation of EbA options. Secondly, this project further worked to determine the vulnerabilities of the area. During this period of determining vulnerabilities through systematic methodology of Vulnerability Impact Assessment (VIA), the implementers employed the no-regret[1] approach as an important part of EbA implementation. And finally actions were identified in the form of EbA options, collaboratively with the affected communities and stakeholders.

This paper will provide only one case study from the whole area, clearly articulating the layered approach.

[1] The no-regret actions under the EbA Mount Elgon project were defined as those including autonomous measures by communities which do not worsen vulnerabilities to climate change or which increase adaptive capacities; and measures that will always have a positive impact on livelihoods and ecosystems regardless of how the climate changes.

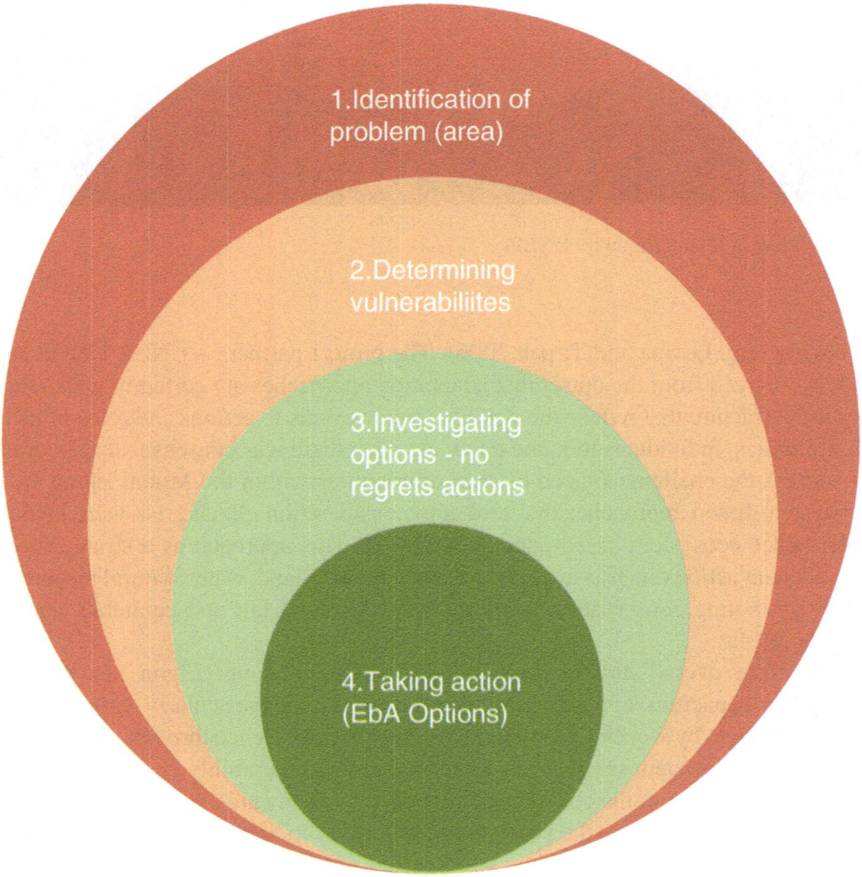

Fig. 3 A layered approach to undertaking EbA in the Mount Elgon ecosystem

6 Implementation of EbA in the Sanzara Parish of Kapchorwa District Through the Four-Layered Approach

6.1 Identification of the Problem Area

Lessons from this project demonstrated that EbA though framed as a comprehensive approach and is usually rooted in both its view on ecosystems services and climate change adaptation (Chong 2014) the reality was such that activities were implemented independently. Within the overall context of the project the activities in the different locales were motivated *either* by changes in ecosystem services *or* climate change adaptation. Therefore for the purposes of this paper the focus will be on Sanzara Parish an area whose ecosystem services had been altered.

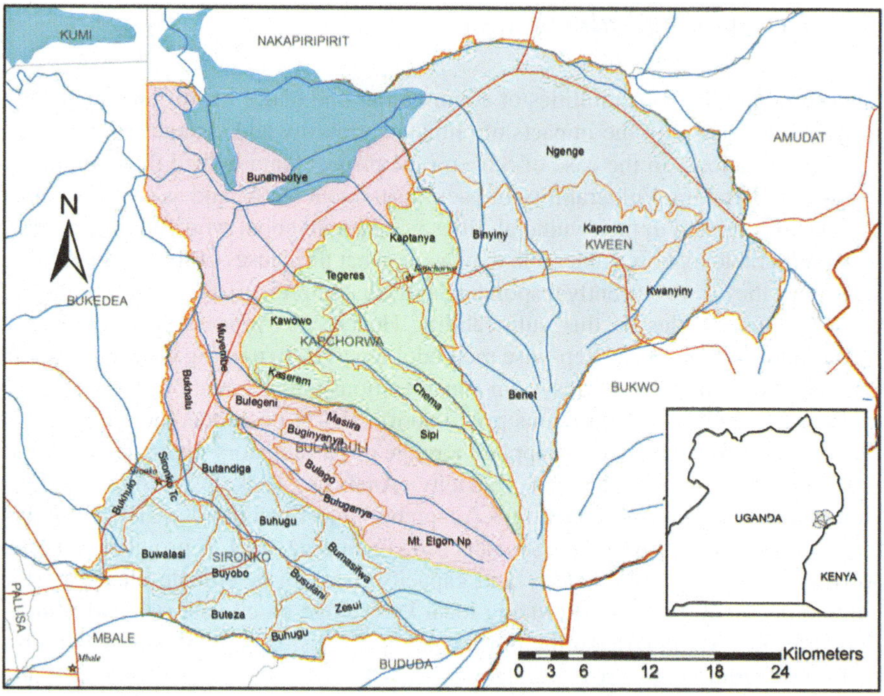

Fig. 4 Map of the project area in Mount Elgon – Uganda

Selection of Sanzara Parish as a problem area was based on earlier mapping[2] which highlighted the uniqueness of the parish due to its location in the rain shadow and the projected threats to the ecosystem and the livelihoods that would increase as a result. The communities in Sanzara parish[3] are purely subsistence farmers with systems ranging from food and cash crops (both perennial and annual) as well as animal rearing. Just as in the case of the entire district, the parish has two seasons, wet and dry which were observed to have changed in frequency hence hardly sustaining vegetation to healthy maturity. This as well as other poor agronomic practices has rendered the place almost bear of forest. This uniqueness made Sanzara a problem area which required deeper and concentrated understanding of the vulnerabilities, planning and demonstrating the probable EbA options to make a case.

[2] IUCN, in partnership with UNDP and IISD undertook a study in 2012 on climate risk management for sustainable crop production in Uganda and the mapping identified Sanzara as one of the most vulnerable parishes.

[3] According to the Sub-county Development Plan (2011–2014/2015–2016), the population is projected at 11,270 people.

6.2 Determining Vulnerabilities

Understanding the vulnerabilities of communities and hence identifying adaptation measures to cope with the impacts of climate variability and change require localized investigations. In the case of Sanzara the project team helped the community through undertaking Vulnerability Impact Assessment (VIA) and as such a number of questions related to their vulnerabilities were posed about who and to what they were vulnerable to, how vulnerable they were, what the causes of their vulnerability were, how they were currently responding and what improvements would strengthen their responses to lessen this vulnerability. Hence VIA process was based on a rather iterative than linear step-wise methodology as shown in Fig. 5 (Munroe et al. 2015) that took integrated ecosystem considerations into account.

Hence to address the questions posed above, IUCN positioned communities at the center of their analysis and applied a range of participatory approaches such as the Climate Vulnerability and Capacity Analysis Took (CVCA),[4] and the Community-based Risk Screening – Adaptation and Livelihoods (CRiSTAL) and also Geographical Information Systems (GIS) mapping. CVCA provides a framework for analyzing vulnerability and capacity to adapt to climate change at the community level through prioritizing local knowledge on climate risk and adaptation strategies. CRISTAL on the other hand is a decision support tool which promotes integration of risk reduction and climate change adaptation into community-level projects. CRISTAL uses information from the vulnerability assessments. Building on the information generated by the CVCA and CRISTAL, IUCN then applied the GIS mapping to further visualise, interpret and understand the relationships, trends and patterns from a scientific point of view. This in effect facilitated the community to analyse their information (Fig. 5). These approaches were instrumental and critical in building community awareness, consensus and certainly capacity of both the local communities and also local government leaders and partners. They also contributed to the stakeholders having the ability to map the ecosystem goods and services within their areas, identify the risks at hand and also the EbA options that needed to be undertaken to address the issues (Fig. 6).

Ultimately the community themselves, through the comprehensive VIA's and also rapid assessment methodologies, prioritised floods and droughts as major climate risks, which interfered and reduced their capability and capacity to use the land to meet their livelihood needs. The community further highlighted and identified the perennial River Sippi as a vital and main opportunity for them to adapt to climate change.

However further interrogation and analysis showed that this major source of water for Sanzara area actually originated from the slopes the Mount Elgon National Park in the upper reaches of the mountain. Furthermore, it was evident that the river traversed an intensely used landscape that undermines the water quality. These findings further demonstrated how the community's adaptive capacity would be threatened

[4]CVCA is a tool that is used as a guide to gather information using key questions at different levels: individual/Household, Local government/community and national levels.

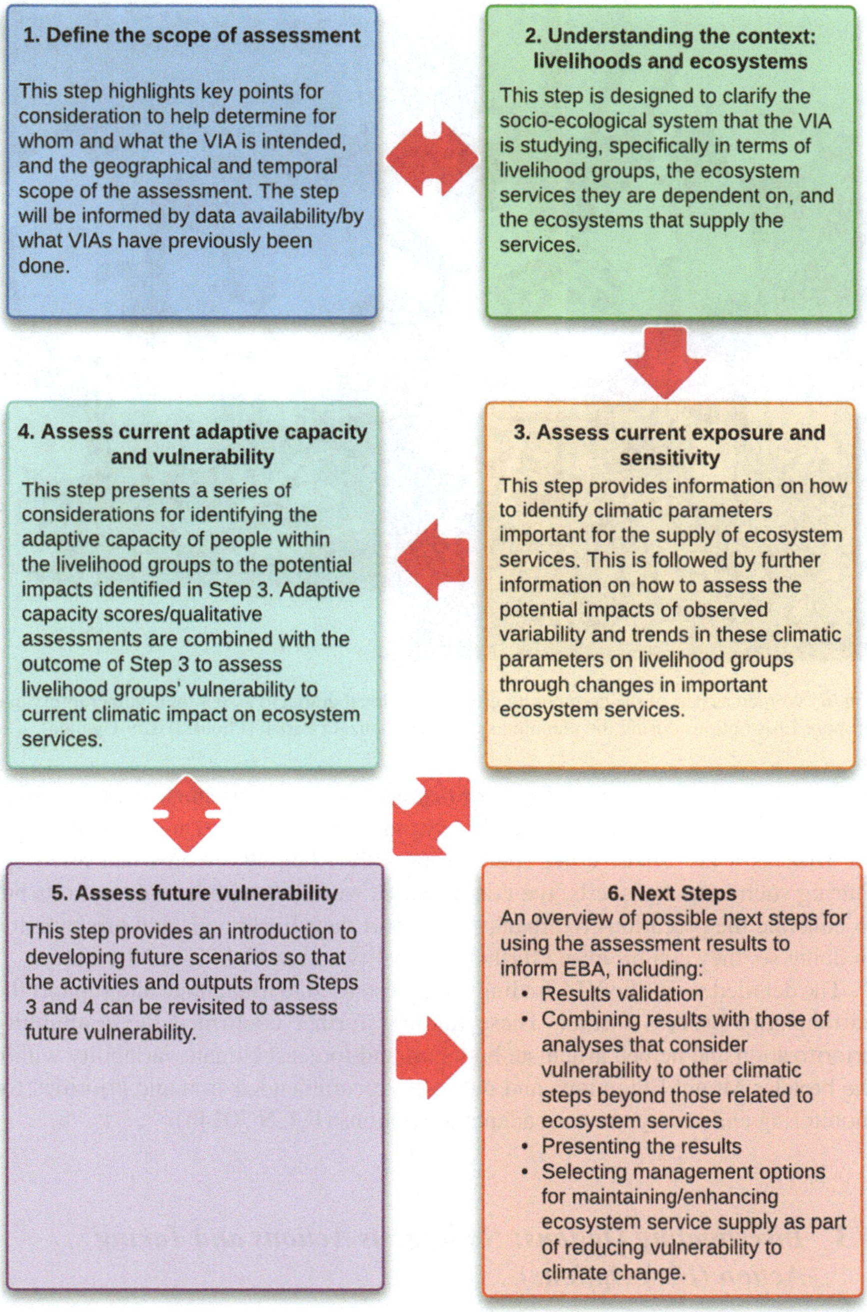

Fig. 5 Framework of VIA guidance steps (Munroe et al. 2015)

Fig. 6 Women as part of the community group using local drama to assess their vulnerability and propose EbA options during the planning sessions in Sanzara Parish (Photo: IUCN-Uganda)

due to the water sources high levels of degradation and contamination. In addition the rivers deep river banks were a risk to women and children who frequently used the river as a last resort water source especially when all streams had dried up. During such long dry spells, the communities were left with limited options but to cultivate the precarious riverbanks to support their livelihoods and productivity. In doing so, they compromised further the quality and availability of the water.

The detailed scientific VIA findings were also used to triangulate and support the participatory process further. These findings further confirmed the community prioritization and helped articulate both past and forecast climate variability within the broader Mount Elgon area and thereafter recommended strategic priorities for monitoring and management of adaptation options (IUCN 2014b).

6.3 Investigating Options: No Regrets Actions and Taking Action (EbA Options)

Water provisioning as an ecosystem service was identified, and River Sippi as d the ecosystem providing it. Within the overall project area it was clear that the best approach was to protect River Sippi and as such address the challenges it was facing. It was therefore realistic and crucial to start with the most hard-pressed and vulnerable community – Sanzara Parish. Furthermore, the project needed a short-term intervention that would in effect mobilise and incentivise the community to appreciate the value of the River Sippi and hence no-regrets actions were identified. The project partners agreed on the working definition of no-regrets actions as those that included "*autonomous measures by communities which do not worsen vulnerabilities to climate change or which increase adaptive capacities*; *and measures that will always have a positive impact on livelihoods and ecosystem services regardless of the climate changes*." The no-regret approach was therefore seen as an important part of the long-term EbA actions for the area.

Thus this provided a strategic opportunity and also entry point for the project to commence its work and hence sustain the ecosystem service to manage the entire ecosystem for the community in its proximity. The added value was that upstream communities would also benefit even though the floods did not affect them. In essence the Sanzara community had been facilitated to be at the forefront of requesting the upstream communities to reduce the degradation of the riverine area.

Henceforth, a partnership was forged between the downstream (Sanzara Parish) and the upstream community – Kapchorwa District Water Office and their communities. Through the discussions that ensued and with the support from this project, the communities supported the tapping of water from River Sippi through a gravity flow scheme – a form of gray infrastructure (e.g. Palmer et al. 2015) (Fig. 7). The historical experiences evidenced by the disasters and hence vulnerabilities as a

Fig. 7 Gravity water scheme at Sanzara Parish after completion (Photo by IUCN – Uganda)

result of the variability within the area, convinced communities on the agreed plan of action. To emphasise the EbA approach, an Environmental Impact Assessment (EIA) was carried out and this confirmed the feasibility of the intervention lest the project further undermined the security of the people, and the ecological integrity. The findings of the EIA confirmed that the abstraction of the water would not compromise the regular supply of water and mitigation measures were recommended. The bigger picture of the scheme within the context of EbA was that the scheme would enhance the water provisioning service, through access to water for food production through irrigation. The scheme would also enhance the supply of clean water for domestic use and thereby increasing the resilience (e.g. Epple and Dunning 2013) and adaptive capacities of over 1000 households in Sanzara Parish area. It was also envisaged that this initial step would catalyse the Sanzara community to take lead in influencing the upper catchment communities to sustain the initiative through catchment restoration and ensure the continuous flow of this water (Fig. 8).

Through thorough analysis, consultations and discussions with the communities, the project settled for a "semi-gray" infrastructure option in the form of the gravity flow scheme as an entry point for engaging with both upstream and downstream water users (UNEP 2014). The strategy therefore was to use this type of infrastructure to restore the integrity of river from its source while ensuring water supply downstream.

The Gravity water scheme (UNEP 2014) demonstrated additionality through its ability to use an existing river to address water stress and mobilise both upstream and downstream communities in integrated water catchment management. The scheme provided a platform for planning and demonstration the value of catchment

Fig. 8 Irrigated field (downstream) using running water in trench after the completion of Gravity Scheme (upstream) (Photo by IUCN – Uganda)

management in enhancing social and ecosystem resilience. Through this platform, communities developed a 10 year catchment plan with a vision (Fig. 9) for resilient people and ecosystems within Sanzara area. Key interventions which have been implemented as part of this plan include river bank rehabilitation, soil and water conservation structures and tree planting to restore the entire degraded landscape on which River Sippi depends. In the face of climate change, the integrated watershed management practices are expected to control soil erosion, reduce water pollution risk and increase crop productivity. Agro-forestry systems will be integrated in the farming systems for reduced pollution loading and reduced pressure through provision of domestic energy. Establishment of a buffer zone along the river was expected to support natural regeneration hence stabilising the soils and making them more resilient to floods, in addition to providing water for production during the dry season, and clean water for domestic use.

These actions, therefore set the premise for EbA in Sanzara and in effect the project applied most of the principles of EbA as indicated in IUCN (2014b).

What is evident through literature and experience is that EbA is a relatively new concept and approach whose premise is to systematically harness the services of ecosystems to cushion local communities from extreme events therefore facilitating

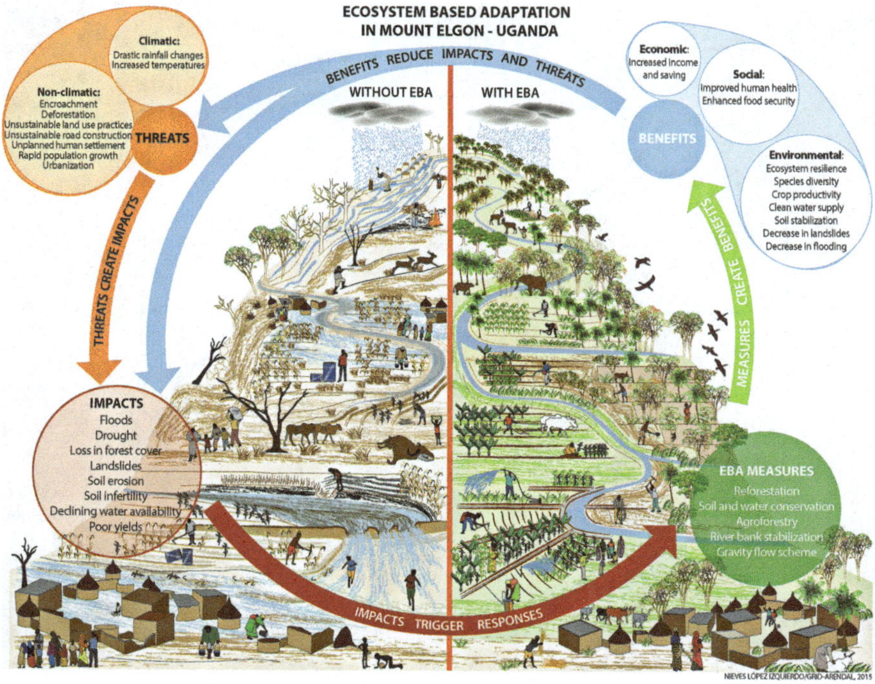

Fig. 9 Sanzara Vision as developed by local communities – Impression by local artist (Source: Nieves Lopes Izquierdo/GRID-Arendal, based on technical guidance from T. Rossing, P. Dourojeanni, C. Petersen, N.I. Nyman & P. Nteza)

adaptation to the severe effects of climate change (Wamsler et al. 2014; Munang et al. 2013). As indicated earlier the provisioning service identified by the community was one through which the natural ecosystem sustained and fulfilled the human life. It is on this premise that after the initial step of enabling the Sanzara community to access water, the subsequent process involved mobilising both the downstream and upstream communities in catchment restoration (Fig. 8). Evidence from many parts of the world has shown that the ecosystems concept is considered as an effective avenue to advance sustainable local government planning (Ahern et al. 2014; Wamsler et al. 2014). As such in order for this project to advance the sustainable planning, the role of the local government of the area was emphasised. Thereafter the project also established an incentive scheme for the upstream communities, as it was evident that the degradation emanated from their socio-economic activities. On their part this was not intentional as this was the only option and resort they had (Fig. 10).

Many factors were taken into consideration during the implementation of this project component. Notably an Environmental Impact Assessment (EIA) conducted and approved by the National Environment Management Authority (NEMA) demonstrated the leadership, commitment and support of government to address the communities' challenges.

Fig. 10 Map developed by the community identifying the priority restoration areas within Mount Elgon

The involvement of various stakeholders to understand the EbA, in participatory planning and implementation, ensured that leadership and the community's awareness, committed, to take lead. The technical local government leaders saw this as delivering on their mandate. While the political leaders show this as being responsive to their constituency. It was an opportunity to engage their people and mobilise them towards a critical goal. The awareness and tools provided to the local leaders enabled them lobby the extension of the project through government support to the villages not reached. Generally a process was established that strengthened social adaptive capacity of the leaders to respond through their institution, ecosystem adaptive capacity and the communities. The project facilitated regular community planning and reflection meetings on progress of planned target. This enabled communities to engage and take charge of the EbA actions, including demanding for actions from their leaders. In fact, community members based on the awareness and the engagement platforms that had been created through the project, regularly summoned leaders. Some village chiefs were changed, the district chairman always moved with technical officers to respond to people's interest and demands. As a result the district had to budget for the extension of the project to the neighbouring villages that had not been reached. Further, they took lead in convening council meetings to discuss actions in the upstream that would affect the lower stream action learning groups and agree on working modalities and management structures for better and coordinated implementation of activities on the ground.

It should be noted that the community contributed 30 % of the cost of the gravity flow scheme through their own labour. The participatory communal work of constructing trenches to lay the initial pipes within community land was particularly relevant taking into account the land tenure dynamic. Participatory construction of the community gravity flow was officially launched on 11 April 2012 by the project in partnership with Kapchorwa District Local Government. Construction of the trenches was eventually completed in September 2012 with a total of 6 km (kilometer) trenches constructed and pipes laid.

Capacity building was seen as a necessary component of the approach. As already mentioned the aim was to place communities at the center of the project, to assess situations and determine best way forward in using the ecosystem to respond to the effect of climate change. However, the project ensured that a scientific analysis were undertaken to triangulate and inform options for the community. Climate change scenarios for the ecosystem were mapped and these were used to raise awareness and facilitate discussion with the communities on the possibilities based on the changing climate.

Communities were trained and introduced to an incentive mechanism as a strategy for promoting better ecosystem management. This was tagged to better management of River Sippi to ensure sustainable flow of quality water in the right quantities, through better land management practices which the communities decided on indicators. The indicators helped determine the best contributors and the mode of payment to motivate and encourage better performance.

7 Conclusion

It's clearly evident that climate change is increasingly threatening local livelihoods in the Mount Elgon area and the very ecosystems they are dependent on. Ecosystem-based adaptation (EbA) provides an opportunity as a robust and flexible strategy that can at the same time cope with the uncertainties that climate change poses. The experience from Sanzara also showed that involvement of all the different stakeholders coupled with local and scientific knowledge decreased the likelihood of maladaptation and also promoted ownership and sustainability (e.g. IUCN 2014a). In addition as demonstrated at the site level in the Mount Elgon area, EbA has the potential for synergies with other adaptation options as its part of an overall adaptation strategy. What is equally evident from this work is that EbA still remains under-utilized by policy makers and associated stakeholders.

Acknowledgments This paper is based on work undertaken as part of a wider project the "Ecosystem based adaptation in Mountain Ecosystems of Nepal, Peru and Uganda" supported by the German Government (BMUB). We would like to particularly thank the project team members based in Uganda (for Mount Elgon), local communities, local governments and all the partners involved in this work (UNEP, UNDP and IUCN). The authors of this paper would like to acknowledge various contributors mainly colleagues working on this project who provided valuable input to this chapter.

References

Articles

Ahern, J., Cilliers, S., Niemela, J. (2014). The concept of ecosystem services in adaptive urban planning and design: A framework for supporting innovation. *Landscape and Urban Planning, 125*, 254–256.

Banana, A. Y., Byakagaba, P., Russell, A. J. M., Waiswa, D., & Bomuhangi, A. (2014). *A review of Uganda's national policies relevant to climate change adaptation and mitigation: Insights from Mount Elgon. Working paper 157*. Bogor: CIFOR.

CBD. (2009). Connecting biodiversity and climate change mitigation and adaptation: Report of the Second Ad Hoc Technical Expert Group on Biodiversity and Climate Change.

Chong, J. (2014). Ecosystem-based approaches to climate change adaptation: Progress and challenges. *International Environmental Agreements Political Law Economy, 15*, 17–41.

CIGI. (2007). International risk report: The Center for International Governance (CIGI).

Dudley, N., Stolton, S., Belokurov, A., Krueger, L., Lopoukhine, N., MacKinnon, K., et al. (2010). *Natural solutions: Protected areas helping peoplecope with climate change*. Gland: IUCN-WCPA, TNC, UNDP, WCS, The World Bank and WWF.

Epple, C., & Dunning, E. (2013). *Ecosystem resilience to climate change: What is it and how can it be addressed in the context of climate change adaptation? Technical report for the Mountain EbA Project*. Cambridge: UNEP World Conservation Monitoring Center, UNEP.

FAO. (2010). *Global forest resources assessment. FAO Forestry paper 163*. Rome: FAO of the United Nations.

Government of Uganda. (2007). National Adaptation Programmes of Action *Kampala.*
Government of Uganda. (2009a). National development plan 2010/11–2014/15.
Government of Uganda. (2009b). The state of Uganda population report, 2009: Addressing the effects of climate change on migration patterns and women (Funded by UNFPA Uganda).
Hepworth, N. D. (2010). *Climate change vulnerability and adaptation preparedness in Uganda.* Nairobi: Heinrich Böll Foundation.
IUCN. (2014a). Technical paper – Ecosystem based adaptation: Building on no regret adaptation measures. UNFCCC COP20, Lima.
Kazoora, C. (2001). *Uganda Country Report.* Compiled for IUCN and WWF.
Knapen, A., Kitutu, M. G., Poesen, J., Bruegelsmans, W., Deckers, J., & Muwanga, A. (2006). Landslides in a densely populated county at the slopes of Mount Elgon (Uganda): Characteristics and causal factors. *Geomorphology, 73*, 149–165.
Levina, E., & Tirpak, D. (2006). *Adaptation to climate change: Key terms.* Paris: OECD.
Levine, et al. (2012). *The Relevance of 'resilience'?.* ODI, 2012.
Moyini, Y. (2007). *Pro-poor and pro-conservation policies and operational procedures in the Mount Elgon Ecosystem.* Nairobi: IUCN, Eastern Africa Regional Office.
Munang, R., Thiaw, I., Alverson, K., Mumba, M., Liu, J., & Rivington, R. (2013). Climate change and ecosystem-based adaptation: A new pragmatic approach to buffering climate change impacts. *Current Opinion in Environmental Sustainability, 5*, 1–5.
Munroe, R., Hicks, C., Doswald, N., Bubb, P., Epple, C., Woroniecki, S., et al. (2015). *Guidance on integrating ecosystem considerations into climate change vulnerability and impact assessments to inform ecosystem - based adaptation.* Cambridge: UNEP-WCMC.
Mutekanga, F. P., Kessler, A., Leber, K., & Visser, S. (2013). The use of stakeholder analysis in integrated watershed management: Experiences from the Ngenge watershed, Uganda. *Mountain Research and Development, 33*(2), 122–131.
Nakakaawa, C., Moll, R., Vedeld, P., Sjaastad, E., & Cavanagh, J. (2015). Collaborative resource management and rural livelihoods around protected areas: A case study of Mount Elgon National Park, Uganda, Forest Policy and Economics (2015). http://dx.doi.org/10.1016/j.forpol.2015.04.002
NEMA. (2008). *State of environment report for Uganda.* National Environment Management Authority Publications.
Palmer, M. A., Liu, J., Matthews, J., Mumba, M., & D'Odorico, P. (2015). *Science, 349*(6248), 584–585.
Petursson, J. G., Vedeld, P., & Sassen, M. (2013). An institutional analysis of deforestation processes in protected areas: The case of the transboundary Mount Elgon, Uganda and Kenya. *Forest Policy and Economics, 26*, 22–33.
Sassen, M., Sheil, D., & Giller, K. (2015). Fuelwood collection and its impacts on a protected tropical mountain forest in Uganda. *Forest Ecology and Management, 354*, 56–67.
Sassen, M., Sheil, D., Giller, K. E., & ter Braak, C. J. (2013). Complex contexts and dynamic drivers: Understanding four decades of forest loss and recovery in an East African protected area. *Biological Conservation, 159*, 257–268.
Scott, P. (1998). *From conflict to collaboration: People and forests at Mount Elgon, Uganda.* Gland: The World Conservation Union (IUCN).
UNEP. (2014). Fast tracking ecosystem-based adaptation: Building resilience to climate change in Africa. UNEP Technical report.
UNFCCC. (2007). The Nairobi work programme on impacts, vulnerability and adaptation to climate change. UNFCCC Secretariat, Accessed from: https://unfccc.int/files/adaptation/sbsta_agenda_item_adaptation/application/pdf/nwp_brochure.pdf
United Nations Development Programme (UNDP), & Bureau for Crisis Prevention and Recovery (BCPR). (2013). *Climate risk management for sustainable crop production in Uganda: Rakai and Kapchorwa districts.* New York: UNDP BCPR.

Wamsler, C., Lenderitz, C., & Brink, E. (2014). Local levers for change: Mainstreaming ecosystem-based adaptation into municipal planning to foster sustainability transitions. *Global Environmental Change, 29*, 189–201.

Books

IPCC. (2007). In M. L. Parry, O. F. Canziani, J. Palutikof, P. J. van der Linden, & C. E. Hanson (Eds.), *Climate change 2007: Impacts, adaptation and vulnerability. Contribution of Working Group II to the Fourth Assessment Report of the Intergovernmental Panel on Climate Change*. Cambridge: Cambridge University Press. 976 pp.

Millennium Ecosystem Assessment. (2005). *Ecosystems and human well-being: Synthesis*. Washington, DC: Island Press.

Internet Resources

IUCN. (2014b). *Ecosystem based adaptation – An approach responding to climate hazards*. Fact sheet. https://cmsdata.iucn.org/downloads/eba_english.pdf

UNFCCC. (2010). Dec 1/CP.16, para 14 (d), http://unfccc.int/resource/docs/2010/cop16/eng/07a01.pdf

United Nations. (2012). The future we want. Shared at Rio+20. http://daccess-dds-ny.un.org/doc/UNDOC/GEN/N11/476/10/PDF/N1147610.pdf?OpenElement

Vulnerability Assessments for Ecosystem-based Adaptation: Lessons from the Nor Yauyos Cochas Landscape Reserve in Peru

Pablo Dourojeanni, Edith Fernandez-Baca, Silvia Giada, James Leslie, Karen Podvin, and Florencia Zapata

Abstract The development of Vulnerability Assessments (VA) to climate change is a rapidly evolving activity within the broader climate adaptation planning process. As such it is receiving significant attention from the communities of adaptation researchers and practitioners. It is uncommon to carry out more than one VA in the same place and at the same time thus this case study presents a unique opportunity to compare the application of three different Vulnerability Assessment approaches that were carried out simultaneously in the same location: the Nor Yauyos Cochas Landscape Reserve in Peru, during the period of 2012 through 2013. All three approaches shared the goal of identifying Ecosystem-based Adaptation (EbA) measures based on the ecological and social vulnerabilities the VAs helped to identify in the target area. Each approach, however, was different in terms of methodologies and conceptual foundation. The following case study describes the application of a participatory VA approach, a model-based VA approach and a deductive VA approach, using a set of descriptors in a custom designed matrix. We also present a narrative description of each approach to explain in more detail the process undertaken by each Vulnerability Assessment. Key lessons learned are that EbA measures require abundant information (pertaining to climate, ecosystems, biodiversity, land use practices, livelihoods, etc). As a result, interaction between scientific knowledge and traditional (local) knowledge is vital. Of importance, all three approaches rendered useful and pertinent results and surprisingly recommended very similar adaptation measures. Nevertheless, the participatory approach was the only one that did

P. Dourojeanni (✉) • E. Fernandez-Baca • J. Leslie
UNDP, Lima, Peru
e-mail: pablo.dourojeanni@undp.org; edith.fernandez-baca@undp.org

S. Giada
UNEP-ROLAC, Panama City, Panama

K. Podvin
IUCN, Lima, Peru

F. Zapata
TMI Andean Program, Lima, Peru

N. Salzmann et al. (eds.), *Climate Change Adaptation Strategies – An Upstream-downstream Perspective*, DOI 10.1007/978-3-319-40773-9_8

not require additional studies to implement measures following the Vulnerability Assessment. The three approaches also proved to be advantageous for application at different scales. While the participatory approach turned out to be most useful at the community level, the model-based approach and the deductive approach delivered information at a broader scope that served to better understand vulnerability for the entire ecosystem target area.

Keywords Ecosystem-based Adaptation (EbA) • Vulnerability Assessment • Peru

1 Introduction

Identifying current climate change adaptation measures is usually done through assessing vulnerability for a selected site. Yet, there are many different approaches to assess vulnerability, making it difficult to evaluate the approaches comparatively. Furthermore studies that examine approaches to Vulnerability Assessments exist, but these assessments are often undertaken at different sites (for an example see: Hammill et al. 2013). Vulnerability as a context-specific phenomenon makes it difficult to reach overarching conclusions on how the different approaches perform (Hammill et al. 2013; Munroe et al. 2012). Moreover, there are not many documented cases that allow for an examination of different approaches for selecting adaptation measures in the same site. Given that the three approaches examined in this case study vary in many aspects (see descriptions further below), we use a broad definition of VA to be able to encompass them: VA is a process that helps identify adaptation needs and options (adapted from PROVIA 2013) (Fig. 1).

Ecosystem-based Adaptation (EbA) is gaining traction as an approach that makes valuable contributions to and fills important gaps in broader climate change adaptation strategies (Travers et al. 2012; Munroe et al. 2012; Reid 2011; CBD 2009) and in the past few years there is a growing number of cases that show that EbA measures are being implemented across a wide range of ecosystems. Ecosystem-based Adaptation is defined by the Convention on Biological Diversity (CBD) as ***"the use of biodiversity and ecosystem services to help people adapt to the adverse effects of climate change"*** (CDB 2009). As further elaborated by Decision X/33 on Climate Change and Biodiversity, this definition also includes the ***"sustainable management, conservation and restoration of ecosystems, as part of an overall adaptation strategy that takes into account the multiple social, economic and cultural co-benefits for local communities"*** (CBD 2010)

Some EbA measures are based on traditional knowledge of local communities, while others are based on know-how from elsewhere and/or adapted to current environmental problems. The selection of which EbA measures to apply in a given place or time is shaped by the three aspects: the local context; how one frames the problem that needs to be addressed (a combination of defining the threats, hazards,

Fig. 1 The Nor Yauyos Cochas Landscape Reserve (RPNYC); located in the high and middle basin of the Cañete River and in the Cochas Pachacayo basin (photograph by Pablo Dourojeanni)

risk factors, sensitivity and adaptive capacity of the system under scrutiny) and how to go about identifying possible solutions (the assessment process itself) (GIZ 2014; Hammill et al. 2013). It is then interesting to pose the following question: Would the use of different approaches for assessing vulnerability, in a given place and time, arrive at different solutions? In this case study we have the unique opportunity to evaluate three different approaches that were applied simultaneously within the same study area, the Nor Yauyos Cochas Landscape Reserve in the Peruvian Andes.

2 Overview of Institutional Settings and Case Study Site

This case study derives from activities executed to date by the Ecosystem-based Adaptation (EbA) in Mountain Ecosystems Project. This is a collaborative initiative of the United Nations Environment Programme (UNEP), the International Union for Conservation of Nature (IUCN) and the United Nations Development Programme (UNDP), funded by Germany's Federal Ministry for the Environment, Nature Conservation, Building and Nuclear Safety (BMUB). In Peru, the Project is

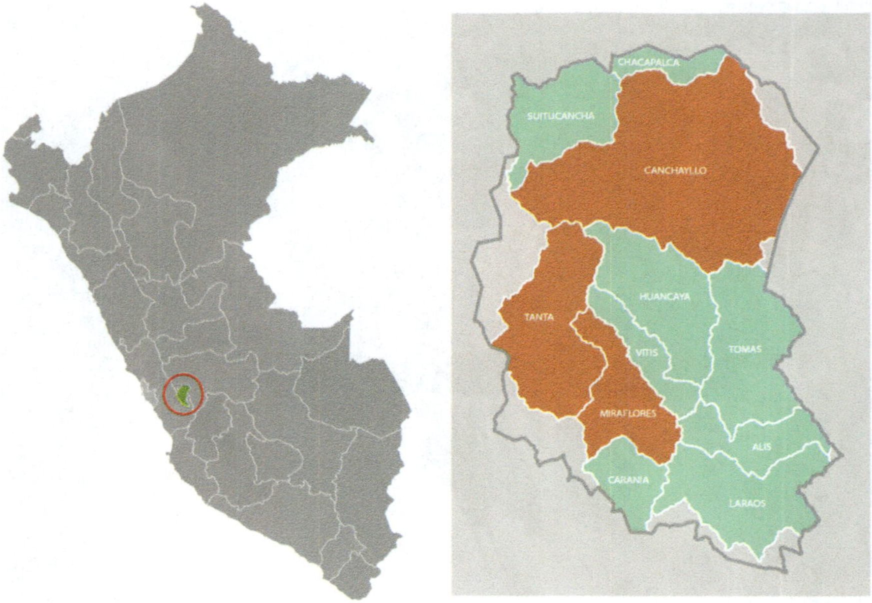

Fig. 2 Map of Peru and case study site showing the three districts where adaptations measures are being implemented

commissioned by the Ministry of Environment of Peru (MINAM for its Spanish acronym) and is implemented in the Nor Yauyos Cochas Landscape Reserve (NYCLR) with the support of the National Service of Natural Protected Areas (SERNANP for its Spanish acronym). The activities under IUCN's responsibility are implemented in partnership with The Mountain Institute (TMI).

The NYCLR is one of only two landscape Reserves in Peru (SERNANP 2014). Direct use of the natural resources by the local population is encouraged, but closely regulated by park authorities. The area holds great diversity of high Andean flora and fauna and is a source of many ecosystem services both for the local population as well as for populations located outside the Reserve (INRENA 2006). Water provision from the Reserve to downstream cities and highly productive agricultural valleys is a key conservation objective. It is estimated that more than 11 million Peruvians, including the inhabitants of Lima City downstream, depend on water that origins in this Reserve (Fig. 2).

It is widely accepted that current climate change impacts in the Andean mountains include changing rainfall patterns, glacier melt and reduction of downstream flows, increased temperatures, an increase in occurrence of frost and extreme weather events and the upward movement of flora and agriculture. In the future it is expected that these impacts will be exacerbated across the region (MINAM 2010). The three assessments showcased here focused on assessing climate vulnerability. Nevertheless other (local) drivers of ecosystem change were also explicitly taken into account (such as overgrazing, land use change, national and local policy).

The EbA Mountain project in Peru decided to undertake three approaches for assessing vulnerability in order to generate knowledge about ecosystem-based VA methodologies. UNEP undertook two parallel Vas (the model-based and deductive approaches) while the collaboration between UICN and TMI opted for the participatory approach. This provided the opportunity to assess vulnerability in three different ways within the same site (the NYCLR). Thus we avoided examining methods and results of VAs from different sites and different contexts, thereby allowing us to focus on studying the approaches themselves. The overall shared aim among the Vas was to identify EbA measures. Nevertheless, as an agreed implementation strategy between project partners, a "no-regrets measures" strategy was also conducted. These no regret measures were led by TMI and IUCN within the participatory approach and served as an entry point into communities with the explicit intention of being based on ecosystems.

To evaluate the approaches in a systematic way, a set of descriptors used for describing variables was created. In light of this evaluation, we propose an integrated VA approach that derives from the project experience in order for climate adaptation practitioners and government officials to learn from this experience.

3 Definition of Descriptors Used to Homologize Approaches

The designs of the three different Vulnerability Assessments respond to different conceptual approaches of how to identify adaptation measures, and are described as "deductive approach", "participatory approach" and "model-based approach". Through each approach, data was collected and analyzed. After completing the Vas, the project team identified lessons learned. Quantitative and qualitative data derived from the approaches were classified by a set of custom made descriptors to homologize the three different approaches, and were documented in a matrix as part of the case study. The descriptors included in the matrix are: scope, resolution, duration, expense, inputs (data and information requirements), outputs (kinds of results and recommendations of adaptation measures) and level of local participation in adaptation measure construction and selection (see Table 1). Definitions of descriptors are listed below.

<u>Scope of VA area:</u> refers to the geographical area covered by the VA. It can range from large to small, such as the entire project site (the NYCLR including its buffer zone), a watershed within the project site, an ecosystem, a district or a village. In this case study, the scope is measured in geographical units (square kilometers).

<u>Resolution of VA:</u> refers to the degree of detail of the inputs and outputs of the assessment. It can range from being a fine resolution where both input information and output information have great detail or a coarse resolution where outputs are coarse (even though inputs might have a fine resolution). Examples of fine and coarse resolution can be information on vulnerability at village level (fine

Table 1 Matrix of descriptors, attributes and units of measurements

Descriptor	Attribute	Unit of measurement	Value of attribute
Scope	Size of area covered by assessment	Square kilometer (Km^2)	Depends on the approach
Resolution Input information	(input) resolution of source data and information	Relative (Fine or coarse)	Fine: use of Primary data (surveys, interviews, focal groups, field observations, participatory methods)
			Coarse: secondary data (literature review or other sources of data) (Or a combination of both (primary and secondary data) (fine)
Resolution Output information	(output) resolution of results and EbA measure recommendation	Relative (Fine or coarse)	Fine: village, community
			Coarse: district, ecosystem, watershed or entire project site
Duration	Time spent conducting assessment	Months	User defined value
Expense	Resources spent doing assessment	US Dollars	User defined value
Inputs	Data and information required for assessment	Relative (low, medium, high, very high)	User defined value
Outputs (products and recommendations)	(products) types and number of different products	Count number and type	User defined value
	(recommendations) types and number of proposed EbA measures	Count number and type	As defined in the variable SCALE of output resolution information

Participation level	Local participation level	Local participation level according to forms of participation	Very low: Contractual participation form
			One social actor has sole decision-making power over most of the decisions taken during the process
			Low: Consultative participation form
			Most of the key decisions are made by one social actor, but emphasis is put on consultation and gathering information from others
			High: Collaborative participation form,
			Different actors collaborate and are put on a more equal footing, emphasizing linkage through an exchange of knowledge, different contributions and a sharing of decision-making power during the process
			Very High: Image Collegiate Participation
			Different actors work together as colleagues or partners. “Ownership” and responsibility are equally distributed among the partners, and decisions are made by agreement or consensus among all actors
			(Adapted from Vernooy (2005)
Beneficiaries of the information	Stakeholders directly interested in the results of assessment	# of institutions or stakeholders	User defined value
Following steps	Readiness for implementing EbA measures after end of assessment	List of steps and/or time in months to undertake those steps	User defined value

resolution) or at entire project site (coarse resolution). In this case study the resolution is measured twice, distinguishing between input and output information.

Duration of VA: refers to how long the VA took to reach final selection of proposed EbA measures. The beginning of the duration is set by the start of the assessment. The start of the assessment is the signature of the consultant contract of the team who leads the approach. The end of the duration of assessment is marked by the delivery of the final report with the proposed EbA measures to be implemented at project site. The duration of assessment is measured in time units of months.

Expense of VA: refers to how much it cost to conduct the assessment. The cost includes personnel who conduct the assessment, field activities associated with data collection and validation, data procurement and materials. The expense of the assessment is measured in currency units of US Dollars.

Inputs necessary for generating VA: refers to the amount of data and information needed to characterize the social and ecological system(s) under scrutiny. Inputs may vary from literature review, meteorological data, future climate scenarios and projections, census information, surveys, field observations, etc. By default any approach may use any given amount of data and information that is available or that it is possible to generate. Nevertheless, by design, the three approaches featured in this case study also have different demands of input information that is needed for generating results (outputs). Some are more information-demanding than others. To measure the inputs necessary for each of the assessments, we needed to categorize each approach and list the inputs needed for generating results (see section on description of approaches below). A scale of four classes was established to measure the amount of inputs necessary for generating the three assessments featured in this case study. The scale is divided into low, medium, high and very high demand of inputs.

Outputs generated by VA: refers to the two types of possible outputs that are of interest in an adaptation project. These refer to products that synthesize the Vulnerability Assessment and the final recommendations (EbA measures to be selected for implementation). The most common types of results can be a wide range of products, such as maps, reports, or indicator-based vulnerability scales. The final recommendations refer to the EbA measures that are proposed on the basis of the results of the assessment. As well, as explained above (see resolution of assessment), the output information resolution is an important aspect to be taken into account and is in direct relation with the EbA measure. For example, there is a difference in the degree of certainty in recommending an adaptation measure suited for one village and the same recommendation for the entire project site. This can also be viewed as a correlation between the variables of scope, resolution and output information. The design of each assessment has a different scope and resolution. The output information should, therefore, be in accordance with these variables.

Level of local participation during VA duration: refers to the degree of involvement the local beneficiaries of the proposed EbA measures have had in the development of the VA. This is in direct relation to the design of the assessment approach (deductive, participatory or model-based). It is measured in terms of local participation level in the selection process. Local participation level is referred to as the degree of community control over the process, the stage of the process when local people participate, and the level of representation of different stakeholders and community groups in the process (Vernooy 2005). In this sense, key questions for analyzing level of local participation are: who controls and makes decisions, who undertakes activities, and who benefits from the results (Ibid.). There is usually an inverse correlation between the level of local participation with the scope of the assessment and a positive correlation with resolution of assessment. In other words, the greater the scope the lower level of local participation and the larger the resolution, the higher level of local participation is implied.

Beneficiaries of the information: refers to the stakeholders that have a direct interest in the results of the assessments. They can include local communities and project partners, as well as external organizations and public and private institutions.

Following steps: refers to the immediate actions that are required prior to the implementation of the EbA measures (under the premise that VAs should always lead towards implementation of EbA measures). This can be measured in two ways. The first is a list of steps that need to be taken, for example, further research or consultations; and second, in the amount of time that these steps would take to accomplish.

3.1 Matrix for the Descriptors and Units of Measurements

A matrix has been custom created for this case study to tabulate results of the variables used to evaluate the performance of the approaches used for the selection of EbA measures. In Table 1 there is an explanation of the attributes being evaluated, the scale used to value them and the units of measurements used for scoring.

Some of the attributes of the variables have a user-defined value in the sense that it is the user that defines the value according to their expected outcomes or context. For example, a time consuming and expensive Vulnerability Assessment can be valued low because it takes too much time to deliver results. On the other hand, if there are no time or budget constraints, a long and expensive Vulnerability Assessment can be valued highly in accordance with the number of outputs, scope and resolution of results.

4 Narrative Description of Assessments and Matrix of Descriptors

A narrative description of each of the three approaches is added in this section to present complementary information that is not captured by the descriptors. A recount of background information on the decision to choose a given approach, or the methodological process and information requirements has been added here to better understand each of the Vulnerability Assessments. At the end of this section we present the completed matrix in Table 2.

Table 2 Matrix of descriptors and their values according to approach taken for selecting EbA measures

Descriptor	Participatory approach	Model-based approach	Deductive approach
Scope	Canchayllo, area of 76.5 Km^2	All of the NYCLR. Including buffer area 3307 Km^2	All of the NYCLR. Including buffer area 3307 Km^2
	Miraflores, area of 202 Km^2		
Resolution	Both fine and coarse	Both fine and coarse	Coarse
Input information			
Resolution	Fine	Coarse	Coarse
Output information			
Duration	8	16	3–4 months (in a 12 month period)
Expense	80,000	130,000	50,000
Inputs	High	Very high	low
Outputs	Products generated:	4 reports including:	1 set of criteria
	2 memoirs of the IPRA	40+ maps; complete list of species; ecosystem distribution; species distribution; future climate change exposure scenarios; ecosystem services list and distribution; demand and supply of ecosystem services maps; sensitivity of population to climate change impacts and	3 versions of possible EbA measures
		11 radial diagrams depicting vulnerability (1 per district)	
	16 memoirs of field visits, meetings and others.		
	10 Technical reports		
	6 Maps		
Products	2 Documents with the detailed design of two no-regrets measures package (see description in recommendations).		

Outputs	(1) Community-based sustainable water management of upper river micro-basins	(1) Pasture management associated with wildlife management scheme	(1) Vicuña management
			(2) pasture management
Recommendations of EbA measures (no regret measures in the case of the participatory approach)	(2) Community-based sustainable communal native grassland management.	(2) water sources conservation and schemes for ecosystem payments	(3) conservation of watersheds and schemes for ecosystem payments
			(4) MAPs collection
		(3) Improvement and maintenance of the water infrastructure	(5) Agroforestry
		(4) sustainable tourism activities	
Local participation level	High.	Low.	Low to very low.
	Collaborative level of participation is reached. Local co-researchers are key members of the assessing team. EbA measures are selected and designed through an exchange of knowledge between local and external partners and a sharing of decision-making power during the process	Consultative level of participation due to the application of different tools for consulting and discussion of issues (interviews and focus groups)	Consultative level of participation to key informants at local level. More consultation was done at higher levels of government (regional, national)
Beneficiaries of the information	Local population at community level; park authorities; local authorities; SERNANP; EbA project organizations	SERNANP, park authorities,	SERNANP, park authorities, EbA Project organizations; MINAM
		EbA project organizations;	
		Local communities, MINAM	
Following steps	Capacity building at community level; implementation of EbA measures on the basis of specific technical studies that need to be developed	Use of the information coming from the study to identify specific ecosystem-based adaptation measures. This will imply a consultative process with the communities identified as most vulnerable.	Couples with the results of the VIA to pinpoint site location for intervention. Further detailed studies at site level needed for measure design and implementation

4.1 Participatory Approach: Vulnerability and Community Planning[1]

A "participatory approach" manages to assemble adaptation measures based on the many observations and opinions compiled from the local community in a process of extensive consultation in which local partners increase their decision-making power (see for example CRISTAL tool or CVCA tool). TMI, IUCN's implementing partner in the NYCLR, developed a methodological process to select, design and implement no-regrets measures in two pilot communities of the NYCLR. We used the following definition of no-regrets measures to guide our work: *"No-regrets measures maximizing positive and minimizing negative aspects of nature-based adaptation strategies and options. No-regrets actions include … measures taken by communities [and/or facilitated by organizations] which do not worsen vulnerabilities to climate change or which increase adaptive capacities and measures that will always have a positive impact on livelihoods and ecosystems regardless of how the climate changes"* (Raza et al. 2014).

Canchayllo and Miraflores communities, both located inside the NYCLR, were initially selected for assessment based on environmental, social, ecological, political and operational criteria (the selection process was done by the Mountain EbA project and the NYCLR Head and staff) (TMI 2013). In Canchayllo the main livelihood is cattle farming, although many families complement their income with other activities. The community of Miraflores, on the other hand, also depends on cattle farming; however agriculture is likewise an important livelihood strategy. There is a high level of out-migration and a low birth rate in Miraflores.

Field trips and workshops were carried out to identify vulnerabilities based on local perceptions, communities' needs and priorities, and generate ideas on how to address such vulnerabilities. Local communities and key stakeholders including the Reserve staff and district municipality authorities participated in such activities (Podvin et al. 2014).

An Integrated Participatory Rural Appraisal (IPRA) followed, where environmental and social impacts of pre-selected no-regrets measures were analyzed by a group of local stakeholders with experts' support, to make the final selection and design. Results of the IPRA were presented and validated by local stakeholders (and presented in the assemblies of both local communities), NYCL Reserve staff, and the EbA partners.

The methodological process that was carried out in Canchayllo and Miraflores took into account existing data and information, but was largely based on new information generated through the participatory process. As a result, two no-regrets measures were selected in each community. Canchayllo prioritized community-based native grassland management and improvement of ancestral hydrological infrastructure. Miraflores, on the other hand, selected community-based native grassland management as well, while also selecting conservation and management of upper micro-watersheds, wetlands and water courses.

[1] For a complete description of this approach refer to Podvin et al. 2014.

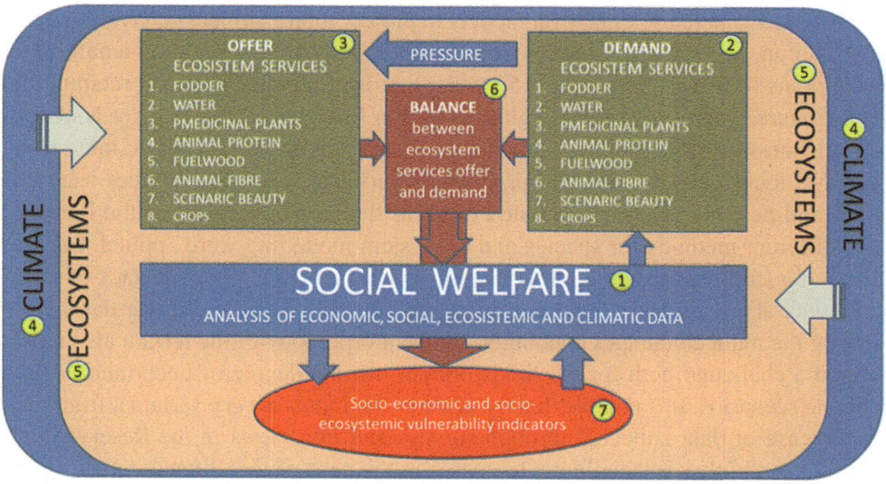

Fig. 3 Social welfare (*1*) depends on the economic activities in order to meet people livelihood. People activities (welfare) generate a demand for ecosystem services (*2*), and a pressure on the supply of ecosystem services (*3*). Supply of ecosystem services depends on both climate (*4*) and the existence of the ecosystems themselves (*5*). The interaction between demand (*2*) and supply (*3*) of ecosystem services generates a balance of supply and demand (*6*) of the ecosystem services which impacts on social welfare (*1*), which is also affected by the climate (*4*) and the presence of ecosystems (*5*). These interactions generate socio-economic and socio-ecosystemic vulnerability (*7*) which ultimately depends on social welfare (*1*)

4.2 *Model-Based Approach: Vulnerability and Impact Assessment (VIA)*[2]

A VA (FDA 2013), based on a set of statistical and spatial models for different aspects of the entire project site (the NYCLR) was developed and is here termed as the "model-based approach" (for example see Munroe et al. 2015). Data and information inputs were required to generate past, present and future climate scenarios, ecological processes, hydrology, economic processes, livelihoods and local people's perceptions.

The main goal of the "model-based approach" applied to the NYCLR and its buffer zone was to analyze the vulnerability to climate change of the territory of the Reserve and identify which areas would be most vulnerable to climate change, according to future emissions scenarios. It also resulted in a list of possible adaptation measures with indication of the districts where they could be implemented. This characterizes and differentiates the model-based approach from the other methods described in this chapter by the development of future climatic scenarios as one of the main methodological characteristics as well as by its focus on how the provision of ecosystem services and people's vulnerability would vary according to the climate scenarios. A conceptual scheme of the VIA is provided in Fig. 3.

[2] For a complete description of this approach refer to FDA, 2014 and Pablo Dourojeanni et al. 2014.

The study made use of several models to project future climate behavior, species distribution, as well as future ecosystem services provision. It required a number of data inputs from various sources (50-year records of climatic data on precipitation, temperature, numerical grids, Digital Elevation Models or DEMs, records of species richness and distribution, population censuses, agricultural census, interviews, surveys, focus groups). In addition, both primary and secondary sources were utilized and new information generated (such as the land cover map of the Reserve) and different methods for species and ecosystem modelling were applied. Lack of chronologically consistent data series (e.g. water quantity), as well as lack of available data at the proper scale (e.g. water quantity and quality, climate data) were among the main challenges encountered. The morphology of the terrain also represented a challenge both for the fieldwork and for the degree of uncertainty of the climate scenarios. It is notable that the scale of the inputs is very variable, from fine, in the case of data collected through surveys and interviews in the Reserve communities and other stakeholders, to coarse in the case of the climate data – downscaled later on. The scale of the results has generally been homogenized to the district level.

To develop such a study, the participation of an integrated and interdisciplinary group of specialized experts was needed in order to properly analyze all components that merged into the final results (see Fig. 1) and to develop the methodology used which considered changes in ecosystem services as one of the main elements. Consultations with external specialists were also occasionally required.

4.3 Deductive Approach: Literature Research and Expert Knowledge

Another VA, a "deductive approach", was conducted based on literature review, consultations with experts, field observations and consultations with local stakeholders. Results were reached using general rules (criteria) designed to delineate the boundaries of what could be possible EbA measures (Dourojeanni 2012a). These criteria worked as a rule-based envelope model into which information from literature and expert opinions was fed, and the output was a set of EbA measures that were based on the context of the entire NYCLR and watershed level (Dourojeanni 2012b). Conclusions (identified EbA measures) were only made as a reference to test the conclusions of the other approaches applied in the project.

The aim of this approach was to identify possible EbA measures that could be suitable to implement in the project site (the NYCLR), and therefore the first step was to define EbA. A set of criteria was derived from the literature and field visits were performed to gather information on site. A first set of 19 possible EbA measures (Dourojeanni 2012b) was discussed and prioritized into five options

(Dourojeanni 2013). These options were reviewed and further documented with the support of field visits and expert consultations.

It is worth noting that information generated through the model-based and the deductive approaches was used as inputs for the identification of a third intervention site (in addition to Miraflores and Canchayllo), namely the community of Tanta, as well as the EbA measures to be implemented there. As with Canchayllo, the community's main livelihood activity is livestock production. EbA measures selected for Tanta included community-based grassland management and domestic livestock husbandry, but also associated with management of vicuñas in the wild.

5 Lessons for Nor Yauyos Cochas Landscape Reserve and Beyond

The participatory methodologies applied in the planning, design, validation and implementation phases of the participatory approach have been key to deliver bottom-up activities that empower and enhance the involved local communities' ownership of project activities and results (TMI 2014). Even though the time extension (duration) of this approach has been shorter than the other two approaches, it has managed to produce a valuable amount of site-specific information (at community level). The added value of the participatory nature of this approach has been that this information has helped both communities and researchers reach a common understanding of local vulnerabilities to climate change and how they can be addressed. Also, in order to define the no-regrets measures, it was crucial to have a multidisciplinary team comprised of local researchers and external experts to analyze pre-selected no-regrets measures and their net potential social and environmental impacts. Another added value of the participatory approach was that at its conclusion, the adaptation measures were almost fully designed and only needed minor technical studies for the start of implementation, thus saving time from having to carry out further in depth studies. A key difference between the participatory approach and the other two approaches is that it focused its assessment at a very fine-scale—only two communities—and analyzed in the field pre-selected adaptation measures.

The model-based approach was a time-consuming, expensive and complex study. Nevertheless, the large amount of useful information that was produced rendered its execution valuable for understanding the complicated relations between supply and demand of ecosystem services in the entire project area (NYCLR), regardless of the uncertainty associated with modelling (resolution of data, amount of data and assumptions used in models). The model-based approach also made a complete review of existing statistical information on population and agriculture and compared it with primary data gathered through a survey. The compilation of all this information is highly important, not only for the selection and implementation of EbA measures. Park authorities and local communities will also use it to update local planning schemes for conservation and development and national authorities will further expand their knowledge on how to conduct Vulnerability Assessments in protected areas.

The deductive approach only used existing information, expert consultations, field site visits and key interviews to produce a potential list of possible EbA measures. The main value of this approach was the conceptual exploration of EbA and the production of a general framework for classifying possible EbA activities regardless of the site (Dourojeanni 2012a). Nevertheless, the lack of involvement of the local population in the selection process of possible measures renders the exercise insufficiently grounded in the target territory and in a need for a second phase to design specific activities together with local communities.

The model-based approach demanded the largest amount of data, but also produced the largest amount of information for the entire project site (NYCLR). Nevertheless, as in the case of the deductive approach, the recommendations of EbA measures made by this approach were loosely grounded in the territory and required further design and consultation with local communities. Due to its participatory nature, the participatory approach resulted in recommendations that were highly grounded in the two local communities where the analysis was performed. However, the scope of the assessment was small and only rendered valuable information for the two selected communities. In contrast, the other two approaches produced information for the entire NYCLR which contributed to assessing and comparing vulnerability at a broader scope.

All three approaches used research, involving external (non-local) scientists to design a specific intervention at a selected site. The participatory approach did this through the IPRA and the other two approaches were led by an off-site scientist. This means that all approaches demanded a solid scientific basis for the design and implementation of EbA or no-regrets measures.

5.1 Towards an Integrated Approach

For future interventions, a combination of the three approaches for VA could be useful to reduce the amount of time, financial costs and effort involved in selecting and designing EbA or no-regrets measures. Once a set of criteria for identifying and classifying EbA measures (deductive approach) is created, the amount of time for pre-selecting possible measures has been reduced. Pre-selecting possible measures helps to narrow down plausible options (at community level or for a larger segment of territory), thus helping to allocate resources to the in-depth study of fewer subjects (as was the case of the participatory approach that also used a set of criteria for pre-selecting measures). As a next step to pre-select EbA measures, the development of an IPRA, as was conducted by the participatory approach, will help to develop together with the community the final selection and design of adaptation measures, while strengthening local capacities. Nevertheless, concerning the challenge of choosing where to perform (scope: what specific community or site at broader scale), the implementation of measures, the VA is still reliant on a broader basis of information or on a finely defined project that targets a specific site.

Many of the specific methods for modeling and mapping ecosystem services applied in the model-based approach can be incorporated into the IPRA to help refine the identification of vulnerabilities of the population, ecosystems and ecosystem services. This is especially useful when the measures are already selected and the specific sites are determined, so that the research can have a great amount of detail (fine resolution).

6 Conclusion

The application of the three approaches described in this paper offer valuable lessons learned about how to design and conduct future Vulnerability Assessments for selecting measures in the context of Ecosystem-based Adaptation. These lessons led to the generation of several key recommendations to ensure the effectiveness and desired results of future Vulnerability Assessments.

Before initiating the design of an assessment, it is important to frame it appropriately, within the context of the overall objectives and desired outcomes of the given EBA initiative. This framing includes an early identification of hazards, the limits and units of analysis of the affected socio-ecological system, the specific purpose(s) of the assessment, available inputs (i.e. data, financial resources, technical capacities), and the target audience(s) of the outputs as well as other relevant stakeholders who should be engaged in the assessments development. In first completing this framing exercise, one can then determine, adjust and even package different methodologies to ensure the production of the desired outputs.

The 'why' of the assessment provides the foundation upon which to select and even integrate approaches. In the specific case of the Ecosystem-based Adaptation in Mountain Ecosystems project in Peru, two primary objectives shaped the multiple approaches to the assessment: (i) reduce the vulnerability of critical mountain ecosystems in Peru through Ecosystem-based Adaptation, and (ii) reduce the vulnerability to climate change of the local communities living within the Nor Yauyos Cochas Landscape Reserve. Both objectives required the engagement and in some cases active participation, of multiple stakeholders to ensure the effectiveness, legitimacy and credibility of the assessment results. Furthermore, both objectives demanded differentiated outputs from the assessment process in order for it to be relevant.

Therefore, stakeholder participation is critical at all stages of the assessment development. For example, the deductive approach, as described in this paper, engaged scientific experts, national and subnational authorities, and to a more limited extent, local actors. The participation of these multiple stakeholders served to gather the necessary information inputs, and also to periodically validate draft results. In the case of the participatory approach, the local communities were fundamental as co-researchers, providing their perceptions on risks, as well as needs, priorities and deep knowledge that could guide the selection and design of the adaptation measures. Additionally, participation of local communities increased the likelihood of ownership and use of the information to decide on what, how and where adaptation measures should be implemented.

The decision about which approach to apply, or a combination of approaches, is dependent ultimately on the end uses or applications of the analysis that should be defined during the initial framing exercise. If the emphasis is on mobilizing local, community-level action, it is important to follow a methodology that facilitates active community participation in all stages of the analysis. However, if there is also an explicit objective to bring to scale the adaptation response, through policy, planning and accompanying financial mechanisms, it is important to ensure that stakeholders responsible for these processes at district and national levels are effectively engaged, in order to ensure their ownership of the results.

Given the EBA Mountain project's two-tier objectives stated above, it is in this light recommended that an integrated, blended approach to the assessment be employed. Community input to the participatory approach ensures social relevance and local ownership of the results, while the participation by policy-level decision makers in the deductive and model-based approaches ensures that results will have a certain level of ownership and credibility among these key stakeholders needed to inform policy and planning decisions at a larger systems-wide scale.

In order to maximize the benefits derived from applying multiple approaches, care should be taken to conceptualize, map and plan them under a single process. Doing so will allow that intermediate outputs be validated through the other approaches, in effect spurring a reiterative and adaptive learning process with the input of multiple methodologies, and the collective participation of diverse stakeholder groups. For example, local community engagement in the initial stages could then provide focus to the deductive approach, e.g. in identification of priority ecosystems and the services they provide, while local actors could be re-engaged to validate initial findings from the "top-down" assessments. The articulation of the bottom-up and deductive models will further ensure a more robust Vulnerability Assessment with a multidimensional and multilevel structure built with the iterative contributions of all key actors.

In both cases, communications and packaging of the results are key. While potentially technically similar in nature, the results need to be translated into the cultural and technical language of each key stakeholder group to facilitate the appropriation and application at the relevant scale.

Acknowledgements The authors would like to acknowledge the valuable contributions of all the persons that contributed directly or indirectly to the culmination of this research. In this sense it is our duty to deeply thank the researcher of the three original vulnerability assessments (VA) described in the chapter. For the participatory VA we would like to thank the scientific coordinator of the research team, Jorge Recharte and his team, comprised both of local campesinos and academic researchers. For the Model based VA we want to thank the Fundación para el Desarrollo Agrario (FDA) of the Universidad Agraria la Molina and the subsidiary research units that undertook the research. For the Deductive VA we want to thank Cordula Epple of UNEP-WCMC and her colleagues for guiding that work. We also want to acknowledge the contribution of Aneli

Gomez and Woodro Andia, field coordinators of the EbA Mountain project. Last but not least we want to express our heartfelt gratitude to Tine Rossing for the invaluable and detailed comments to the final draft of this chapter.

References

CBD (Secretariat of the Convention on Biological Diversity). (2009). *Connecting biodiversity and climate change mitigation and adaptation: Report of the second ad hoc technical expert group on biodiversity and climate change* (Montreal, technical series No. 41, p. 126). Montréal: Secretariat of the Convention on Biological Diversity.

CBD (Secretariat of the Convention on Biological Diversity). (2010). *COP 10 Decision, X/33.* http://www.cbd.int/decision/cop/?id=12299

Dourojeanni, P. (2012a). *Ejercicio de elaboración de criterios para la selección de medidas de adaptación basada en ecosistemas en la RPNYC, Perú.* Documento de trabajo interno.

Dourojeanni, P. (2012b). *Lista preliminar de posibles medidas de adaptación basada en ecosistemas para la RPNYC.* Documento de trabajo interno.

Dourojeanni, P. (2013). *Documentación de la lista corta de medidas de adaptación basada en ecosistemas para la RPNYC.* Documento trabajo interno.

Dourojeanni, P., Giada, S., & Leclerc, M. (2014). *Evaluación de Vulnerabilidad e Impacto del Cambio Climático en la Reserva Paisajísitca Nor Yauyos-Cochas y su Zona de Amortiguamiento: Resumen Técnico.*

FDA (Fundación para el Desarrollo Agrario). (2013). *Evaluación del Impacto y Vulnerabilidad del Cambio Climático de la Reserva Paisajística Nor Yauyos Cochas y áreas de amortiguamiento – VIA RPNYC.* Elaborado en el marco de la colaboración interinstitucional CDC-FEP-Universidad Nacional Agraria La Molina, Escuela de Ingeniería de Antioquía y IRI-EICES-Columbia University. Lima, PNUMA.

GIZ. (2014). *The vulnerability sourcebook concept and guidelines for standardized vulnerability assessments.* Deutsche Gesellschaft für Internationale Zusammenarbeit (GIZ) GmbH. https://gc21.giz.de/ibt/var/app/wp342deP/1443/wp-content/uploads/filebase/va/vulnerability-guides-manuals-reports/Vulnerability_Sourcebook_-_Guidelines_for_Assessments_-_GIZ_2014.pdf

Hammill, A., Bizikova, L., Dekens, J. & McCandless, M. (2013). *Comparative analysis of climate change vulnerability assessments: Lessons from Tunisia and Indonesia.* Deutsche Gesellschaft für Internationale Zusammenarbeit (GIZ) GmbH. https://gc21.giz.de/ibt/var/app/wp342deP/1443/wp-content/uploads/filebase/va/vulnerability-guides-manuals-reports/Comperative-analysis-of-climate-change-vulnerability-assessments.pdf

INRENA. (2006). *Reserva Paisajística Nor Yauyos Cochas, Plan Maestro 2006–2011.* Instituto Nacional de Recursos Naturales – INRENA Intendencia de Áreas Naturales Protegidas – IANP

MINAM. (2010). *Segunda Comunicación Nacional del Perú a la Convención Marco de las Naciones Unidas sobre Cambio Climático (SCNCC).*

Munroe, R., Dilys, R., Doswald, N., Spencer, T., Möller, I., Vira, B., Reid, H., Kontoleon, A., Giuliani, A., Castelli, I., & Stephens, J. (2012). *Review of the evidence base for ecosystem-based approaches for adaptation to climate change.* http://www.environmentalevidencejournal.org/content/pdf/2047-2382-1-13.pdf

Munroe, R., Hicks, C., Doswald, N., Bubb, P., Epple, C., Woroniecki, S., Bodin, B., & Osti, M. (2015). *Guidance on integrating ecosystem considerations into climate change vulnerability and impact assessments to inform ecosystem-based adaptation.* Cambridge: UNEP-WCMC.

Podvin, K., Cordero, D., & Gómez, A. (2014). Climate change adaptation in the peruvian andes: Implementing no-regrets measures in the Nor Yauyos-Cochas landscape reserve. In R. Murti & C. Buyck (Eds.), *Safe havens: Protected areas for disaster risk reduction and climate change adaptation.* Gland: IUCN.

PROVIA. (2013). *PROVIA guidance on assessing vulnerability, impacts and adaptation to climate change*. Consultation document, United Nations Environment Programme, Nairobi, Kenya. 198 p.

Raza, A., Barrow, E., Zapata, F., Cordero, D., Podvin, K., Kutegeka, S., Gafabusa, R., Khanal, R., & Adhikari, A. (2014). *Ecosystem based adaptation: Building on no regret adaptation measures*. Technical paper. 20th session of the Conference of the Parties to the UNFCCC and the 10th session of the Conference of the Parties to the Kyoto Protocol, Lima, Peru, 1–12 December 2014.

Reid, H. (2011). *Improving the evidence for ecosystem-based adaptation. Sustainable development opinion paper*. London: IIED.

SERNANP. (2014). Lista oficial de ANPs. http://www.sernanp.gob.pe/sernanp/archivos/biblioteca/mapas/ListaAnps_12112014.pdf

TMI (The Mountain Institute). (2013). *Memoria de la reunión de trabajo para selección de sitios del Proyecto EbA Montaña*. Internal report. 12 p.

TMI (The Mountain Institute). (2014). *The mountain institute report on action learning for mountain EbApProject, Perú/RPNYC*. First cycle of Action Learning. Internal Report. 10 p.

Travers, A., Elrick, C., Kay, R., & Vestergaard, O. (2012). *Ecosystem-based adaptation guidance: Moving from principles to practice*. UNEP Working document, April 2012.

Vernooy, R. (2005). The quality of participation: Critical reflections on decision making, context and goals. In J. Gonsalves, et al (Eds.), Participatory research and development for sustainable agriculture and natural resource management: A sourcebook. Volume 1: Understanding participatory research and development. Laguna/Ottawa: International Potato Center-Users' Perspectives With Agricultural Research and Development/International Development Research Centre.

The Role of Ecosystem-based Adaptation in the Swiss Mountains

Veruska Muccione and Ben Daley

Abstract Ecosystem-based Adaptation (EbA) to climate change addresses the links between ecosystem services, climate change adaptation and sustainable resource management. This study explores the role of EbA in the mountain areas of Switzerland by looking at existing and potential EbA interventions, their effectiveness, opportunities and challenges. It analyses the Swiss policy context and how this can be conductive to EbA. EbA interventions in the Swiss mountains are identified in the area of disaster risk management, water management and agriculture. The research highlights some characteristics of these interventions. Challenges and opportunities of EbA are attributed in general to knowledge, acceptance and socio-economic factors. The Swiss policy context appears to be poorly conductive to EbA, with the Swiss adaptation strategy promoting sectoral approaches at the expense of more integrative interventions. The role of new cross-sectoral institutions in the form of boundary organisations is suggested as a way to better integrate EbA into Swiss policy and practice.

Keywords Climate change • Disaster risk reduction • Ecosystem-based adaptation • Swiss mountains

1 Introduction

Since the publication of the Millennium Ecosystem Assessment Report (MEA 2005a, b), a growing body of knowledge has discussed the links between healthy, well-managed ecosystems and human wellbeing (MEA 2005a; Carpenter et al. 2012; Wang et al. 2013), not least in terms of providing goods and services to cope

V. Muccione (✉)
Department of Geography, University of Zurich,
Winterthurerstrasse 190, CH-8057 Zürich, Switzerland
e-mail: veruska.muccione@geo.uzh.ch

B. Daley
Centre for Development, Environment and Policy (CeDEP), SOAS, University of London,
Russell Square, London WC1H 0XG, UK
e-mail: bd9@soas.ac.uk

N. Salzmann et al. (eds.), *Climate Change Adaptation Strategies – An Upstream-downstream Perspective*, DOI 10.1007/978-3-319-40773-9_9

with a changing climate (Jones et al. 2012; Munang et al. 2013a, b). The term ecosystem-based approaches was first adopted at the second Conference of Parties (COP) of the Convention on Biological Diversity (CBD) held in Jakarta in 1995 and refers to the integrated management of land, water and living resources that promotes conservation and sustainable use in an equitable way.[1] Applied to climate change, ecosystem approaches refer to a strategy integrating the use of biodiversity and ecosystem services to help people adapt to climate change (CBD COP X 33[2], World Bank 2009; Munroe et al. 2012). Examples of EbA include:

- Conservation and restoration of forests to stabilise slopes and protect against landslides, debris flow and avalanches (CBD 2009).
- Sustainable management of grasslands and rangelands, which contribute to increasing the resilience against floods and droughts (Colls et al. 2009).
- Establishment of diverse and resilient agriculture and agro-forestry systems to help cope with changing climatic conditions (CBD 2009).
- Sustainable water management of aquatic ecosystems in order to provide resilient water storage, flood regulation and reduction of erosion/siltation (Opperman et al. 2009; Midgley et al. 2012).

Evidence of the broad ability of ecosystem-based approaches to help communities adapt to climate change have been showcased in Pérez et al. (2010). The book reports 11 case studies on the effectiveness of EbA in adaptation planning ranging from EbA interventions in the high mountain ecosystems of Colombia to integrated water resources management in the island ecosystems of Fiji. Jones et al. (2012) assert that although most of the documented EbA interventions to date are in developing countries, there is great potential for them also in developed countries. For example, Batker et al. (2010) discuss the value of restoring the Mississippi Delta, which can be an important tool in coastal defence, and Tengö and Belfrage (2004) examine the use of alternative agriculture practices such as intercropping and crop rotation in smallholder farming in Sweden.

The World Bank (2009) recognises that ecosystem-based approaches should constitute an important pillar in national and international climate change adaptation strategies for their ability to deliver solutions, which are cost-effective, flexible and provide a wealth of co-benefits. For example, protecting forests, swamps and mangroves also contributes to the maintenance of existing carbon stocks and the enhancement of carbon sinks, which are important strategies for climate change mitigation (World Bank 2009; Naumann et al. 2011). Jones et al. (2012) discuss the important nexus between EbA and traditional conservation interventions, in the sense that where ecosystems are protected for their adaptation properties they also provide significant opportunities for conservation management. In turn, protecting ecosystems through improved management increases their resilience and thus reduces the risks of degradation and potential negative impacts of climate change, such as the crossing of tipping points (Mumby et al. 2007; Game et al. 2009). EbA

[1] https://www.cbd.int/ecosystem/description.shtml.

[2] http://www.cbd.int/climate/doc/cop-10-dec-33-en.pdf.

also seems to promote local and traditional knowledge; hence it can be adapted to the local context in which the adaptation and ecosystem management are undertaken (Vignola et al. 2009).

There are nonetheless some caveats to observe when promoting the EbA concept. As Oliver et al. (2012) observed, there is a long tradition of using ecosystem-based approaches for conservation and ecosystem management practices and only recently the potential for adaptation embedded in such initiatives has been realised. The risk is that promoting such interventions as adaptation bears the possibility that they might not be effective under changing climatic conditions and lead to maladaptation. Indeed, there is considerable uncertainty in the level of climate change that ecosystems can bear to be able to continue providing their services and this is particularly true for EbA (Jones et al. 2012). In a systematic review aimed at assessing the evidence base for the effectiveness of EbA worldwide, Doswald et al. (2014) show that there is evidence that the use of ecosystem-based approaches can be effective in reducing vulnerability to climate related impacts globally and in several different types of ecosystem. In spite of this, they find it difficult to draw conclusions on the long-term effectiveness of EbA interventions in a changing climate. Thus, whether existing ecosystem-based approaches can be classified as EbA depends first on the specific context in which they are embedded and second on the ability of the interventions to cope with climate change. These necessitate proper local vulnerability and impact assessments (Oliver et al. 2012). Finally, EbA appears to be more difficult to implement compared to technical solutions because it requires considerable institutional cooperation as well as cooperation across communities and sectors and its co-benefits are usually very broadly spread amongst different sectors and communities (IPCC 2014).

As outlined in Doswald and Osti (2011), since the concept of EbA started emerging many case studies and good practice examples were developed around the world, with the large majority in developing countries. Following this, two exhaustive assessments on EbA good practice and potentials for climate change adaptation and mitigation in Europe were published in 2011 (Doswald and Osti 2011; Naumann et al. 2011). Naumann et al. (2011) also provided a first thorough assessment of the policy context for EbA in some European countries. However, to date, EbAs are often overlooked in overall national or sub-national adaptation plans (Campbell et al. 2009). Doswald et al. (2014) find a number of research gaps that need to be addressed in order to translate EbA research into policy relevant advice and on the implementation of EbAs in certain ecosystems, such as mountain ecosystems. This chapter addresses explicitly one of such gap by examining the role of EbA in the Swiss mountains. Semi-structured in-depth interviews are used to answer the following research questions: (a) what type of EbA measures are preferred in the Swiss mountains; (b) what are the challenges and opportunities for EbA in the Swiss mountains and (c) is the current policy environment is favourable to the integration of EbA as an overall strategy to deal with climate change?

We focus on the mountain regions because mountain ecosystems and their services are increasingly important for the Swiss economy and national identity as well as in relation to their ability to provide interventions to cope with global change

(Grêt-Regamey et al. 2012). Conversely, mountain ecosystems are increasingly threatened by global change (Löffler et al. 2011; Gobiet et al. 2013) and little progress has been made to halt the loss of biodiversity and to adapt to climate change (SCNAT 2012). The protective function of ecosystems against a backdrop of increasing hazards in mountain areas is already well recognized in Switzerland (Briner et al. 2013; Grêt-Regamey et al. 2013; Huber et al. 2013) together with the ability of ecosystems to provide goods and services that are socially and economically relevant for mountain and downstream communities (Körner and Ohsawa 2006; OcCC 2007; Koellner 2009; Grêt-Regamey et al. 2012). In addition, EbA solutions can be found in a strategic challenge of the Swiss federal government[3] until Rio+ 30, precisely "Adaptation to climate change: elaborate a new risk culture with regard to natural hazards; reinvent Alpine tourism; develop strategies for biodiversity conservation" (SCNAT 2012). Nevertheless, the links between ecosystem services and climate change adaptation are yet to be established in Switzerland and the reasons behind this knowledge gap are worth investigating.

2 Methods

Qualitative research methods were used for the data collection and analysis. Semi-structured interviews with key individuals were carried out between April and June 2014. Key informants included policy makers at the federal and cantonal level, researchers, private sector managers, consultants and representatives of civil society. Such categories represent the full breadth of knowledge and opinions concerning climate change adaptation and EbA in Switzerland, from academic and applied research to policy formulation and social perspectives. The interviews were divided in three different sections corresponding to three main research questions. Namely, the first part was dedicated to explore the understanding of EbA concept as well as the type of interventions preferred in the Swiss mountains; the second part explored the opportunities and challenges of EbA through questions addressing co-benefits of EbA, challenges and evidence of their effectiveness. Finally, the third part was concerned with the Swiss policy context for EbA.

We selected the informants from the different categories by means of purposive sampling (Bryman 2012). In addition a snowballing technique was added to the initial purposive sampling. Informants were contacted exclusively by email. In the email it was stated the thematic areas of the research, main objectives as well as some practical information on the length of the interview, suggested dates and locations of the interview. Informed consent was obtained to record the interview and make use of the information provided during the interview. An interview sheet was prepared before each interview in order to insure that the important topics were

[3] The Swiss political system is organised as a federal parlamentary republic consisting of 26 Cantons (e.g. equivalent of states).

covered; in some cases, it was sent to informants in advance of the scheduled interview. In total thirteen expert interviews were carried out. Once the interviews were recorded, they were transcribed and coded by using NVivo Software.[4] The use of coding for content analysis is commonly employed in qualitative data analysis (Corbin and Strauss 2008). Codes are constructs of words that synthesise concepts appearing in individual pieces of language-based data with the scope to identify patterns and categories (Saldaña 2013). The coding followed a procedure common in qualitative research where categories were created in accordance to the research objectives, different categories were eventually recoded and finally codes grouped under overarching categories.

3 Results and Discussions

3.1 The EbA Concept and EbA Interventions

A comparison between existing interventions and interventions potentially effective to cope with climate change impacts is provided in Fig. 1. The quantitative data in Fig. 1 were extracted from the qualitative interviews by looking for key words, proxies of keywords and synthesis from the transcripts.

Informants identified interventions in the areas of disaster risk reduction, natural resources management (mainly associated with agriculture, forestry and biodiversity) and water regulation and storage. The majority of informants agreed on (i) the protection/management of forests to protect against gravitational hazards as being also effective to cope with climate change, and (ii) river revitalisation to protect against floods. The intervention "forest to protect against gravitational hazards" emerged as one of the favourite ecosystem-based approaches, because it combines both acceptability by local people and authorities and consequently institutional capability. Other interventions related to natural resource management (e.g. sustainable management of grasslands and mixed agriculture exploitation) were mentioned as already existing in Switzerland and likely to work also under a changing climate. The broader agreement on protective forests and river restoration could be due to a general tendency for people to think of climate change adaptation in terms of coping with extreme events and less in terms of slow onset events related to changes in precipitation, phenology, droughts and increased erosion. Several informants mentioned that many EbA interventions are already in use and would be used anyway even without climate change. This is a key opportunity for EbA, as it implies that adaptation to climate change can be realised with small changes in existing approaches. However, it also bears the risks that the interventions might be inadequate under changing climatic conditions.

[4] More details about NVivo and its features can be found here: http://www.qsrinternational.com/.

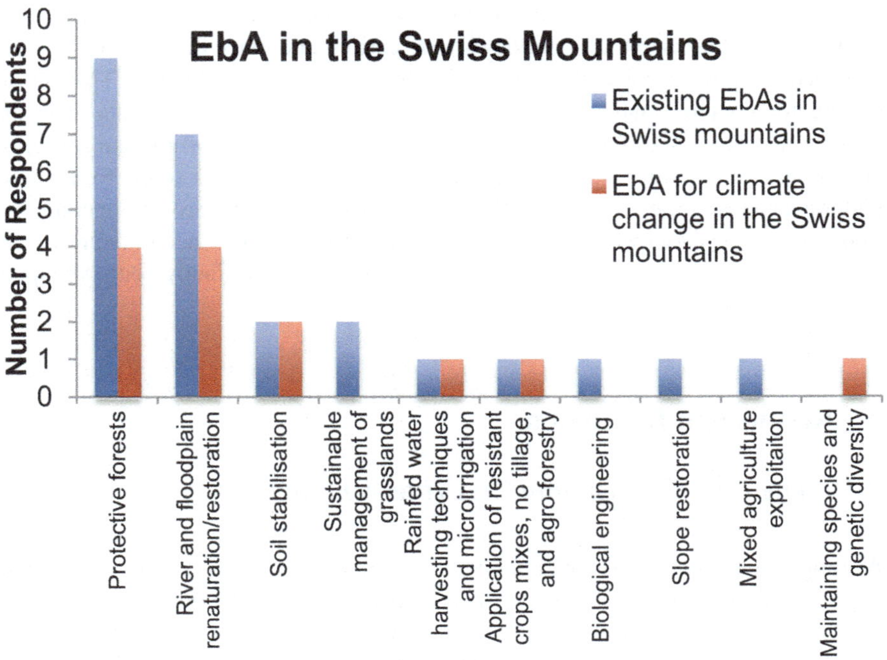

Fig. 1 EbA in the Swiss mountains according to the informants. The *blue bars* indicate EbA relevant interventions (possibly developed for climate extremes and variability) and *red bars* represent foreseen EbA interventions to cope with climate change impacts

3.2 Challenges and Opportunities of EbA

We present the challenges and opportunities of EbA ordered according to three main categories developed during the coding of the interviews, namely knowledge, acceptance and socio-economic opportunities and challenges. The categories as well as corresponding main topics are schematically provided in Fig. 2 (challenges) and Fig. 3 (opportunities). In addition to the diagrams, complementary information is also provided in narrative form to better contextualise our findings.

One policy maker, one academic and one consultant agreed that climate change and biodiversity management are perceived by mountain communities as secondary issues when compared to short term economic development. Lay people have difficulties in seeing the long-term purposes of climate change adaptation and biodiversity conservation (both associated with EbA), since their contributions to societal well-being are not readily evident. At the institutional level, the fact that the Swiss social organisations are mainly based on sectorial distribution of powers (reflected at all level of social organisation from local to cantonal to federal) is perceived as a strong challenge for EbA by several informants.

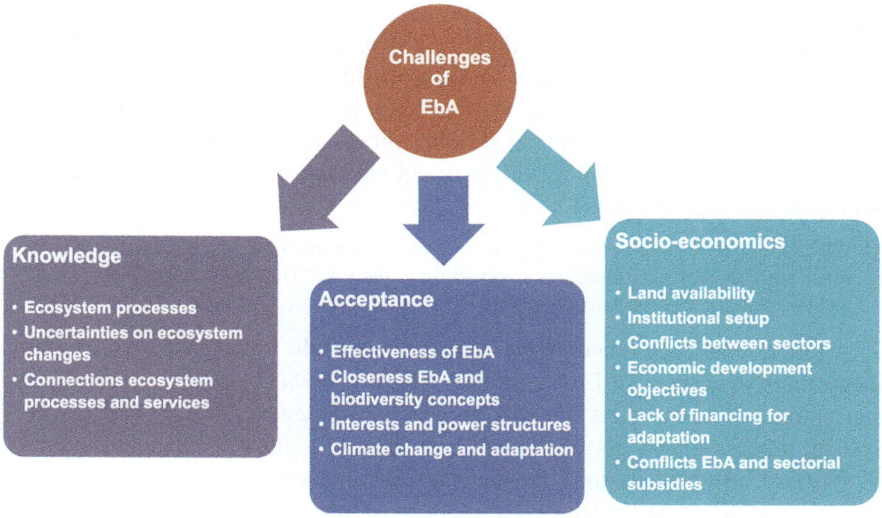

Fig. 2 Challenges of EbA. The *boxes* depict the main categories of knowledge, acceptance and socio-economics challenges

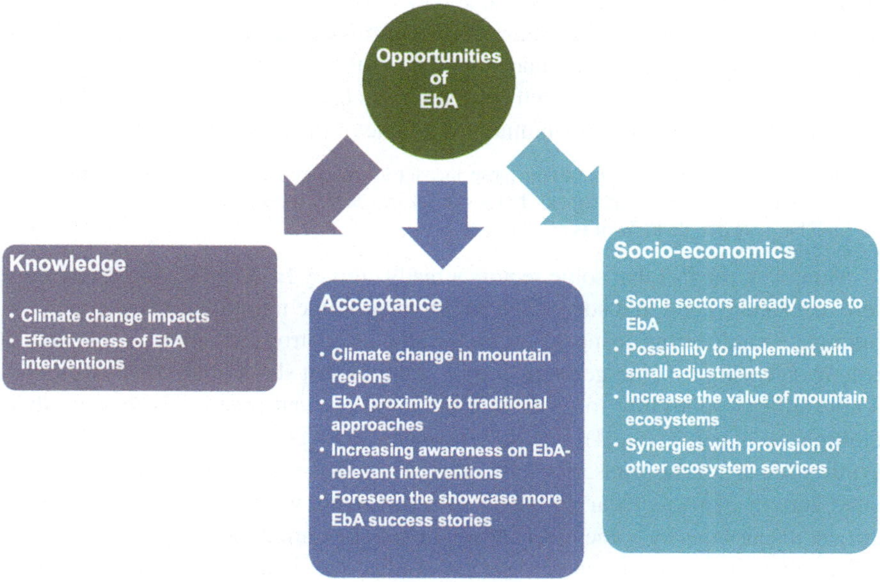

Fig. 3 Opportunities of EbA. The *boxes* depict the main categories of knowledge, acceptance and socio-economics opportunities

Next to this, recurrent challenges throughout the interviews concern land use issues. In mountain areas land availability is considered at its limit and EbA interventions are perceived as needing more space in comparison to more technical interventions. An example of land appropriation issues reported by more than half of the respondents was the conflict between river enlargement and farming in alpine valleys. One researcher dealing with issues of land use change and climate change summarised this problem as follow:

> What I know for example is that agriculture is very much opposing this re-naturalisation of rivers since usually agriculture goes very close to the river. Thus the farmers lose money and they don't like it. They really oppose it.

A further challenge relates to the EbA concept itself. It emerged from the interviews that the concept of EbA is poorly known in Switzerland. A reason for this conceptual gap seems to be related to limited knowledge on the links between ecosystem processes and services and their changes under a changing climate. Moreover, EbA demands good knowledge of the interactions and feedback between natural and human systems, which is per se a new and evolving research field (Carpenter et al. 2012). Some informants suggested that showcasing more EbA pilot initiatives would indeed increase informal knowledge and acceptability at the community and cantonal level. This is in line with international efforts to mainstream EbA into decision-making frameworks (Munang et al. 2013a).

In terms of opportunities in general, it was stressed that policy makers and civil society need to acquire a better understanding of the effectiveness of EbA interventions and that this can be achieved by continuous exchange amongst research, policy and practice. This is well summarised by one of the researchers:

> One aspect is whether people think these measures are effective. If they think so, then I think they will be easily accepted. Let's say for protective forests the acceptance is very high because people know it is effective.

Moreover, the fact that some sectors already employ EbA solutions (e.g. agriculture and forests) and that some EbA solutions promote traditional knowledge (e.g. biological engineering using wood and plants to control torrential run-off and to restore steep slopes) emerge as great opportunities that should be further exploited.

A further opportunity to promote EbA in policy and practice is through their multiple co-benefits (Fig. 4). Figure 4 shows the main co-benefits of EbA that were synthesised from the transcripts. It should be added that while some informants cited certain items as clear co-benefits, others items were accompanied by more critical discussions by some other informants. For example, while a slight majority agreed that EbAs are in general more cost-effective than technical interventions, others considered cost effectiveness of EbA an uncertain co-benefit because: (i) cost-effectiveness is measure and context dependent and cannot be generalised to the whole EbA; (ii) cost-effectiveness depends on the services (or disservices) addressed; and (iii) cost-effectiveness depends on the physical boundaries, timescales and stakeholders included in the assessment. A certain number of informants

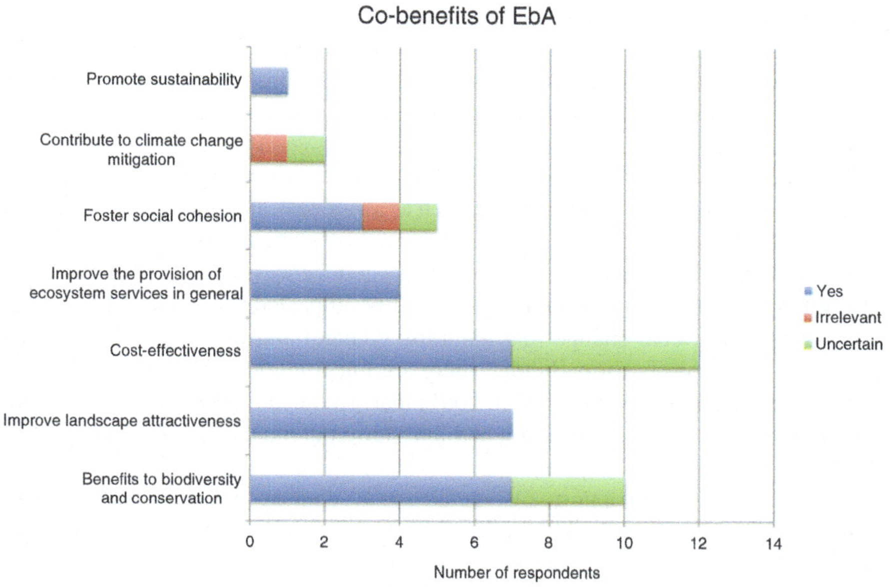

Fig. 4 Co-benefits of EbA. The *blue bars* represent the number of informants who discussed the item as a straight co-benefit; the *green bars* indicate an uncertain opinion regarding to co-benefit function; and the *red bars* indicate irrelevance of the item as a co-benefit. The *bars* do not add up to the whole number of informants, since not every informant discussed the same set of co-benefits

reported uncertainties in the broader benefits of EbA to biodiversity, mainly due to poor familiarity with biodiversity functions and processes in general. It should be noted that some co-benefits were cited by some informants and some were not, thus the number of informants per each co-benefits does not add up to the total number of informants. There is also no apparent relation between the different co-benefits. As an example, promote sustainability was cited only in very few cases although the Fig. 4 suggests a tendency towards social, economic and environmental sustainability. Co-benefits can also promote spill over from one economic sector to another. As an example, improved resilience of forests can be beneficial for the provisioning of other services such as timber productivity. The fact that EbA increases landscape attractiveness can have positive impacts on tourism, which is one of the most important sources of income in mountain communities. It was surprising that most of the informants did not see any benefits for climate change mitigation. In the literature, ecosystem-based approaches are also broadly discussed in the context of climate change mitigation through the improvement of carbon sinks (Naumann et al. 2011). However, this was not the case in the interviews, possibly due to a tendency in seeing mitigation as mainly associated to technical solutions. Furthermore, there is a general tendency in the EbA literature to promote this type of adaptation as being more cost effective in comparison to technical solutions (Jones et al. 2012).

However, this was not perceived to be the case by the informants. Indeed, the discussion on cost-effectiveness as well as issues surrounding distribution of costs and benefits over space and time confirmed the lack of established methods to reliably calculate economic, social and environmental costs and benefits for EbA, as also reported in Jones et al. (2012).

In general it can be said that several EbA interventions appear to bear great potential for adaptation in Switzerland, such as awareness, flexibility and associated co-benefits. However, they are also accompanied by some important challenges of different nature. Being mainly based on ecosystem processes and functions, EbA interventions depend on how ecosystems will be impacted by climate change and other drivers of global change (e.g. land-use change) and the knowledge we can gather on such aspects. In particular, there are uncertainties and tipping points in ecosystem behaviours (Scheffer et al. 2001) that will need to be addressed in future research to better frame the medium and long-term potential of EbA (Munroe et al. 2011).

Other challenges and opportunities related to acceptance and socioeconomic prerequisites were also identified throughout the interviews. Acceptance and socioeconomic challenges can be strongly correlated, meaning that EbA-relevant interventions can be better accepted by different stakeholders (e.g. farmers, foresters, communities) if they are aligned with the socio-economic objectives of the cantons or the communities. Biodiversity, adaptation and climate change are not priority issues for the cantons and the communities in comparison to for instance economic development. This means that creating the opportunities for EbA is challenging unless the strategy of an economic sector is already close to EbA (as is the case for agriculture, forestry and to a lesser extent water and disaster risk management) or EbA interventions are possible with only small adjustments (streamlining existing ecosystem management practices into adaptation).

3.3 The Institutional Framework

The third part of the interviews looked at the Swiss institutional framework in relation to EbA. The entry question was the relationship between the federal adaptation strategy and EbA. As highlighted above, there is no direct mention of ecosystem-based approaches to adaptation in the Swiss Federal Adaptation strategy. The reasons identified throughout discussions with the informants can be summarised as follows:

- Knowledge on EbA in Switzerland is still limited.
- Lack of enough tangible evidence of the effectiveness of EbA.
- EbA-relevant interventions not referred to as EbA.
- Tendency to think of adaptation mainly in terms of technical interventions.

- The people drafting the strategy being too technical and inexperienced with the discourse around ecosystems and ecosystem services.
- The lack of integrative approaches in adaptation strategies.
- The sectoral ownership of the different adaptation interventions in the strategy and possible conflicts with the integrative character of EbA.

At the cantonal level, it emerged that there is not yet any overarching coordinated strategy for adaptation, let alone any discussion about different types of adaptation. Some cantons are about to start aligning their respective adaptation strategies with the federal strategy, while the majority still lag behind. It was highlighted that there might be sectorial agencies aware of interventions such as protective forests and river restoration, but possibly these offices do not deal with climate change. Only one canton was very positive on the potential effectiveness of EbA.

The final question explored sectorial policies more in details (mainly at the federal level, unless otherwise stated) that could either favour or hinder the integration of EbA interventions into a portfolio of climate change adaptation interventions and the reasons for their inclusion/exclusion.

There was good agreement amongst the informants that tourism policies in mountain areas could hinder EbA both at the federal and cantonal levels due to a tendency to develop (and subsidise) large-scale technical infrastructure. One policy maker believed that a shift of focus from winter to summer tourism (cited as an adaptation intervention in the sectoral adaptation strategy) could instead profit from EbA interventions (e.g. improvement in landscape attractiveness). Spatial development is second in respondents' order by that could hinder EbA due to a general tendency in mountain communities to want to keep expanding and the construction of critical infrastructures. However, spatial development is also seen as offering opportunities because it provides new guidelines for the appropriate re-naturalisation of steep mountain slopes. The agricultural policy is believed (by almost an equal number of informants) to hold both great potential for EbA, such as:

- Increase in surface area dedicated to the promotion of biodiversity.
- Subsides for biodiversity conservation and no-tillage techniques.
- A lot of potential because the allocated budget is very high.
- Investments in landscape services and quality.

as well as great obstacles:

- Subsides of high alpine intensive farming activities.
- Promote the use of agricultural land exclusively for agricultural production.
- General conflict with guidelines for river restoration.

The two sectoral policies most conductive to EbA are believed to be the water and forestry management strategies. For water management, there is agreement amongst the informants that river enlargement and restoration are the two most promising EbA relevant interventions. For forestry management, the opinion of the informants converge on the strong policy support for the enhancement of the protective function of forests against gravitational hazards. However, some conflicts were also reported, namely conflicts between river enlargement objectives and ground water resources as well as conflicts with the energy policy (i.e. hydropower).

An analysis of the Swiss policy context for promoting and mainstreaming EbA at different levels of social organisation (from local, cantonal to federal) has revealed several important institutional challenges and some good opportunities. The EbA concept and associated interventions could be successfully embedded in the federal adaptation strategy and consequently in the cantonal strategies, because from the interviews it emerged that EbA promotes sustainability (economic, social and environmental), working with nature (concept), social cohesion (co-benefits) and no-regret measures (interventions that would be there anyway, even without climate change). These are all aspects in line with principles in the Swiss federal adaptation strategy as outlined in FOEN (2012):

- Preferred adaptation interventions are those that allow the use of natural processes and regulating services.
- Preferred adaptation interventions are those that will be valuable independently of the rate of climate change and that come with additional co-benefits.
- Preferred adaptation interventions are those that promote cooperation and social cohesion.

Nevertheless, there is no dedicated discussion in the adaptation strategy on EbA. A first possible reason for this is again the lack of adequate knowledge on the links between ecosystem processes, services and climate change adaptation, followed by an excessive focus of adaptation on infrastructural and technical solutions. Based on the expert opinion of the authors and following Naumann et al. (2011), more nuanced reasons are:

- Institutional arrangements not in line with the issue to be resolved
- Different governmental offices claiming authority over a specific issue

Swiss policies are coordinated by different economic sectors, e.g. the federal office for agriculture is responsible for agricultural policy, the federal office for energy is responsible for the energy policy, etc. Moreover, the cantons and communities (*Gemeinden*) have considerable freedom in implementing their policies. In contrast, adaptation demands more cross-sectoral actions since climate change is perceived as a problem encompassing almost all economic sectors and different levels of social organisation. According to key individuals from civil society, consultancies and academic research, sectors and cantons are neither inclined nor open to renegotiating their policy competences, because competences are interest driven and power laden. As a consequence, the Swiss federal adaptation strategy appears to have neither the authority nor the scope to claim jurisdiction over the "climate change adaptation" issue, but rather acts as a coordinating instruments for the "sectoral adaptation strategies". The result is that of 63 adaptation interventions, 54 are sectoral interventions and only 9 are more overarching and mainly focus on knowledge and capacity building (FOEN 2014). EbA interventions can hardly be defined at the sectorial level, because several interventions encompass more than one sector and the services provided are spatially and administratively separated (IPCC 2014). River restoration provides a good example of this issue. River restoration done upstream to protect downstream communities creates benefits for downstream communities since they will be better protected against floods. However, upstream farmers have difficulties in accepting the need to forgo part of their land, as they will not benefit directly from the service. Thus, developing such interventions implies a good knowledge of the characteristics and effectiveness of the services, where the services are located and which co-benefits can be exploited. This requires first of all scientific knowledge; second a spatially explicit allocation of ecosystem services in order to identify distributions of costs and benefits (not only economic but also social and environmental) over time and space; and third it has to be locally negotiated by all involved stakeholders. Hence, EbA demands collective action from different actors (policy makers, researchers, civil society) as well as at different scales of jurisdiction (national, cantonal, local). Collective action may be achieved by creating institutions to bridge and strengthen vertical (across jurisdictional scales) and horizontal (across sectors) integration for climate change adaptation and ecosystem service management. A similar conclusion was also achieved by Vignola et al. (2009) in an analysis of the role of EbA in policy and society, and by Huber et al. (2013) in discussion of the provisioning of ecosystem goods and services in a Swiss local mountain community under climate change and land-use change. The need for integrative approaches in adaptation policy formulation was further supported by the analysis on sectoral policy. Policy integration may allow synergies to be exploited and trade-offs to be reduced between those policies that seem more inclined to favour (e.g. forestry, water management and to a certain extent agriculture) and those policies more inclined to disfavour EbAs (tourism, spatial planning and to a certain extent agriculture) (Wertz-Kanounnikoff et al. 2011). Moreover, the interviews highlighted that the cantons lag well behind the federal jurisdiction on adaptation issues in general, necessitating more horizontal dialogues. Creating cross-sectoral institutions would foster information exchange amongst researchers, policy makers (federal and cantonal) and local communities for effective adaptation responses to climate change (Duit and Galaz 2008). Cross-sectoral institutions could take the form of boundary organisations, which are organisations brokering the dialogues in climate change between different actors (Agrawala et al. 2001; Turton et al. 2007). Some informants suggested that such institutions could indeed promote information exchange and knowledge across different scales and actors (researchers, policy makers, civil society) for the purpose of supporting multilevel (federal, cantonal, and local) collective action for effective ecosystem-based adaptation to climate change. However, none of the informants provided clear examples of boundary organisations neither in Switzerland nor abroad.

4 Conclusions and Recommendation

In this study we looked at the status of knowledge on and implementation of ecosystem based adaptation (EbA) interventions in the Swiss mountains. We found out that EbA interventions in the Swiss mountains are mainly in the areas of disaster risk management, natural resources management and water management. Forest management to protect against gravitational hazards emerged as the first most discussed and studied EbA related intervention, whereas the potentials for using river restoration to adapt to climate change have so far not been explored extensively by the research community. River restoration is nonetheless widely discussed and its acceptance is increasing at the community level as the informants have stressed, albeit some important conflicts over land availability and water resources. Interventions in natural resources management mainly from agriculture are widely researched and ownership is high in Switzerland. Answers to our second research question (i.e. what are the challenges and opportunities for EbA in the swiss mountains?) have highlighted that research efforts should be geared towards a better understanding of the relationships between mountain ecosystem processes, ecosystem services and the ability of such ecosystems to cope with changing climatic and socio-economic conditions. This kind of research can strenghten the case for EbA and contribute to improve its acceptance. Opportunities exist for some interventions to be easily mainstreamed in current policy trends (e.g. forestry policy and water management policy) and traditional knowledge. In relation to the third research question (i.e. is the current policy environment is favourable to the integration of EbA as an overall strategy to deal with climate change?), we observed that the current sectorial policy context for climate change adaptation (both at the federal and cantonal level), converging into the Federal Adaptation strategy, appears to be poorly conductive to integrative adaptation approaches such as EbA. The development of new transversal institutions in form of boundary organisations could on one side create a more diffuse dialogue amongst adaptation stakeholders (researchers, decision makers, civil society) and on the other side improve vertical and horizontal policy integration, so fundamental for EbA and also for adaptation in general.

Acknowledgments The elaboration of this article has been supported by the University Research Priority Program on Global Change and Biodiversity (URPP GCB) of the University of Zurich and by the Swiss Federal Office for the Environment (FOEN). We acknowledge contribution of all our informants for the interesting information they provided.

References

Agrawala, S., Broad, K., & Guston, D. H. (2001). Integrating climate forecasts and societal decision making: Challenges to an emergent boundary organization. *Science, Technology & Human Values, 26*(4), 454–477.

Batker, D., de la Torre, I., Costanza, R., Swedeen, P., Day, J., Boumans, R., & Bagstad, K. (2010). Gaining ground: Wetlands, hurricanes, and the economy: The value of restoring the Mississippi river delta. *Environmental Law Report: News & Analysis, 40*(11), 11106–11110.

Briner, S., Elkin, C., & Huber, R. (2013). Evaluating the relative impact of climate and economic changes on forest and agricultural ecosystem services in mountain regions. *Journal of Environmental Management, 129*, 414–422. doi:10.1016/j.jenvman.2013.07.018.

Bryman, A. (2012). *Social research methods* (3rd ed.). Oxford: Oxford University Press, 427 pp.

Campbell, A., Kapos, V., Scharlemann, J. P. W., Bubb, P., Chenery, A., Coad, L., Dickson, B., Doswald, N., Khan, M. S. I., Kershaw, F., & Rashid, M. (2009). *Review of the literature on the links between biodiversity and climate change: Impacts, adaptation and mitigation* (Technical series no. 42). Montreal: Secretariat of the Convention on Biological Diversity.

Carpenter, S. R., Folke, C., Norstrom, A., Olsson, O., Schultz, L., Agarwal, B., Balvanera, P., Campbell, B., Castilla, J. C., Cramer, W., DeFries, R., Eyzaguirre, P., Hughes, T. P., Polasky, S., Sanusi, Z., Scholes, R., & Spierenburg, M. (2012). Program on ecosystem change and society: An international research strategy for integrated social–ecological systems. *Current Opinion in Environmental Sustainability, 4*, 134–138. doi:10.1016/j.cosust.2012.01.001.

CBD. (2009). *Connecting biodiversity and climate change mitigation and adaptation: Report of the second ad hoc technical expert group on biodiversity and climate change* (Technical series no. 41). Montreal, 126 pages.

Colls, A., Ash, N., & Ikkala, N. (2009). *Ecosystem-based adaptation: A natural response to climate change*. Gland: IUCN. 16 pp.

Corbin, J., & Strauss, A. (2008). *Basics of qualitative research: Techniques and procedures for developing grounded theory*. Thousand Oaks: Sage Publications.

Doswald, N., & Osti, M. (2011). *Ecosystem-based approaches to adaptation and mitigation – Good practice examples and lessons learned in Europe*. Bonn: Federal Agency for Nature Conservation.

Doswald, N., Munroe, R., Roe, D., Giuliani, A., Castelli, I., Stephens, J., Möller, I., Spencer, T., Vira, B., & Reid, H. (2014). Effectiveness of ecosystem-based approaches for adaptation: Review of the evidence-base. *Climate and Development, 6*(2), 185–201. doi:10.1080/17565529.2013.867247.

Duit, A., & Galaz, V. (2008). Governance and complexity: Emerging issues for governance theory. *Governance: An International Journal of Policy, Administration and Institutions, 21*(3), 311–335. doi:10.1111/j.1468-0491.2008.00402.x.

FOEN. (2012). *Adaptation aux changement climatiques en Suisse. Objectifs, défis et champs d'action. Premier volet de la stratégie du Conseil Fédéral*. Bern: Swiss Federal Office for the Environment.

FOEN. (2014). *Adaptation aux changements climatiques en Suisse Plan d'action 2014–2019. Deuxième volet de la stratégie du Conseil Fédéral*. Bern: Swiss Federal Office for the Environment.

Game, E. T., Bode, M., McDonald-Madden, M., Grantham, H. S., & Possingham, H. P. (2009). Dynamic marine protected areas can improve the resilience of coral reef systems. *Ecology Letters, 12*(12), 1336–1346. doi:10.1111/j.1461-0248.2009.01384.x.

Gobiet, A., Kotlarski, S., Beniston, M., Heinrich, G., Rajczak, J., & Stoffel, M. (2013). 21st century climate change in the European Alps – A review. *Science of the Total Environment, 493*, 1138–1151. doi:10.1016/j.scitotenv.2013.07.050.

Grêt-Regamey, A., Brunner, S. H., & Kienast, F. (2012). Mountain ecosystem services: Who cares? *Mountain Research and Development, 32*(S1), 23–34. doi:10.1659/MRD-JOURNAL-D-10-00115.S1.

Grêt-Regamey, A., Brunner, S. H., Altwegg, J., Christen, M., & Bebi, P. (2013). Integrating expert knowledge into mapping ecosystem services trade-offs for sustainable forest management. *Ecology and Society, 18*(3), 599–619. doi:10.5751/ES-05800-180334.

Huber, R., Rigling, A., Bebi, P., Brand, F. S., Briner, S., Buttler, A., Elkin, C., Gillet, F., Grêt-Regamey, A., Hirschi, C., Lischke, H., Scholz, R. W., Seidl, R., Spiegelberger, T., Walz, A.,

Zimmermann, W., & Bugmann, H. (2013). Sustainable land use in mountain regions under global change: Synthesis across scales and disciplines. *Ecology and Society, 18*(3), 115–136. doi:10.5751/ES-05499-180336.

IPCC. (2014). In C. B. Field, V. R. Barros, D. J. Dokken, K. J. Mach, M. D. Mastrandrea, T. E. Bilir, M. Chatterjee, K. L. Ebi, Y. O. Estrada, R. C. Genova, B. Girma, E. S. Kissel, A. N. Levy, S. MacCracken, P. R. Mastrandrea, & L. L. White (Eds.), *Climate change 2014: Impacts, adaptation, and vulnerability. Chapter 14: Adaptation needs and option. Contribution of Working Group II to the Fifth Assessment Report of the Intergovernmental Panel on Climate Change* (pp. 1–51). Cambridge: Cambridge University Press.

Jones, H. P., Hole, D. G., & Zavaleta, E. S. (2012). Harnessing nature to help people adapt to climate change. *Nature Climate Change, 2*, 504–509. doi:10.1038/nclimate1463.

Koellner, T. (2009). Supply and demand for ecosystem services in mountainous regions. In *Global change and sustainable development in mountain regions* (Alpine space e man & environment, 7, pp. 61–70).

Körner, C., & Ohsawa, M. (2006). Mountain systems. In R. Hassan, R. Scholes, & N. Ash (Eds.), *Ecosystem and human well-being: Current state and trends. Millennium ecosystem assessment* (Vol. 1, pp. 681–716). Washington, DC: Island Press.

Löffler, J., Anschlag, K., Baker, B., Finch, O., Diekkrüger, B., Wundram, D., Schröder, B., Pape, R., & Lundberg, A. (2011). Mountain ecosystem response to global change. *Erdkunde, 65*(2), 189–213.

MEA. (2005a). *Ecosystems and human well-being: Synthesis. Millennium ecosystem assessment.* Washington, DC: Island Press.

MEA. (2005b). *Ecosystems and human well-being: Summary for decision makers. Millennium ecosystem assessment.* Washington, DC: Island Press.

Midgley, G. S. M., Barnett, M., & Wågsæther, K. (2012). *Biodiversity, climate change and sustainable development – harnessing synergies and celebrating successes.* Final Technical Report, Jan 2012.

Mumby, P. J., Hastings, A., & Edwards, H. J. (2007). Thresholds and the resilience of Caribbean coral reefs. *Nature, 450*(7166), 98–101.

Munang, R., Thiaw, I., Alverson, K., Mumba, M., Liu, J., & Zhen, H. (2013a). The role of ecosystem services in climate change adaptation and disaster risk reduction. *Current Opinion in Environmental Sustainability, 5*(1), 47–52. doi:10.1016/j.cosust.2013.02.002.

Munang, R., Thiaw, I., Alverson, K., Mumba, M., Liu, & Rivington, M. (2013b). Climate change and ecosystem-based adaptation: A new pragmatic approach to buffering climate change impacts. *Current Opinion in Environmental Sustainability, 5*(1), 67–71. doi:10.1016/j.cosust.2012.12.001.

Munroe, R., Doswald, N., Roe, D., Reid, A., Giuliani, A., Castelli, I., & Mölleret, I. (2011). *Does EbA work ? A review of the evidence on the effectiveness of ecosystem-based approaches to adaptation.* Cambridge, UK: Cambridge Conservation Initiative.

Munroe, R., Roe, D., Doswald, N., Spencer, T., Möller, I., Bhaskar, V., Kontoleon, A., Reid, H., Giuliani, A., Castelli, I., & Stephens, J. (2012). Review of the evidence base for ecosystem-based approaches for adaptation to climate change. *Environmental Evidence*, 1–13. Available from DOI: 10.1186/2047-2382-1-13.

Naumann, S., Anzaldua, G., Berry, P., Burch, S., Davis, M., Frelih-Larsen, A., Gerdes, H., & Sanders, M. (2011). *Assessment of the potential of ecosystem-based approaches to climate change adaptation and mitigation in Europe.* Final report to the European Commission, DG Environment, Ecologic institute and Environmental Change Institute, Oxford University Centre for the Environment.

OcCC. (2007). *Climate change and Switzerland 2050 – Expected impacts on environment, society and economy. Umwelt, Gesellschaft und Wirtschaft.* Bern: OcCC/ProClim.

Oliver, J., Probst, K., Renner, I., & Klemens, R. (2012). *Ecosystem-based Adaptation (EbA) – A new approach to advance natural solutions for climate change adaptation across different sec-*

tors. Eschborn, Germany: Deutsche Gesellschaft für Internationale Zusammenarbeit (GIZ) GmbH.

Opperman, J. J., Galloway, G. E., Fargione, J., Mount, J. F., Richter, B. D., & Secchi, S. (2009). Sustainable floodplains through large-scale reconnection to rivers. *Science, 326*(5959), 1487–1488. doi:10.1126/science.1178256.

Pérez, A. A., Herrera, F. B., & Cazzolla Gatti, R. (2010). In IUCN (Ed.), *Building resilience to climate change building resilience to climate change ecosystem-based adaptation and lessons from the field*. Gland: IUCN.

Saldaña, J. (2013). *The coding manual for qualitative researchers*. Los Angeles: SAGE Publications.

Scheffer, M., Carpenter, S., Foley, J. A., Folke, C., & Walker, B. (2001). Catastrophic shifts in ecosystems. *Nature, 413*(6856), 591–596. doi:10.1038/35098000.

SCNAT. (2012). *Mountains, a priority for a planet under pressure and for Switzerland*. Fact Sheet n°2 "Rio +20".

Soussan, J., & Springate-Baginski, O. (2001). *Livelihood-policy relationships in South Asia: A methodology for policy process analysis*. Working Paper 9 prepared for DFID, UK: DFID.

Tengö, M., & Belfrage, K. (2004). Local management practices for dealing with change and uncertainty: A cross-scale comparison of cases in Sweden and Tanzania. *Ecology & Society, 9*(3), 1.

Turton, A. R., Hattingh, H. J., Maree, G. A., Roux, D. J., Claassen, M., & Strydom, W. F. (2007). *Governance as a trialogue: Government–society–science in transition*. Germany: Springer. Berlin, Springer-Verlag.

Vignola, R., Locatelli, B., Martinez, C., & Imbach, P. (2009). Ecosystem-based adaptation to climate change: What role for policy-makers, society and scientists? *Mitigation and Adaptation Strategies for Global Change, 14*, 691–696. doi:10.1007/s11027-009-9193-6.

Wang, S., Fu, B., Wei, Y., & Lyle, C. (2013). Ecosystem services management: An integrated approach. *Current Opinion in Environmental Sustainability, 5*(1), 11–15. doi:10.1016/j.cosust.2013.01.003.

Wertz-Kanounnikoff, S., Locatelli, B., Wunder, S., & Brockhaus, M. (2011). Ecosystem-based adaptation to climate change: What scope for payments for environmental services? *Climate and Development, 3*(2), 143–158. doi:10.1080/17565529.2011.582277.

World Bank. (2009). *Convenient solutions to an inconvenient truth: Ecosystem based approaches to climate change*. Washington, DC: The World Bank.

Community Perceptions and Responses to Climate Variability: Insights from the Himalayas

Anju Pandit, Anmol Jain, Randhir Singha, Augustus Suting, Senti Jamir, Neera Shresthra Pradhan, and Dhrupad Choudhury

Abstract A prerequisite for the formulation of effective adaptation strategies and plans is an in-depth understanding of impacts resultant of climate variability, the measures adopted by communities as a response to such stress and the support needs to reduce vulnerabilities arising out of such challenges. A satisfactory, updated information base covering all these aspects is difficult to come by, particularly in the Hindu Kush Himalayan countries, posing serious challenges for any agency tasked with the responsibility of formulating climate change adaptation strategies and plans. To bridge this knowledge gap, an extensive participatory assessment was undertaken in selected districts of Bhutan, India and Nepal, covering 90 villages spanning an altitudinal range of 50–3500 MSL. The results of this extensive survey are reported in this chapter, with special focus on perceptions of mountain communities on climate variability, their impacts and the responses of the communities to overcome the resultant stress.

Keywords Community perception • Climate variability • Impact assessment • Coping mechanism • Adaptive mechanism

---

A. Pandit • N.S. Pradhan • D. Choudhury (✉)
International Centre for Integrated Mountain Development (ICIMOD), Kathmandu, Nepal
e-mail: Dhrupad.Choudhury@icimod.org

A. Jain
Independent Consultant, Mussourie, India

R. Singha
Resources Centre for Sustainable Development,
Byelane 12, Kundilnagar, Guwahati, Assam, India

A. Suting
Meghalaya Rural Development Society, Shillong, Meghalaya, India

S. Jamir
West Garo Hills Community Resource Management Project, west Garo Hills, Tura, India

N. Salzmann et al. (eds.), *Climate Change Adaptation Strategies – An Upstream-downstream Perspective*, DOI 10.1007/978-3-319-40773-9_10

1 Introduction

Climate related disasters have been on the increase in the past few years across the Hindu Kush Himalayas (Haque 2003; ICIMOD 2010; Xu et al. 2009) – the long drawn dry spells in many parts of western Nepal and the Indian Himalayan states between 2007 and 2010, followed by the devastating flash floods in Uttarakhand in India (Das 2013; Uniyal 2013), Chitral in Pakistan (Rahman and Khan 2013) and across major parts of western Myanmar in the immediate past. Severe rainfall induced landslides in 2015 in Bandarban, Bangladesh and the Chin state in Myanmar are a few examples of climate induced disasters in the region. These events, together with the increasingly erratic patterns of rainfall in the region, underscore the urgent need for formulating effective adaptation action plans in order to enhance adaptive capacities of mountain communities to cope with stress (Pradhan et al. 2014, 2015).

A pre-requisite to effective adaptation strategies and plans, however, is a sufficiently robust information base describing the local impacts of climate variability and the responses of communities to such change – an understanding of impacts and community capabilities to deal with such challenges. Yet, the lack of adequate and reliable information is a singular, fundamental challenge facing decision makers while attempting to formulate locally relevant adaptation plans and strategies. The IPCC Fourth Assessment Report pointed out the lack of information and research on the Himalayan region. This was reaffirmed by a review of recent published literature on the subject. The only studies reported from the region on community perceptions and impacts on livelihoods are those of Byg and Salick (2009) for Tibet, Chaudhury et al. (2011) and Piya et al. (2012) for Nepal who document the observations of selected communities on the changing weather patterns and perceptions of changed weather conditions on livelihood. The paucity on information, therefore, clearly suggests an information (and knowledge) gap which could seriously hinder the formulation of effective adaptation strategies and plans.

Against this backdrop of the need to generate information on the impacts of climate variability as well as the need to document responses of communities to such changes in order to improve our understanding on these issues, the International Centre for Integrated Mountain Development (ICIMOD) initiated an extensive participatory assessment in 2010 and 2011, to document community perceptions on climate (weather) variability and change, community responses to such changes and the institutional dependency of such communities to deal with the stress arising out of increasing weather variability. An extensive participatory qualitative assessments covering 90 villages across 15 districts of Bhutan, India and Nepal and spanning an altitudinal range varying from 50 MSL to 3500 MSL, provided a better understanding of community perceptions, climate change impacts on their livelihood and an insight into the capacities of mountain communities to respond to such change. In addition, the assessments also sought to provide an understanding of the institutional dependencies of the mountain communities and the support that such institutions were required to extend in order to reduce the vulnerabilities of the communities arising out of such change. This chapter highlights some of the results from this assessment, focusing on the impacts and responses of the mountain communities from selected districts in the three Himalayan countries. The results on institutional dependency is being reported elsewhere, and hence is not included in this chapter, except for a brief summary.

2 Methodology

An extensive participatory assessment was designed, covering 90 villages across 15 districts (6 villages per district) of Bhutan, India and Nepal and spanning an altitudinal range varying from 50 MSL to 3500 MSL (Table 1). Out of 15 districts, from west to east, 3 were selected in Uttarakhand (western India), 5 in Nepal, 2 in Bhutan and 5 in North East (NE) India. A Participatory Rural Appraisal (PRA) toolkit, comprising of five tools was developed for the purpose. The PRA Toolkit comprised: (i) Seasonal Calendar – modified to capture present and past weather trends and hence document changes, (ii) Hazard Ranking (Spider Web) – ranked on a scale of 0–5, the tool helped to identify the weather events (and weather induced events) that a particular community perceived as severely hazardous to livelihood pursuits, particularly for food and income security, (iii) Seasonal Dependency Matrix – a modified Seasonal Calendar developed specifically to map dependencies of communities on different support systems (natural production systems as well as non-land based

Table 1 The study districts with the altitudinal range of study villages

S.N.	Districts	Altitudinal range of study villages (m)
Bhutan		
1.	Pemagatshel	1100–2000
2.	Trashi yangtse	850–2350
Nepal		
3.	Bajhang	950–2400
4.	Humla	2000–3100
5.	Dailekh	650–3500
6.	Tanahu	750–1650
7.	Terathum	750–1870
India		
8.	Almora	1100–2100
9.	Bageshwar	900–1790
10.	Tehri	Na
11.	West Garo Hills	50–1040
12.	East Garo Hills	275–750
13.	Ri Bhoi	69–1020
14.	Karbi Anglong	406–669
15.	Ukhrul	1600–2200

options); the tool helps in mapping diversity of support systems and resources within each system as well as their seasonal availability; (iv) Seasonal Activity Calendar – to document all farm and non-farm activities undertaken through different seasons in a year, with focus on livelihood security pursuits, (v) Institutional Dependency (Venn Diagram) – to document institutional dependencies of households at times of stress: this tool helps to document dependencies on institutions across the spectrum (formal, informal, statutory, private and public bodies) as well as the purposes for which communities turn to such institutions (hence, support services provided).

To document impacts of change and community responses to deal with impacts resultant to such change, an iteration, based on results of tool (ii) was initiated focusing on the findings of tools (i) and (iv). The combined application of the results of these three tools helped to succinctly bring out impacts as well as the coping and adaptive responses of communities at each location. The PRA exercises were complimented with Focused Group Discussions and intensive iterative consultations with smaller groups and key informants conducted in selected villages across the three countries subsequent to the PRA assessments.

3 Results and Discussions

3.1 Community Perceptions on Changes in Weather Patterns

Communities across the study areas perceived a change in weather patterns, particularly in regard to precipitation. With the exception of Bhutan, communities in all locations indicated a reduction in the annual duration of rainfall. This reduction was most perceptible in Uttarakhand, but communities in Nepal also perceived a similar reduction, significantly more than that perceived in Bhutan and NE India. A trend analysis, plotting the perceived reduction in duration of rainfall across the longitudinal spread for all locations showed a significant trend, with the reduction in duration being maximum in western locations, progressively reducing towards eastward. This shows that communities in western locations (Uttarakhand and Nepal) perceived significant reductions in the annual duration of rainfall, but this reduction was less dramatic as one moved eastwards through Bhutan and into NE India (Fig. 1).

Communities also perceived a reduction in the duration of annual snowfall in the study locations in Uttarakhand, Nepal and Bhutan – in some cases reporting the complete absence of snowfall in recent years (particularly in Uttarakhand). Communities in all locations also reported a delay in the onset of snowfall, on an average by about a month; the same trend was also perceived in regard to completion of snowfall.

The reduction in annual duration of precipitation events were perceived, quite expectedly, to have influenced the annual duration of dry periods. Communities

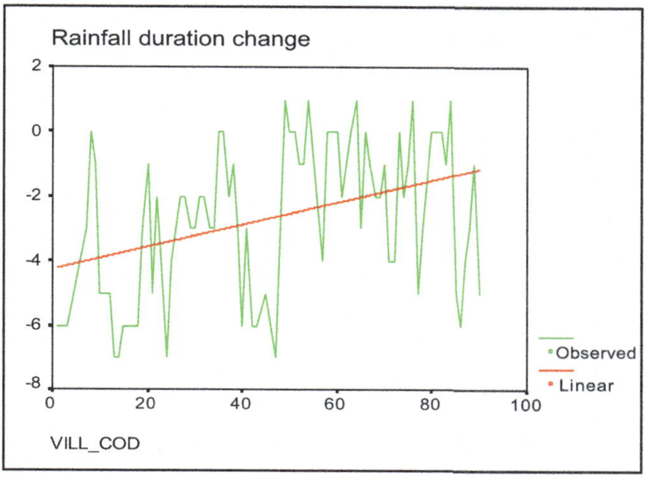

Fig. 1 Trend analysis of the perceived reduction in duration of annual rainfall across the study areas (west to east) (*x-axis*: village codes listed from western most village in Uttarakhand to eastern most in Ukhrul; *y-axis*: in months) (Based on participatory assessments carried out in the study villages, 2010–2011)

across the study location perceived an increase in the duration of dry periods, the notable exception being Bhutan. As in the case of annual duration of rainfall, an increase in the duration of dry spells were perceptibly significant in Uttarakhand. The duration of dry spells were also perceived to have increased in Nepal, being significantly more than that in NE India or Bhutan; this increase was also seen in NE India. A trend analysis of the perceived duration of dry spells along a longitudinal spread showed a significant trend of prolonged dry spells in the western locations which progressively decreased towards eastwards (Fig. 2). This trend was also reflected in the case of early onset and prolongation (delayed completion) of dry spells, being statistically significant in both cases. On an altitudinal gradient, an increase in the duration of dry spells was positively correlated with elevation suggesting an increase in dry spells with a corresponding increase in elevation. This was also true in case of early onset of dry spells, suggesting that the onset of dry spells commence earlier as elevation increases.

Most areas of the HKH region lack sufficient observational records to draw conclusions about trends in annual precipitation over the past century (Shrestha and Aryal 2010). Overall, in South Asia, the frequency of heavy precipitation events is increasing, while light rain events are decreasing (Hijioka et al. 2014). Observational data seem to validate community perceptions as verified by the observational records suggesting a reduction of rainfall duration and increase in intensity. The perceived trends in temperature generally agree with observational record and show significant warming in last decades (Shrestha and Aryal 2010). As indicated by observational and model results (Shrestha and Aryal 2010; Piya et al. 2012), this study shows that communities perceive a clear change in weather patterns, reduced rainy season and increased dry weather.

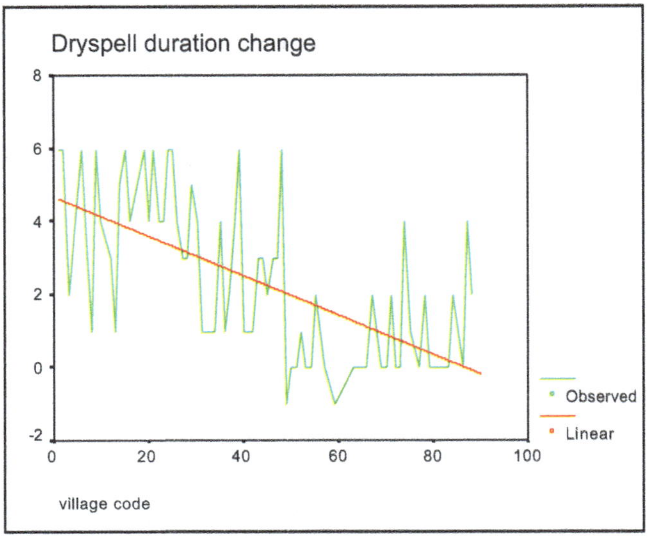

Fig. 2 Trend analysis of the perceived increase in annual dry spells (Based on participatory assessments carried out in the study villages, 2010–2011)

3.2 Impacts of Weather Variability on Food and Income Security

The iteration using the results of Seasonal Weather and Activity Calendars (Tools I and IV) (Fig. 3) drew out sufficient information to assess the vulnerabilities consequent to the weather hazards, triggered discussions on community adaptation to impacts of weather hazards, and allowed the documentation of coping and adaptive responses while providing insights on the external support required.

The iteration with communities suggest changed precipitation patterns– delayed onset and early completion of rainfall and snowfall, intermittent spells of rainless days, extreme events of very heavy rainfall, hailstorms and winds and the prolonged dry spells– with enhanced the risks of poor crop performance and at times, even crop failure, rendering most rainfed dependent agrarian households extremely vulnerable. Farmers across the study districts perceived enhanced risks of poor crop performance – and crop failures in some areas – as a result of increased exposure to changes in precipitation patterns and prolonged spells of dry weather. Farmers across the region reported severe impacts on crops, particularly cereal and vegetable crops due to the changes in weather conditions, especially in precipitation patterns (Fig. 4a, b). The prolongation of dry spells, depletion of soil moisture and the resultant water scarcity in Uttarakhand and Nepal had severe consequences for women as they had to spend long hours to collect water or trudge long distances to fetch clean, potable water and collect fodder and forage. The increased drudgery, particularly in the hot, dry weather also increased women's vulnerability to health issues. In many villages in the Garo Hills districts of NE India, the cumulative effects of dry

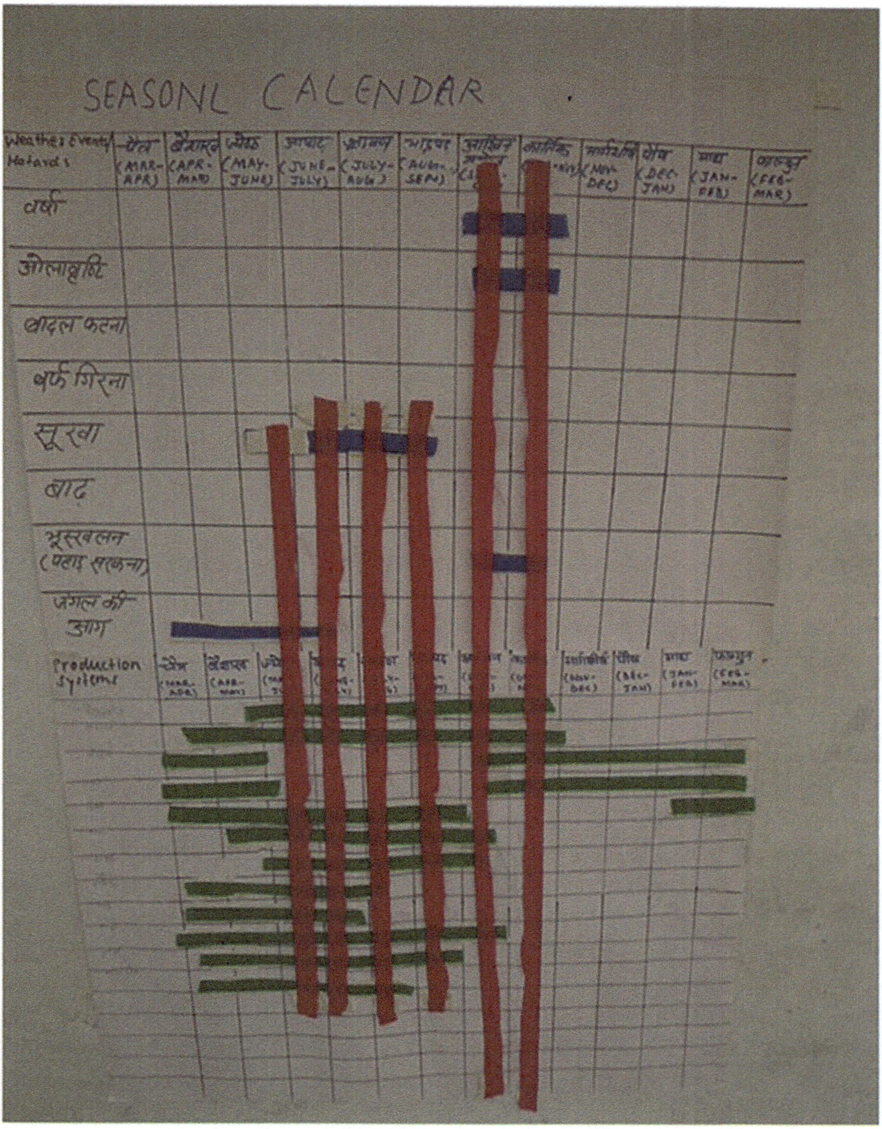

Fig. 3 Combination of the Seasonal weather and Seasonal Activity Calendars for conducting a participatory assessment of risks, vulnerabilities arising out of weather hazards and document community responses to address the impacts

weather, erratic rainfall and storms resulted in drastic decline in agricultural yields. For shifting cultivators inhabiting these districts, particularly those with short-fallow cycles, the drastically reduced yields forced men to move out of the villages to look for seasonal wage employment in coal mines nearby. With the men engaged in mines, the women were left with the added burden of looking after their

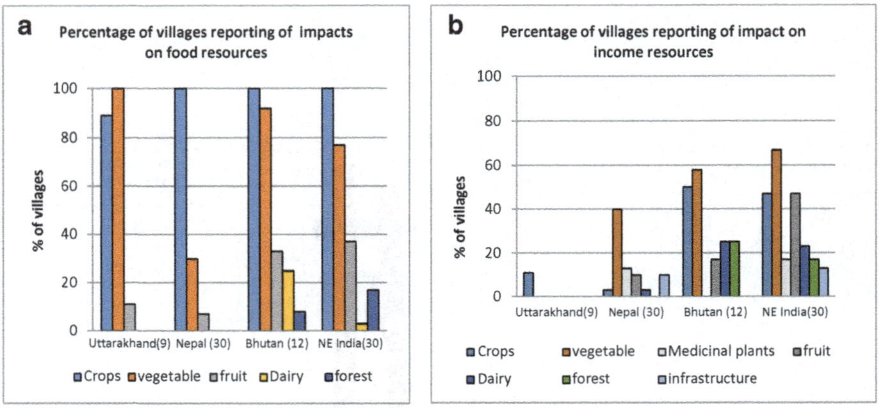

Fig. 4 (**a**, **b**) The impact of weather changes on agriculture and other land-based livelihood pursuits in the study districts (Based on participatory assessments carried out in the study villages, 2010)

agricultural fields which now included the demanding task of jungle clearing. For many households with limited labour, families had to abandon shifting cultivation and the women forced to take up wage employment as domestic helps with better off families in the village or nearby urban centres. With limited opportunities for wage employment and the seasonal availability of employment opportunities in mines, the vulnerability of such families had increased manifold and they precariously verge on slipping into chronic poverty.

For many farmers elsewhere, particularly in Bhutan and parts of NE India, hailstorms and winds damaged early season vegetable crops and destroyed blooms of many fruit crops. While these events affected both the poor and the well off, for the smallholding farming families of Ri Bhoi district, the early season hailstorms and heavy rainfall destroyed both vegetable and paddy crops while the late season events damaged whatever standing crops of paddy that escaped damage from earlier events, thus depriving them of the little food and income security they would otherwise have, rendering such households extremely vulnerable to food scarcity and burdened with debt. The findings suggested increased vulnerability of smallholder agrarian households across the region from impacts of climate variability, especially those that are dependent on rainfed agriculture. This seems to be consistent with trends seen elsewhere (Cooper et al. 2008).

3.3 Coping and Adaptive Mechanisms

The coping responses from the field survey are reactive, temporal and of immediate nature, essentially to cope with a particular stress immediately, whereas, adaptive responses are anticipatory, precautionary and long-term (Smit et al. 2000). Impacts of climate change were generally perceived to be negative with reports commonly highlighting responses to stress and shocks; however, under adverse situations, it can also be turned into opportunities. Communities across the study villages reported different coping and adaptive mechanisms to deal with the change in weather patterns (Tables 2 and 3).

Table 2 A summary of coping responses adopted by communities of the study villages

Particulars	Uttarakhand	Nepal	NE India	Bhutan
Cropping pattern	Cropping delayed (15 days – 1 month)	Early sowing	Early sowing	Harvesting and weeding done in subsidized rain
		Cropping delayed (1 month)	Cropping delayed	
	Covering sown seeds with litter;		Repeat sowing	
		Repeat sowing	Seeds dibbled (deep sown); seeds soaked, broadcast	Mulching done with oak leaves
	Polyhouses for vegetables	Wet maize seeds prior to sowing; deep sow seeds		
			Early harvest (to avoid storms/ hailstorm)	
			Cropping season shift (summer to winter)	
			Bamboo matting mesh to avoid siltation of fields due to flash floods	
		Dry seed beds (rice) – seeds germinated in moist pouches, transplanted to seed beds only after first rains, beds covered with straw, mulch	Use of pomelo (citrus) peals, *Samsneng, Magvit*, bamboo shoot extract, *So-ik*, *jarman* and *l-upo* against pests; hang crab carcass against rice bug	
		Crop rotation – Horsegram/ blackgram/ sesame		
		Mixed cropping – maize, beans; millet, blackgram		
Alternative crops	Rice – pulses, soya, sesame, coarse grains (*mandua/ragi*) or *madira* (fodder)	Millet – maize, upland rice, black gram, chino, mustard, *Philunge* (beans)	Sesame, soyabeans, rice beans, cowpea	
		Wheat, barley – buckwheat		
	Mandua – madira or ginger, soya, urad (pulse) or potato	Buckwheat – turnip, mustard, green leafy vegetables,		

(continued)

Table 2 (continued)

Particulars	Uttarakhand	Nepal	NE India	Bhutan
Seasonal occupational shifts			Farming to sand mining (rivers), weaving	
			Seasonal wage earning in mines, plantations, road building	
			Wage earning from Mahatma Gandhi National Rural Employment Guarantee Programme (MNREGA)	

Source: Based on participatory assessments carried out in the study villages, 2010

Although farmers claimed that crop replacements were in response to weather-related stress, this, however, could be debatable as most of these changes seem also to be a response to growing market opportunities. The responses do indicate an undeniable shift in crops and suggested a move towards transformative adaptation, whatever the stimuli. This is evident from the fact that the shifts reflect a replacement of food grains with commodity crops that were commercially important and hence reflected a transformation in the intended outcome of crop replacements – attempts to enhance returns to agriculture and increase cash incomes of households instead of focusing only on food self-sufficiency – a characteristic of subsistence farming. A reflection of this intent was also evident in the increased dependency on forest products such as NTFPs in NE India.

While replacement of food crops with cash crops seemed a widespread adaptive response across Uttarakhand and Nepal, crop varietal replacement, particularly with traditional varieties of food crops, was more frequently reported in Nepal, Bhutan and NE India. Varietal replacements of wheat and paddy seemed quite widespread in the villages in Nepal and NE India, with villagers opting for early maturing, water-stress tolerant, pest resistant or dwarf varieties as an adaptation to prolonged dryspells, pest outbreaks and crop lodging due to late season heavy rains. In specific cases in NE India, when crops in irrigated systems wilted after germination due to delay in the onset of monsoons, upland farmers resorted to growing these varieties in traditional shifting cultivation fields with encouraging outcomes. In the same areas, farmers have also introduced rainfed crop varieties into irrigated systems with positive results.

While farmers in Uttarakhand predominantly chose new varieties introduced by agricultural extension agencies, those in Bhutan and NE India seemed to prefer traditional varieties to new varieties. This could possibly be a consequence of the relatively easier access to improved varieties in Uttarakhand while access to new varieties could be more difficult in Bhutan and NE India and the degree of appropri-

Table 3 A summary of adaptive responses adopted by communities of the study villages

Particulars	Uttarakhand	Nepal	NE India	Bhutan
Crop replacements	Change from wheat to ginger, tumeric, mustard	Change from wheat, barley to potato		
	Change from maize to cauliflower, peas, soya, vegetable climbers	Change from apple to potato		
	Change from cereals to potato			
	Groundnut to ginger			
	Change from cereals to horticulture, aloe vera	Change from groundnut to blackgram and horsegram		
		Change from maize to banana		
		Change from rice to maize/ginger		
Livestock	Reduced numbers (fodder, forage shortage)	Reduced numbers (fodder, forage shortage)	Preference for smaller ruminants (goats), piggery, poultry	Stall feeding, agro pasture, improve pasture land , reduced unproductive cattle heads, introduced improved cattle breeds, grass supplement with kitchen waste and dry hay
	Change from large ruminants to small animals (goats)	Change from large ruminants to small animals (goats)		
Varietal replacements		Wheat: *Daudkhani* by *Jhuse*	Rice -*Sapa, Methungia, Amosa, Chuibok, Soksu, Naka*	Change in crop variety {*Sawn II/Yangtsepa, Yangtsip* (short heighted)}
		Rice: *Chhiti/Jwali* by *Laidiya/Rui/Anjana*		
		Marso, Darnali by *Munyasonali/Jhyali*	Short duration and dwarf varieties grown to avoid loss from hailstorms	Wind resistant varieties are preferred
			Millet – *Alika*	

(continued)

Table 3 (continued)

Particulars	Uttarakhand	Nepal	NE India	Bhutan
Mixed cropping			Banana intercropping with areca and litchi (diversification in horticulture plantations)	Potato, beans inter cropping with maize
NRM and water management	Water-harvesting Oak regeneration, catchment protection Institutional arrangements Water sharing Labour sharing arrangements			Controlled watering to improve the growth, Light irrigation
Occupational shifts	Farming to carpentry, masonry Wage earning (roadside eateries, wage labour) Migration: seasonal or long-term	Farming to carpet weaving, carpentry Wage earning Seasonal migration (lean months – to India) Long-term – Gulf, SE Asia or Korea		Increasing preference for indoor labor work
Systems shift			Change to shifting cultivation Increased dependency on forest produces (Non Timber Forest Products, wild edibles)	
New opportunities (due to warmer weather)	Groundnut, beans (early season) Peas, cauliflower (after potato harvest – additional income)			

Source: Based on participatory assessments carried out in the study villages, 2010

ateness of the improved varieties introduced by extension agencies to the local conditions – suitability to local agricultural systems, preference, taste – reflecting the degree to which agricultural research was linked to ground realities in each area, and thereby, the effectiveness of agricultural research and extension in these areas.

Communities have also demonstrated a diversification of 'portfolios' in response to change – intensifying home gardens and integrating this with small livestock rearing. A clear trend in animal husbandry was the rapid replacement of larger livestock with smaller ruminants, especially goats. While growing scarcity of fodder resultant to reduced rainfall and inadequate labour are reasons put forth in Uttarakhand and Nepal for this shift, the preference for smaller ruminants, swine and poultry are also attributed to latent economic opportunities and the ability of these animals to independently forage for food, thus reducing drudgery. Households across the areas perceived investments in smaller animals not only as an affordable saving to be encashed with handsome returns especially during festivals (or special occasions such as marriages), but an investment that could also be quickly liquidated at times of distress arising out of shocks; they were fully aware, however, of the high risks such investments could hold, given the susceptibility to diseases and epidemics and the lack of access to veterinary healthcare services.

As water scarcity emerged as a serious concern across Uttarakhand, villagers in some parts of the state have adopted long-term measures, reflecting their determination to overcome the challenge. Several villages constructed water harvesting structures to improve access to water for domestic and irrigation purposes, and importantly, to reduce the drudgery of women. Drawing on traditional wisdom and the experiences of the famous Chipko movement, several villages took pro-active steps towards the regeneration of oak forests and encouraged catchment protection, recognizing the positive consequences of such action in improving the hydrological cycle and long-term water recharge. In several areas, the rejuvenation of traditional springs attained central focus and dedicated action. Revival of water sharing practices and traditional institutional arrangements to manage this was another action adopted by the villagers as a long-term response to the growing water stress.

Where land-based mechanisms have not been a promising option, households shifted to non-farm opportunities – forest resources, carpentry, masonry, weaving, mining; they have also actively sought daily wage earning opportunities in mining, petty business or in government programmes, such as rural road building schemes in Nepal and the Mahatma Gandhi National Rural Employment Guarantee Programme (MNREGA) in India. In extreme cases, members of households were forced to migrate to peri-urban centres and cities to seek seasonal or long-term employment opportunities – a trend highly common and rapidly increasing in Nepal. The underlying purpose, in all these cases, was the singular objective of ensuring income, and through that, a reduction of their vulnerability.

3.4 Assessing Adaptive Capacities: Institutional Dependencies

Informal institutions – friends, families and relatives – are the first institutions households turn to at times of stress. These institutions are immediately accessible for households in remote communities and hence, a crucial source of immediate relief. Formal civic institutions – such as the church or monastery – assume critical importance after the friends and kins, providing much needed relief immediately following the onset of any stress. Formal institutions such government agencies, though important, are difficult to access, and hence not the immediate institutions that households turn to. For immediate access to credit, households find affinity groups such as self-help groups or cooperatives more accessible than formal financial institutions. The results show that while informal civic institutions play a crucial role in providing immediate relief or for addressing basic needs, government agencies still remain the only source for extension and technical services. The situation in Bhutan is unique, where all services – be they access to resources or service support – are provided by the government and no other institution, except the monastery for specific purposes have such an influence in the lives of villagers. However, in all the areas, access to government agencies seemed extremely difficult and villagers in Uttarakhand, Nepal and northeast India uniformly pointed out the lack of information flow from and to the government agencies, creating an impression of a non-responsive governance mechanism that could effectively address their concerns.

4 Conclusions

Adaptive responses to increasing climate variability seem to be aimed at reducing risks and vulnerabilities and enhancing the household's ability to cope with stress. This is done either by enhancing the resilience in production systems and by extension of household food security, or by transformative shifts in crops, resources or occupation that potentially promise to assure and enhance incomes (or both). This increases the household's ability to ensure a higher degree of resilience to cope with stress. It must be pointed out here that the vulnerability faced by villagers do not arise solely because of climate variability. Resource degradation, changing landuse, changes in farming practices and the gradual dilution of traditional practices as 'modern' methodologies in agriculture have been introduced have all contributed to the stressed conditions mountain farming communities face: climate variability and its impacts have acerbated the challenges. Mountain communities are responding to these changes within their limited capabilities; however, this is not adequate to enhance adaptive capacities.

Communities recognize the need and inadequacy of effective support services – extension services, capacity building, veterinary healthcare and access to inputs, credit and markets. They identified the critical need for effective delivery mechanisms and a responsive local governance system as key to consolidation of their initiatives and potentially transforming such initiatives into viable and long term livelihood strategies. Despite the critical dependency on external agencies for support services, access to such services are often beyond the reach of rural communities and the lack of effective delivery mechanisms severely restrict their ability to adapt to the change.

In order to enhance the resilience of mountain communities, the study recommends the need for interventions in on-farm natural resource management, improving access to stress tolerant food crops (including access to seeds), strengthening of extension and support services, providing access to credit, diversifying income generation opportunities and improving the responses of local government agencies to address their needs. Improving support services and responsive mechanisms are essential ingredients of developmental and poverty alleviation programmes (Agarwal and Perrins 2009). Strengthening these efforts will ensure mainstreaming of adaptation into development. Therefore, efforts need to focus on effective and efficient delivery of development and poverty alleviation programmes in order to enhance adaptive capacities of marginalized, rural communities.

Acknowledgement This research was supported by a Regional Grant from the International Fund for Agricultural Development (IFAD)'s Asia and the Pacific Division and partially supported by core funds of ICIMOD contributed by the governments of Afghanistan, Australia, Austria, Bangladesh, Bhutan, China, India, Myanmar, Nepal, Norway, Pakistan, Switzerland, and the United Kingdom. The authors are thankful to the study team in India, Nepal and Bhutan and the local communities and key informants. We acknowledge the inputs provided by ICIMOD colleagues, reviewers and editors to bring this chapter to a final stage.

References

Agarwal, A., & Perrins, N. (2009). Climate adaptation, local institutions and rural livelihoods. In W. N. Adger, I. Lorenzoni, & K. L. O'Brein (Eds.), *Adapting to climate change: Thresholds, values, governance* (pp. 350–367). Cambridge: Cambridge University Press.

Byg, A., & Salick, J. (2009). Local perspectives on a global phenomenon – Climate change in Eastern Tibetan villages. *Global Environmental Change, 19*, 156–166.

Chaudhury, P., Rai, S., Wangdi, S., Mao, A., Rehman, N., Chettri, S., & Bawa, K. S. (2011). Consistency of local perceptions of climate change in the Kangchenjunga Himalaya landscape. *Current Science, 101*(3), 1–10.

Cooper, P. J. M., Dimes, J., Rao, K. P. C., Shapiro, B., Shiferaw, B., & Twomlow, S. (2008). Coping better with current climatic variability in the rain-fed farming systems of sub-Saharan Africa: An essential first step in adapting to future climate change? *Agriculture, Ecosystems and Environment, 126*, 24–35. doi:10.1016/j.agee.2008.01.007.

Das, P. K. (2013). The Himalayan Tsunami " – Cloudburst, flash flood & death toll: A geographical postmortem. *IOSR Journal Environment Science Toxicology Food Technology, 7*, 33–45.

Haque, C. E. (2003). Perspectives of natural disasters in east and south Asia, and the pacific island states: Socio-economic coorelates and needs assessment. *Natural Hazards, 29*, 465–483.

Hijioka, Y., Lin, E., Pereira, J. J., Corlett, R. T., Cui, X., Insarov, G. E., Lasco, R. D., Lindgren, E., & Surjan, A. (2014). Asia. In V. R. Barros, C. B. Field, D. J. Dokken, M. D. Mastrandrea, K. J. Mach, T. E. Bilir, M. Chatterjee, K. L. Ebi, & Y. O. Estrada (Eds.), *Climate change 2014: Impacts, adaptation, and vulnerability. Part B: Regional aspects. Contribution of Working*

Group II to the Fifth Assessment Report of the Intergovernmental Panel on Climate Change (pp. 1327–1370). Cambridge: Cambridge University Press.

ICIMOD. (2010). *Climate change vulnerability of mountain ecosystems in the Eastern Himalayas*. Kathmandu: ICIMOD.

Piya, L., Maharjan, K. L., & Joshi, N. P. (2012). Perceptions and realities of climate change among the Chepang communities in rural mid-hills of Nepal. *Journal Contemporary India Studies: Space and Society, Hiroshima University, 2*, 35–50.

Pradhan, N. S., Khadgi, V. R., & Kaur, N. (2014). The role of policies and institutions in adaptation planning: Experiences from the Hindu Kush Himalaya. In J. Ensor, S. Huq, & R. Berger (Eds.), *Community-based adaptation to climate change: Emerging lessons* (pp. 95–110). United Kingdom: Practical Action Publishing.

Pradhan, N. S., Sijapati, S., & Bajracharya, S. R. (2015). Farmers' responses to climate change impact on water availability: Insights from the Indrawati Basin in Nepal. *International Journal of Water Resources Development, 31*, 269–283. doi:10.1080/07900627.2015.1033514.

Rahman, A., & Khan, A. N. (2013). Chitral flood 2010. *Natural Hazards, 66*, 887–904.

Shrestha, A. B., & Aryal, R. (2010). Climate change in Nepal and its impact on Himalayan glaciers. *Regional Environmental Change, 11*, 65–77. doi:10.1007/s10113-010-0174-9.

Smit, B., Burton, I., Klein, R. J. T., & Wandel, J. (2000). An anatomy of adaptation to climate change and variability. *Climatic Change, 45*, 223–251.

Uniyal, A. (2013). Lessons from Kedarnath tragedy of Uttarakhand Himalaya. *India Current Science, 105*, 1472–1474.

Xu, J., Grumbine, R. E., Shrestha, A., Eriksson, M., Yang, X., Wang, Y., & Wilkes, A. (2009). The melting Himalayas: Cascading effects of climate change on water, biodiversity, and livelihoods. *Conservation Biology, 23*, 520–530. doi:10.1111/j.1523-1739.2009.01237.x.

Drought: In Search of Sustainable Solutions to a Persistent, 'Wicked' Problem in South Africa

Coleen Vogel and Koos van Zyl

Abstract Drought is a persistent, creeping challenge for many countries in southern Africa defying neat definitions. Droughts usually occur somewhere in South Africa during any year. The more recent drought (2014–2016) has harshly reminded people of the need to be more proactive about droughts. Given this repetitive occurrence of drought, South Africa, has a long history and an interesting governance tradition of how droughts are framed and managed. A historical, comparative assessment of the role of the State and other institutional architectures in drought-risk reduction is provided. Attempts in the 1990s, for example, to widen the framing and scope of the drought challenge, with the creation of the National Consultative Drought Forum, resulted in some success in drought risk reduction. Many of the more inclusive approaches and inroads gained by this forum were picked up and continued by the National Disaster Management Act. A sustained central focus on an inclusive drought risk reduction approach has, however, not been fully sustained. Building on the National Consultative Drought Forum's more inclusive approach and vision of holistic drought risk reduction, the case for the creation of more inclusive, permanent and participatory social learning arenas are advocated.

1 Introduction

The world is being buffeted by a range of 'wicked problems or challenges' including poverty, environmental degradation, climate change etc. (Rittel and Webber 1973). Wicked challenges are described as those that are usually linked to public policy issues with one of the key problems being able to precisely 'frame' and design appropriate solutions to address a wicked problem (Rittel and Webber 1973). Wicked problems or challenges (Rittel and Webber 1973) usually have no stopping,

---

C. Vogel (✉)
Global Change and Sustainability Research Institute and Animal, Plant and Environmental Sciences, University of the Witwatersrand, Johannesburg, South Africa
e-mail: Coleen.vogel@wits.ac.za

K. van Zyl
Former: Disaster Risk Management Agri SA, Johannesburg, South Africa
e-mail: kosie@agrisa.co.za

N. Salzmann et al. (eds.), *Climate Change Adaptation Strategies – An Upstream-downstream Perspective*, DOI 10.1007/978-3-319-40773-9_11

have few criteria enabling understanding the problem and causal links in the system are also difficult to delineate. Finding discreet solutions to problems such as climate variability and change, made real in droughts, are thus usually cloaked in a myriad of challenges including that solutions are not 'true' or 'false' but tend to be more complex in nature defying ultimate tests of a solution (Rittel and Webber 1973, 163).

Drought is proving to be a persistent challenge in parts of the USA (e.g. California); parts of Brazil (e.g. Sao Paulo) and also in southern Africa and South Africa. Recent assessments of the pervasive drought impacting South Africa illustrate a wide array of outcomes:

> "One of South Africa's biggest sugar producers, Illovo Sugar, will close its Umzimkulu mill for the year-long 2015 milling season starting in April. "What's worrying is that the drought will continue to affect industry even in 2016" (Business Report, Ntuli, N., 19 Jan 2015, www.iol.co.za).
>
> The impacts of such a pervasive drought include possible seasonal farm labour job losses and wider scale impacts. South Africa is the continent's biggest producer of sugar cane according to FAO (Food and Agricultural Organization (www.bloomber.com/news/articles).
>
> Parched Free State hit by mass exodus ...Farners are leaving in droves and land is lying fallow in the worst drought in 23 years (Kings, Agriculture section, Mail and Gaurdian, October 16–22, 2015.

Part of the difficulty of trying to 'manage' or reduce the risks to a drought and encourage 'coping and adaptation', is that drought being a 'wicked problem and challenge', cannot be neatly described and defined (Wilhite and Glantz 1987). Droughts include biophysical, socio-economic and socio-political attributes. Solutions to one drought situation may thus not be transferable to another drought situation, even if occurring at the same time and/or in the same region and country. Part of the challenge of trying to adapt to droughts lies in the search for a common 'language' particularly when framing the issue at hand. Scientists and policy makers, amongst others, have tried to 'define a drought' (Wilhite et al. 2014). In the recent IPCC report (IPCC 2014, 122) drought is defined as "A period of abnormally dry weather long enough to cause a serious hydrological imbalance". The authors caution however, that drought is a relative term and also refer to a meteorological drought as being that '...with an abnormal precipitation deficit' and a mega drought being that which is '...very lengthy and pervasive' (IPCC 2014, 122). Drought also needs to be related to the specific area that is under discussion e.g. a precipitation deficit impacting crop production would usually be referred to as an 'agricultural drought'. A precipitation deficit impacting soil moisture and runoff would be referred to as a 'hydrological drought' (IPCC 2014, 122).

Attributing changes associated with drought to climate change is also difficult. Droughts as phenomena are usually more clearly coupled to a climate variability discourse (even though climate variability is linked to climate change). Consensus on how droughts may change in the future, for example, is given 'medium confidence' (IPCC 2012). At a country level (e.g. South Africa), recent climate projections are less clear on droughts per se flagging drying trends and rainfall variability rather than drawing attention to periods of extreme drought (Department of Environmental Affairs 2013). Folding a discussion on drought and climate change adaptation into a neat discussion thus becomes very difficult.

Periods of long, protracted drought usually also unveil a range of other 'wicked challenges' usually coupled to various socio-economic and socio-political structural issues e.g. access to resources for farming, access to water etc. that are very pervasive and persistent concerns (Glantz 1994). In north-east Brazil, for example, Nelson and Finan (2009) trace a historical narrative of droughts and their impacts showing that periods of climate variability and severe droughts are tightly coupled to agricultural employment in rural areas. Out migration, for example, is a known livelihood drought coping strategy. Government interventions that are designed to be of 'relief' in a system of strong patronage have resulted, in some cases, in an erosion of vulnerability over time rather than promoting resilience.

The endless dilemma of having to prescribe what is meant by a drought means that a more inclusive approach to addressing the challenge is fraught with issues of: Who are the experts? Whose definition matters? And, then who has the most credibility upon which to base a 'strategy'? (Firman et al. 2014). While many have tried to define a drought, including several actors in South Africa (Bruwer 1989), we refer here to the IPCC definition but acknowledge that this too may be limited in really capturing the range of challenges contained in a 'drought'. The fuzziness of trying to frame a drought and design appropriate interventions, particularly more inclusive and collective interventions, is an example of a sustainable development issue that raises the fundamental and central requirement for more open-minded governance in thinking and flexibility in deriving solutions (Jasanoff 2003; Vob et al. 2009; Leach et al. 2010; Loorbach and Rotmans 2010; Chappells and Medd 2012; Wilhite et al. 2014).

Given this complex array of difficulties when framing a drought the search for appropriate and flexible drought intervention approaches has emerged. Working in Ghana, Fitzgibbon and Mensah (2012) use a model of engagement of participants that seeks to make use of flexible governance approaches and continuous learning predicated on lessons learnt from previous droughts. Issues of equity and power balances using more inclusive and deliberative stakeholder processes are profiled that enable greater and more iterative learning. The authors call for a wide inclusion of actors such as religious groups, extension officers, community communication systems, youth education as well recognition of informal knowledge etc.:

> ...institutional strategies that have proved capable of coping with extreme climate events must be identified and incorporated into regional and national adaptation plans...... Extension workers must be informed and motivated. (Fitzgibbon and Mensah 2012, 11)

Situating this chapter in a suitable 'frame' and approach in the terms of reference given to the authors (upstream and downstream challenges of drought) was difficult. The tendency is to portray drought adaptation as a process that requires an understanding of impacts and also vulnerability. Linked to such a wide understanding of drought is the upstream (State role that sets policy for drought response) and local (or downstream) implications of either failed State interventions or effective State engagement in drought risk reduction. Nelson and Finan (2009), drawing on the work of Lemos and others (2007), however, note that such simplistic notions of drought and drought risk governance may not be very helpful in the long term and observe that:

> As the system comes closer to transformative thresholds, focus must shift from what has promoted persistent vulnerability to ways to direct transformation in a socially desirable way. (Nelson and Finan 2009, 314)

> In our view, sustainable adaptation will occur when the decision-making process becomes truly participatory, when governments make investments in local infrastructure and human capital .... (Nelson and Finan 2009, 314)

These vexations both with what a drought is as well as how best to 'manage' a drought remain a challenge both locally and internationally. By reflecting on the experience of drought risk reduction in South Africa we make a call for a relaxation of what may be termed a 'control mentality' e.g. drought = understanding of vulnerability + impacts + policy response. Such thinking we argue constrains effective adaptation and engagement by a range of actors.

Approaches that constantly try to solve a problem through narrow policy solutions are insufficient (Loorbach and Rotmans 2010). Those using transition management approaches and more pluralistic approaches to complex, wicked challenges suggest that solutions will rather require an expanded group of actor engagements including co-learning approaches (e.g. Fitzgibbon and Mensah 2012), where pluralistic governance and partnerships are used to influence solutions through a design of experimentation rather than trying to 'manage' uncertainty in some clearly defined pathway. Such approaches may be seen by some to be more 'whimsical' in design but they allow for the bringing together of a range of actors and their activities so that 'they can re-inforce to such an extent that they can compete with dominant actors and practices' (Loorbach and Rotmans 2010, 239). As Jasanoff (2003) more eloquently argues, rather than seeking monocausal explanations and solutions it '...would be fruitful to design avenues through which societies can collectively reflect on the ambiguity of their experiences, and to assess strengths and weaknesses of alternative deliberations' thereby enabling learning as a form of civic deliberation (Jasanoff 2003, 242).

Using a historical narrative approach, we reflect on drought risk reduction policy and practice in South Africa drawing attention to the beginnings of a more socially inclusive and 'learning approach' approach that began in the early 1990s (namely the consultative drought forum in South Africa) that required various forms of social learning (Tschakert and Dietrich 2010). We then deliberate the question: What have we learnt about drought risk governance? and Where we are today?

2 Setting the Scene: Drought in South Africa

Managing and 'adapting to drought' in South Africa has been hampered by issues of definition, the science of drought and drought risk governance framing. If one uses biophysical criteria to describe a drought then rainfall and temperature become variables that are often used. Rainfall in the country is characterized by notable variability that overlays a precarious water supply. The average annual rainfall is estimated to be around 500 mm per annum making South Africa vulnerable to situations of drought (rainfall being approximately 60 % of the global average) (Zucchini and Nenadi 2006; Dallas and Rivers-Moore 2014). Indeed more than 65 % of the country receives on average less than 500 mm per annum and includes the arid and semi-arid interior (Dallas and River-Moore 2014). It is only the arguably thinner spatial stretches along the coastal areas that are buffered by the escarpment and the sea, where rainfall has always been more generous in supply.

The classification of 'upstream and downstream' drought adaptation (particularly as some form of holistic approach and or solution) in a country such as South Africa is therefore very difficult to adhere to and for the remainder of this chapter we broaden the sweep of the domain in which we argue for drought to be seen as a persistent and more pervasive reality thereby defying contraction into a 'simple' climate phenomenon. Such an expanded view, we suggest, also has fundamental implications for drought coping and adaptation and 'management' of a drought.

Severe or protracted droughts are those that have 'crept' into the press and have begun to receive some government attention (Glantz 1994). Trying to spatially and meaningfully delimit such a phenomenon, as outlined above, is thus challenging. In the country, a deficit of 25 % from normal rainfall is usually classified as a severe 'meteorological drought' but a deficit of 20 % can produce significant impacts in both social and physical systems and amongst a variety of communities and places (www.weathersa.co.za). During the period between July 1960 and June, 2004 there were eight summer rainfall seasons in which rainfall was less than 80 % of normal (www.weathersa.co.za).

The causes of drought are also difficult to classify with meteorological droughts being caused by, for example, persistent high pressure atmospheric systems and El Niño Southern Oscillation conditions (e.g. Tyson 1986; Vogel et al. 2000). Future projections of drought for the country are also difficult to precisely identify both in terms of timing and spatial extent. Globally, large general circulation models (GCMs) have enabled scientists to begin to make some projections of future droughts and extreme drought periods although these have limitations and can never be seen to be precise. Despite model uncertainty, assessments have shown that droughts have become more frequent and of greater severity in some parts of the world (IPCC 2012). From the Special Report on Extreme Events (SREX) summary for policy makers, for example, it is noted that "...there is medium confidence that some regions of the world have experienced more intense and longer droughts" ... and that "there is medium confidence that droughts will intensify in the 21st century in some seasons and areas, due to reduced precipitation and/or increased evapotranspiration. This applies to regions including ….southern Africa….” (IPCC 2012, 6 and 11). Recent national assessments (e.g. DEA 2013), including projections for six hydrological zones in the country, show that in some areas, notable reductions in rainfall and increases in temperature can be expected in the future.

Climate change projections based on various model capabilities and under various scenarios (e.g. IPCC AR4 A2 and B1 and RCP4.5) are also much better at capturing possible future climate situations than others. There is far less uncertainty for warming trends until the end of this century in South Africa, for example than, for projections based on rainfall, with most showing projections for drying and wetting in almost all parts of the country (DEA 2013, 17). Very significant warming is shown, in some cases greater than 5° centigrade, over the southern interior of the country by the end of this century (DEA 2013, 17). A general pattern of drier conditions to the west and south of the country is also projected (DEA 2013, 17). The projected impacts associated with such changes in various sectors including water, agriculture and health are also provided but determining accurate risks, is not possible.

The science of trying to firstly define a drought, associated impacts and duration, as outlined above, is a therefore clearly a ‘wicked challenge’. In addition, responses to droughts impacting the country at least from the early 1920s to the present, have been dominated by a particular ‘framing’ of drought. Drought has usually been seen as being the outcome of an agricultural problem that requires various forms of State intervention, usually in the form of financial aid to commercial white farmers. The Department of Agriculture has, since its establishment in 1904, been the custodian of drought matters in the country. The financial support from Parliament through annual budget allocations and various financial assistance schemes approved by the National Treasury are all made to the Department of Agriculture. Since the implementation of the Disaster Management Act, 2002, and the establishment of the National Disaster Management Centre, the previous Department of Agriculture, now called The Department of Agriculture, Forestry and Fisheries, a sector department, is appointed as the responsible organ of state for drought acting through a National Drought Committee. The Department of Water Affairs, another sector department, has the responsibility for water and sanitation, with the main disaster risk responsibility, that of flood management. Such a drought framing, built on problems of scientific uncertainty and well-intentioned intervention have been debated including how one defines a ‘risk’:

> ...equilibrium responses - seeking new forms of stable state through a set of interventions, guided by a particular set of knowledge framings, generated by particular practices and institutions. This creates a particular pathway …for socio-technical and governance intervention and change. Yet there is very little attention to the specific challenges presented by long-term, external changes which are not amenable to prediction and control. (adapted from Leach et al. 2010, 86)

The reactive, post-impact response to drought, including various relief measures is also being challenged (Wilhite et al. 2014). Wilhite et al. (2014) call for a drought policy that includes risk and early warning, mitigation and preparedness, awareness and education and policy governance that includes political commitment to a more proactive risk approach. This call for a more proactive drought policy approach, that includes detailed vulnerability assessments, is more aligned to climate change adaptation processes.

Notwithstanding the various challenges of 'framing' drought policy in South Africa, the central question remains are we effective in adapting to drought risks? An initial glance at the facts and figures presented at the outset of this chapter may suggest that we are not doing that well in designing effective process to assist in drought adaptation. One way of trying to better understand drought 'practices' is to reflect and trace past drought experiences through a critically reflexive process, that interrogates past droughts, their nature, impacts, drivers of change and the appropriateness of interventions. As Sainath (1996), Garcia (1981), Davies (2002), and Nelson and Finan (2009) all show, it is by carefully examining past drought experiences that a more informed and sensitive understanding of the mix of causation and context underpinning drought is obtained, thereby enabling better adaptation to such complex climate challenges in the future.

3 Droughts Over Time: Changing Causes and Consequences

A number of droughts of varying spatial and temporal scales have occurred in the southern African region in the period of historical records in the early 1820s, early 1830s, late 1840s and early 1850s, mid-1870s, and mid-1880s. Periods of below-normal rainfall in the meteorological record include the early late 1940s and early 1950s, much of the 1960s and early 1970s, early 1980s and 1990s (Tyson 1996); early 2000s and 2007 (Holloway et al. 2012) and the most severe drought of the past few years.

A brief journey through South African drought history shows that a series of response pathways have been framed and driven by various drought practices that impacted on those trying to farm and derive a livelihood from the land, irrespective of whether you were 'upstream or downstream' of a drought. Droughts during the period 1900–1950, and arguably continuing to today, have been 'framed' as a normal feature of farming. The impacts of a drought are linked to how farmers farm (e.g. using conservation practices of soil and veld management including appropriate farming 'unit' size and carrying capacity of farm) (Vogel et al. 2010). Several drought responses by government were made based on these premises. Notwithstanding the goal of farmers being encouraged to sustainably manage their farms, the government was repeatedly called upon to provide debt relief to farmers culminating in a major drought debt relief injection and crisis in 1992. In summary, it may be argued that drought impacts and particularly interventions, similar to the Brazilian case (Nelson and Finan 2009), have failed to provide 'relief' and rather have increased vulnerability as drought debt roll over from year to year and local livelihoods erode.

To illustrate this argument in some detail, the following brief narrative of drought and State response in South Africa is provided below:

As early as September 1920, the leaders of the country appointed a Commission to inquire into the best means of avoiding losses by drought (Union of South Africa, Interim Report of the Drought Investigation Commission, 1922):

> In the year 1919 the losses in the Union due to drought amounted to over £±16,000,000 a figure approaching the entire contribution of South Africa to the Great War. And a sum of money large enough to construct a 2000 mile of new railway line. (Appendix A of the Final Report, Drought Investigation Commission, 1923, letter Heinrich du Toit, Chairman Drought Investigation Commission)

A commission of enquiry into the drought of the time was called and a report was eventually produced. In April 1922, the Commission presented an interim report (incorporated in the main report) dealing mainly with the position of the small-stock farmer. Based on considerable investigation, which necessitated travelling over large parts of the country, with several public meetings (Union of South Africa 1922, 3) the writers of the report presented the first comprehensive drought assessment in October 1923. While subsequent chapters of the report dealt more specifically with the various sections of drought, the main report highlights that the key factors causing drought losses are the kraaling of stock; inadequacy of drinking water facilities; the destruction of vegetation and resulting soil erosion: "which in turn leads to a diminishing efficiency of the rainfall" (Author unknown 1922, 118):

> "...drought losses can be fully explained without presuming a deterioration in the rainfall... the severe losses of the 1919 drought were caused principally by faulty veld and stock management" (Drought Investigation Commission, 1923, 5, 26).

These assessments of the strong role of people and how they interact with the environment and nature's response to periods of droughts have continued over time. The focus on land use change, soil management and agricultural practices have been the repeated approaches of several other drought commissions and investigations with a persistent response being *financial schemes* that were called on to assist white commercial farmers in times of drought (see for example, the Desert Encroachment Committee, 1951; the Verbeek Committee – Report on Drought Feeding, 1965; Interim Report of the Commission of Enquiry into Agriculture 1968 – Marais Commission; the Jacobs Committee 1979; Drought Aid Scheme of 1982; and the financial rescue bid by the then Minister of Agriculture Van Niekerk in 1992) (Vogel 1994).

In the early 1990s, the convergence of an agricultural debt crisis and a changing political dispensation in the run-up to changing national governance, from an apartheid regime to a more democratic form of governance, created a potentially positive landscape for a change in drought 'thinking'. Various forums contesting the State developed during this period. One such forum, the National Consultative Forum on Drought, heralded in a more proactive and participatory response to official drought risk reduction. This forum, the first of its kind, included inputs from civil society, trade unions, organised agriculture and other sectors and began to focus on the 'other' half of the community, namely the poor, who they suggested had been left out in previous drought considerations:

Table 1 Summary of some of the impacts associated with the early 1990s drought in the southern African region

SADC region	(Estimates total number affected 86 million, 20 million at serious risk of starvation. Cereal output fell from 11.3 to 6.2 million tonnes (Buckland et al. 2000))
Zimbabwe	Manufacturing output declined by 9.3 % with a 25 % reduction in volume of manufacturing output (Benson and Clay 1994)
Mozambique	Impacted more than an estimated 1.3 million people, especially the rural poor of the southern and central zones. Exacerbated by the war food supplies were negatively hampered and it is estimated that the World Food Programme alone spent US$200 million in food aid relief (www.fao.org)
South Africa	An estimated 50,000 jobs were lost in the agricultural sector affecting in terms of wider knock-on impacts an estimated 250,000 people (AFRA 1993). Despite the small contribution by the agricultural sector the loss to the GDP has been estimated as being about US$500 million

> ....we have attempted to view the drought from their perspective. The primary problem is not so much a lack of water as a problem of endemic poverty. (Abrams et al. 1992)

This view was echoed by others and the urgent need to re-examine paradigms of drought 'management' was called for:

> The disastrous drought of 1992, which in many areas of South Africa was the worst recorded since 1922, however, once again brought local drought policy under scrutiny and revealed significant weaknesses in the ability of government structures to respond timeously and effectively to the disaster and reduce the impact thereof. (Walters 1993, 1)

> For the future, drought policy will need to address the social and economic costs of drought induced vulnerability in a far more focused way than hitherto. (Rimmer in, LAPC 1993, 25)

The major drought of the early 1990s thus become a 'benchmark drought period' in the southern African region where a serious reflection on 'relief practices' and reactive drought policy was undertaken. The 1992/1993 drought "undoubtedly one of the most widespread droughts of the last 45 years" (www.weathersa.co.za), resulted in maize imports to South Africa and an array of humanitarian responses (Holloway et al. 2012). The 1992 drought period was triggered by an El Niño but drew into stark focus the plight of the vulnerable as livelihoods were compromised in the region (more than 20 million lives impacted) with widespread livestock losses, problems in urban supplies, regional health impacts and an array of varying challenges. The list of available estimates of the drought for the region at the time presents a chilling picture of loss and devastation (Table 1).

Notwithstanding this more participatory drought response the Forum did not last very long. Obtaining actual reasons for its demise are not easy to find and most suggest that the overwhelming political changes at the time over ran the Forum and its design. Politically-aligned participants of the Forum, however, soon became appointed into positions of government (Pers comm, Mike Walters 2015) e.g. in the Departments of Agriculture and Water.

Following this cathartic drought period, the National Disaster Management Centre (NDMC) was established soon after the change in government in 1994 within the Department of Cooperative Governance and Traditional Affairs (COGTA) co-ordinating all spheres of drought and disaster risk reduction. A number of 'committees' were created including those focussing on early warning and also post-disaster activities (Ngaka 2012). Despite these various tiers of organisation and governance it is questionable whether the paradigm shift to a more proactive approach that began in the mid-1990s through the Drought Forum has been sustained. Currently farmers still seek 'relief' during drought periods and national assessments of the vulnerable and most at risk, although ongoing activities, remains fragmented.

4 The Changing Discourse and Nature of Drought Impacts Over Time

The aforementioned recollections of major droughts in the country and the region are a vivid reminder of the crippling affect a persistent change in rainfall and temperature can have in a place, and indeed across a region, when overlain onto a context of varying poverty and well-being. The true nature of the impacts of drought are, however, very difficult to capture and trying to discreetly identify impacts in separate sectors (water, agriculture) is not really possible. Indeed, trying to capture the political economy of droughts in a region stressed by a range of contributing and concatenating stresses is a 'wicked exercise', not least when the major focus in drought intervention has been historically only on white farmers. Increasingly not only is the deeper socio-economic landscape that underpins a drought requiring more attention but so to the changing nature of disasters across the region (Holloway et al. 2013).

In an assessment of humanitarian risks and challenges from 2000 to 2012, across a range of countries in SADC, an overwhelming conclusion was that it is not only 'multiple stresses' including 'drought' that require attention but a 'reframing is required' in which climate risks, including drought, are seen as "compound and composite" stressors that find expression in the region in "…sequenced and often simultaneous ways" (e.g. drought followed by floods, impacts intersecting such as HIVAIDS, food price shocks, civil conflict etc.) (Holloway et al. 2013, iv–vii). The southern African region is also one that is 'on the move', being shaped by increased mobility and the use of social media, thereby enabling a drought and or a flood and associated impacts to extend their reach and increasingly becoming trans-boundary challenges.

5 Reflecting on Drought Interventions: Are We Even Trying to Learn?

Three other authors of drought policy, practice and drought risk governance have also reflected on drought and the 'wicked' impacts usually associated with them. Between them they have all written extensively on the topic of drought from varying angles and for varying settings. The first, Garcia (1981, 219) in *Nature Pleads not Guilty* argues that the 1970s Sahelian crisis was not only because of failing rains. Indeed he argues that droughts usually 'reveal a pre-existing disequilibrium' much like the argument we have been trying to show in this paper. This disequilibrium is the result of compound and complex 'forces', including policy, regulation, economic change etc. Davies (2002) in a detailed examination of the role of history, politics and nature in *Late Victorian Holocausts El Nino Famines and the making of the Third World* also shows that El Niño is just one part of the story and that tragedies in the last third of the nineteenth century were the result of the mix of 'imperial arrogance' and 'natural incident' (dust jacket review of book, 2002). Finally, Sainath (1996) in his compelling book entitled: *Everybody loves a good drought stories from India's poorest villages*, notes that:

> Drought is, beyond question, among the most serious problems India faces, Drought relief, almost beyond question, is rural India's biggest growth industry. ....That's why some of them (the poor authors insert) call drought relief *teesra fasl* (the third crop). ...the next year the same problems will crop up all over again because the **real issues** were never touched" (Sainath 1996, 255 and 260) (bold emphasis added) and "But at the best of times, the press has viewed drought and scarcity as *events*. And the belief that only events make news, not *processes*, distorts understanding" (Sainath 1996, 267) (italics added).

In this chapter, we have shown that in South Africa, history and past framings and practices, much like in other parts of the world, also matter. Drought framed as an event, for example, usually results in reactive responses including repeated financial bail outs, subsidies and various other reactions. Drought framed as a process, however, as revealed through the National Consultative Drought Forum, calls attention to underlying issues of poverty and other structural concerns that become exposed during a drought (Garcia 1981) and that cannot be addressed with simple response solutions (one of the elements of a 'wicked challenge'). Contestations of power, particularly in heightened and sensitive situations of jockeying for leadership (as occurred in South Africa in 1994) can also high jack well meaning consultative processes such as those begun in the Drought Forum.

Adaptive governance for droughts, as with most other parts of the climate change system, must therefore recognize ..."the interconnectedness of present, past and future human and biophysical systems and attempts to respond to significant challenges" (cited in Hurlbert and Montana 2015, 122). A key remaining issue is, are we ready to begin a serious assessment of droughts that moves beyond a simple risk approach into a more nuanced understanding of drought that may provide some clues as to how to 'better live with drought' (Leach et al. 2010, 92). Why, despite the success of the forum on drought in the 1990s and the establishment of countless committees (Ngaka 2012) do we still struggle to assimilate the challenges of drought in the region and in South Africa?

The pervasive and persistent challenge of droughts, as shown in this chapter, requires that some adaptation to droughts occurs. Treating drought as a *shock event and not a process* is no longer a tenable approach given the developmental challenges that countries such as South Africa faces. Climate change adaptation has increasingly also shifted as a discourse from an events, to more of a *process response and approach*. With such a shift, calls for greater '*learning*' about climate risks and responses (Vob et al. 2009; Tschakert and Dietrich 2010), including assessments on drought using more reflexive approaches, have been made (social learning – e.g. Ison et al. 2015; Lotz-Sisitka et al. 2015), transformative change approaches (Feola 2015) and transgressive learning approaches (Lotz-Sisitka et al. 2015). Examples of such approaches with reference to drought have begun in some cases (e.g. south east England; reflections on adaptive water governance and drought risk governance in Argentina and Canada – see for example, Chappells and Medd 2012; Hurlbert and Montana 2015).

In the South African case, evidence of learning from drought experience has been sporadic but nonetheless evident. The National Consultative Forum on Drought undoubtedly helped to set the scene for a more pluralistic view of drought. Soon after the promulgation of the Act in 2002 (Act no 52 of 2002), the National Disaster Risk Management Framework was published in 2005, providing the detail of Key Performance Areas and Enablers from which disaster management plans and specifically drought management plans are built. From these policy aims a new approach was to be developed towards disaster management i.e. from a reactive dispensation to a more proactive approach. Thus while emergency response is still an essential part of disaster management including essential contingency plans and planning, the need to adopt longer-term, risk-reduction disaster management methods, has become essential (Vogel et al. 2012). The Drought Management Plan (DMP), now known as the Disaster Risk Sectoral Plan, published in August 2005, is yet another addition in the compendia of drought policy documents.

Despite these conceptual realignments, changes in 'framing a drought' and 'stakeholder engagements' we argue that having effective policy also means that relevant actors affected by drought (e.g. actors) need to be engaged in drought risks and the design of drought response (e.g. transgressive and disruptive learning approaches – Lotz-Sisitka et al. 2015). Actors should be engaged ***at the very outset of the design of the intervention*** (Firman et al. 2014) and not called to a 'talk-shop' after a drought has emerged that usually just serves to restrict and delay effective responses. By allocating drought risk reduction only to policy processes and ignoring civic society, we argue, severely constrains and will continue to constrain, effective drought risk reduction and climate change adaptation. The recent drought of several months in 2015, sadly reveals that despite some proactive thinking around drought risks the notion of a more permanent platform is still not mainstreamed into government thinking "In the response to the crisis (food and drought challenges), the government has created a multidepartmental national drought task team" (Kings, Mail and Guardian, 16–22, 2015, p.11) (parentheses added).

Other examples of a weak national drought risk management platform design are evident. A detailed assessment of a local, urban drought and interventions in the Eden and Central Karoo Municipalities in the South Western Cape in South Africa, 2008–2011 (DiMP 2012), for example, shows that drought risks were shown to be amplified by a variety of risk drivers that culminated in a widespread water crisis. The risks were then further aggravated by a lack of a systemic drought risk management plan, despite the presence of legislation in the form of the National Disaster Management Act and legislation that calls for vulnerability assessments and a risk reduction approach. Notwithstanding the huge losses incurred and carried by the municipalities (an estimated R272 million for various relief activities) and five separate disaster declarations (DiMP 2012) the ultimate response was, however, eventually effective. The response was made effective by experienced disaster response managers and the establishment of stakeholder or actor network mechanisms (*a more inclusive design* containing elements of transition and transgressive management approaches). The creation of a 'process' and 'data' system also enabled better informed responses e.g. a 'water crisis risk rating mechanism'.

6 Discussion

What may all this reflection on drought risk mean for policy makers, farmers and others in a pragmatic sense? The framing of droughts as 'once off events' we argue must be reconsidered and where possible, longer-term approaches adopted. This may mean that Early Warning Systems, for example, become integral, ***ongoing parts of information*** that are provided via the appropriate source e.g. SAWS – South African Weather Service on a regular basis and not only when a drought is beginning. While some of this is being undertaken (e.g. by the Department of Agriculture, Forestry and Fisheries) creative approaches that make use of social media, whereby weather and climate information can be provided in easily acceptable modes, may enable more additional risk management tools. Outputs coming from an Early Warning System can then be augmented by inputs coming from users in the form of data and observations using mobile phone applications.

The need for effective monitoring and evaluation (M & E) for droughts is also essential (DiMP 2012). Organised agriculture (e.g. Agri SA) and the Department of Agriculture, Forestry and Fisheries, in collaboration with the University of Free State have, for example, embarked on a study to determine and develop drought indicators as an integral part of drought mapping and to design drought response and recovery measures. Such responses will require constant collection of data and input of such data into computerised programmes for analysis and subsequent decision-making.

Discreet, silo-based science information, however, is not enough. Collaborations between scientists (e.g. 'on-the-ground', social assessments, remotely-sensed data and other reports such as crop moisture indices, the Standardized Precipitation

Index and the Normalized Difference Vegetation Index) and the wider citizenry should be encouraged. The inclusion of citizen approaches, for example, can enrich the 'data capturing' process and also enable citizens to be part of the solution and not treated as mere victims of the problem.

7 Conclusions

Notwithstanding the shift in focus to a more nuanced assessment of droughts in South Africa since the 1990s and the role that government policy and various other forms of 'management' play and have played in heightening and or dampening droughts, the persistent struggle to fully understand and reduce the risks of drought continue to bedevil adaptation to drought. The Disaster Management Act, 2002, designed to strongly profile the vulnerable and the weak, is currently being reviewed to adjust the definitions in the Act to those used internationally and more clearly define the situations when a drought hazard exceeds a threshold meriting a disaster declaration in terms of the Act. The DEA (Department of Environmental Affairs) is also trying to inform its own practice and implementation of drought and other adaptation requirements through development of M&E including reporting obligations to the UNFCCC (United Nations Framework of Climate Change Convention).

Despite the thick pile of policy documents, the *implementation* of a holistic drought plan is still lacking in South Africa. Emergency and event-driven response, following declaration of disasters in terms of the Act, is often still common practice. Response measures are still based on providing animal feed, in this case up to a maximum of 50 large stock units. The response time is also lengthy, that is from the time the drought is acknowledged by Provincial departments of agriculture and, the declaration by the relevant municipal district as a disaster. The recent and growing threat of ENSO in the last few months of 2015 has also re-confirmed that there appears to be no wide and inclusive, 'space' in which to discuss a drought threat as there was in the days of the Forum. The lack of a 'safe, credible and permanent space' to discuss a threatening and severe drought can have damaging long-term consequences. Many farmers, especially small-scale or emerging farmers, are at risk of damaging livelihood impacts occasioned by sporadic drought risk reduction efforts.

Despite the success of a more engaged citizenry as shown in the example of the Western Cape, drought (including municipal engineers and stakeholder groups) and the transformative shift in drought thinking of the National Drought Consultative Forum, **sustained creative** thinking on drought remains patchy at best. Reasons for this are hard to unearth and perhaps it is a fear of 'losing control'. By opening up and making drought risk reduction a more inclusive process, with greater engagement by a broader citizenry, there could be a possible 'fear' of losing centralised control.

The need for 'arenas' (e.g. a drought forum approach) and the identification of 'front runners' (Loorbach and Rotmans 2010), as noted in the Western Cape case, may be useful drought adaptive governance approaches to consider as we face the future of climate and other stresses. Undertaking such experiments, however, will require a serious investigation of current drought risk reduction practice across a variety of levels, examining barriers and identifying opportunities to effective drought risk governance. What may it take to create a more sustained drought risk reduction approach? Perhaps humility.

In the words of an old Nigerian Proverb:

> In the moment of crisis, the wise build bridges and the foolish build dams!

References

Abrams, L., Short, R., & Evans, J. (1992). Root cause and relief restraint report, *National Consultative Report on Drought, Secretarial and Ops Room*. Johannesburg, 8 Oct 1992.

AFRA, Association for Rural Advancement. (1993). *Drought relief and rural communities* (Special report, no. 9). Pietermaritzburg: Association for Rural Advancement.

Benson, C., & Clay, E. (1994). *The impact of drought on sub-Saharan African economies: A preliminary examination* (Working paper, 77). London: ODI.

Bruwer, J. J. (1989). Drought policy in the Republic of South Africa. Part I. *Drought Network News, 1*(3), 14–16.

Buckland, R., Eeele, G., & Mugware, R. (2000). Humanitarian crisis and natural disasters, a SADC perspective. In E. Clay, & O. Stokke (Eds.), *Food and Humanitarian Security*. London: Frank Cass Production, 181–195.

Chappells, H., & Medd, W. (2012). Resilience in practice: The 2006 drought in southeast, England. *Society and Natural Resources: An International Journal, 25*(3), 302–316.

Dallas, H. F., & Rivers-Moore, N. (2014). Ecological consequences of global climate change for freshwater ecosystems in South Africa. *South African Journal of Science, 110*(5/6), 1–11.

Davies, M. (2002). *Late Victorian Holocausts El Nino Famines and the Making of the Third World*. Brooklyn: Verso.

Department of Environmental Affairs, DEA. (2013). *Long-term adaptation scenarios flagship programme (LTAS) for South Africa*. Climate trends and scenarios for South Africa, Pretoria, South Africa.

Disaster Mitigation for Sustainable Livelihoods Programme, DiMP. (2012). *Southern cape drought disaster, "The scramble for water"*. Cape Town: University of Stellenbosch.

Feola, G. (2015). Societal transformation in response to global environmental change: A review of emerging concepts. *Ambio, 44*, (5)376–390.

Firman, C., Roncoli, C., Bartels, W., Boudreau, M., Crockett, H., Gray, H., et al. (2014). Social climate justice in climate services: Engaging African American farmers in the American South. *Climate Risk Management, 2*, 11–25.

Garcia, R. (1981). *Drought and man: The 1972 case history. Vol. 1: Nature pleads not guilty*. New York: Pergamon.

Glantz, M. H. (Ed.). (1994). *Drought follows the plow*. Great Britain: Cambridge University Press.

Holloway, A., Chasi, V., de Waal, J., Drimie, S., Fortune, G., Mafuleka, G., Morojele, M., Penicela Nhambiu, B., Randrianalijaona, M., Vogel, C., & Zweig, P. (2013). *Humanitarian trends in Southern Africa: Challenges and opportunities* (Regional Interagency Standing Committee, Southern Africa). Rome: FAO.

Hurlbert, M. A., & Montana, E. (2015). Dimensions of adaptive water governance and drought in Argentina and Canada. *Journal of Sustainable Development, 8*(1), 120–137.

IPCC. (2012). Summary for policy makers. In C. B. Field, V. Barros, T. F. Stokker, D. Qin, D. J. Dokken, K. L. Ebi, M. D. Mastrandrea, K. J. Mach, G.-K. Plattner, S. K. Allen, M. Tignor, & P. M. Midgley (Eds.), *Managing the risks of extreme events and disasters to advance climate change adaptation* (pp. 1–19). A Special Report of Working Groups I and II of the Intergovernmental Panel on Climate Change. Cambridge/New York: Cambridge University Press.

IPCC. (2014). Annex II: Glossary. In K. J. Mach, S. Planton, & C. von Stechow (Eds.), *Climate change 2014: Synthesis report. Contribution of Working Groups I, II and III to the Fifth Assessment Report of the Intergovernmental Panel on Climate Change* (pp. 117–130) [Core Writing Team, R.K. Pachauri and L.A. Meyer (eds.)]. Geneva: IPCC.

Ison, R. L., Collins, K. B., & Wallis, P. J. (2015). Institutionalising social learning: Towards systemic and adaptive governance. *Environmental Science and Policy.* 53(b), 105–117. http://dx.doi.org/10.1016/j.envsci.2014.11.002.

Jasanoff, S. (2003). Technologies of humility: Citizen participation in governing science. *Minerva, 41*, 223–244.

Land and Agriculture Policy Centre, LAPC. (1993). *Debt relief and the South African drought relief programme: An overview*. Unpublished policy paper no. 1. Johannesburg.

Leach, M., Scoones, I., & Stirling, A. (2010). *Dynamic sustainabilities: Technology, environment, social justice*. London: Earthscan.

Lemos, M. C., Boyd, E., Tompkins, E. L., Osbahr, H., & Liverman, D. (2007). Developing adaptation and adapting development. *Ecology and Society, 12*(2), 26. http://www.ecologyandsociety.org/vol12/iss2/art26/. Accessed 28 Apr 2015.

Loorbach, D., & Rotmans, J. (2010). The practice of transition management: Examples and lessons from four distinct case studies. *Futures, 42*, 237–246.

Lotz-Sisitka, H., Wals, A. E. J., Kronlid, D., & McGarry, D. (2015). Transformative, transgressive social learning: Rethinking higher education pedagogy in times of systemic global dysfunction. *Current Opinion in Environmental Sustainability, 16*, 73–80.

Nelson, D. R., & Finan, T. J. (2009). Praying for drought: Persistent vulnerability and the politics of patronage in Ceara, Northeast Brazil. *American Anthropologist, 111*(3), 302–316.

Ngaka, M. J. (2012). The multi-stakeholder approach: Drought risk reduction in South Africa, In *Disaster risk reduction in Africa, UNISDR informs, special issue on drought risk reduction, United Nations, International Strategy for Disaster Risk Reduction, Regional Office for Africa.*

Rittel, H. W. J., & Webber, M. M. (1973). Dilemmas in a general theory of planning. *Policy Sciences, 4*, 155–169.

Sainath, P. (1996). *Everybody loves a good drought, stories from India's poorest villages*. London: Review, Headline Book Publishing.

Tschakert, P., & Dietrich, K. A. (2010). Anticipatory learning for climate change adaptation and resilience. *Ecology and Society, 15*(2), 11. www.ecology and soceity.org/vol15/iss2/art 11/

Tyson, P. D. (1986). *Climatic change and variability in southern Africa*. Cape Town: Oxford University Press, 220 pp.

Union of South Africa. (1923). *Final Report of the Drought Investigation Commission*, Presented to the House of Parliament by His Royal Highness the Governer-General. Cape Town: Caoe Times Ltd, Government Printers, [U.G. 49-'23].

Vob, J.-P., Smith, A., & Grin, J. (2009). Designing long-term policy:Rethinking transition management. *Policy Science, 42*, 275–302.

Vogel, C. (1994). *Consequences of droughts in southern Africa (1960–1992)*. Unpublished PhD thesis, University of the Witwatersrand, Johanneburg.

Vogel, C., Laing, M., & Munnik. (2000). Drought in South Africa, with special reference to the 1980–94 period. In D. Wilhite (Eds.), *Drought volume 1, a global assessment* (Routledge hazards and disasters series, pp. 348–366), London/New York: Routledge.

Vogel, C., Koch, I., & Van Zyl, K. (2010). "A Persistent Truth"—reflections on drought risk management in Southern Africa. *Weather, Climate and Society., 2*, 9–22.

Walters, M. C. (1993). Present state policy in the RSA and possible areas of adaptation. Paper presented at a seminar on Planning for Drought as a Natural Phenomenon, Mmabhato, former Bophuthatswana, 28/01/1993.

Walters, M. C. (2015). Personal communication.

Wilhite, D. A., & Glantz, M. H. (1987). Understanding the drought phenomenon: The role of definitions. In D. A. Wilhite, W. E. Easterling, & D. A. Woods (Eds.), *Planning for drought: Toward a reduction of societal vulnerability* (pp. 11–27). Boulder: Westview Press.

Wilhite, D. A., Sivakumar, M. V. K., & Pulwarty, R. (2014). Managing drought risk in a changing climate: The role of national drought policy. *Weather and Climate Extremes, 3*, 4–13.

Zucchini, W., & Nenadi, O. (2006). A web-based rainfall atlas for southern Africa. *Environmetrics, 17*(3), 269–283.

Making Climate Resilience a Private Sector Business: Insights from the Agricultural Sector in Nepal

Chiara Trabacchi and Martin Stadelmann

Abstract Public finance for adaptation measures in developing countries falls short of the investment needed to build climate resilience in a world on a 4 °C warming trajectory. It is unlikely to meet these needs on its own. Private finance is an essential complement to set these countries on climate-resilient development pathways.

This study investigates the levers that international public finance can use to involve domestic private actors in the financing and implementation of adaptation efforts. To this end, it takes a case study approach examining a project selected from the portfolio of the Pilot Program for Climate Resilience-backed and run by the International Finance Corporation in the agricultural sector of Nepal. To assess private actors' interest in building climate resilience, barriers to private investment in adaptation measures and how the PPCR project addresses these barriers, the authors conducted semi-structured interviews with project stakeholders, financial modelling, risk assessment and a cost-benefit analysis.

The study identified four key elements of the project strategy that helped to involve and raise the interest of Nepalese private actors in the project: (i) early involvement in project design supported by consultations and in-depth analyses of the country's agricultural sector and related climate change risk; (ii) tailored capacity building measures to build a private delivery model for adaptation; (iii) innovative financing mechanisms to tackle risks and get local financiers on board (iv) a supply-chain approach to leverage the alignment of interest between buyers and suppliers of agricultural products and build the "business case" beyond the project's life.

Keywords Climate finance • Private finance • Climate resilience • Adaptation

---

C. Trabacchi (✉) • M. Stadelmann
Climate Policy Initiative, 20 St Dunstans Hill, EC3R 8HL London, UK
e-mail: chiara.trabacchi@cpiclimatefinance.org

N. Salzmann et al. (eds.), *Climate Change Adaptation Strategies – An Upstream-downstream Perspective*, DOI 10.1007/978-3-319-40773-9_12

1 Introduction

The role of the private sector in adapting to climate change has been increasingly emphasized in international political debates. This is because public financing alone is unlikely to meet the investment needs of a world on a path to a temperature rise of up to 4 °C. Private finance is an essential complement to public finance and, in many instances, it should be the dominant source of finance (Biagini and Miller 2013). The engagement of both international and domestic private actors is thus critical for successful adaptation (Pauw and Pegels 2013).

In developing countries, the domestic private sector in particular can contribute substantially to adaptation through its activities and investments (Pauw and Pegels 2013). In these countries, in fact, private actors – ranging from large businesses and commercial banks to individual farmers – play crucial roles in key climate-relevant sectors like agriculture, which are highly vulnerable to projected climatic changes. These actors' decisions, including the one to provide finance, have the potential to set these countries and communities on a more climate-resilient path.

To provide evidence on the strategies and levers that the public sector can use to involve domestic private actors in adaptation efforts, this study examines a project selected from the portfolio of the Pilot Program for Climate Resilience (PPCR),[1] the "*Promoting Climate-Resilient Agriculture*" in Nepal. This PPCR-backed project, which was developed by the International Finance Corporation (IFC), employs public resources with the aim of strengthening the climate resilience of the agricultural sector. It does so by engaging and developing the capacity of agribusiness firms and local commercial banks to transfer skills and resources to farmers, thereby empowering them to adapt.

The project "*Promoting Climate-Resilient Agriculture*" was chosen as a case study because it was the first within the PPCR portfolio to attract the interest of local businesses and engage them in the delivery of adaptation outcomes. It was also the first to move to the implementation phase. The case study is also relevant as Nepal is among the most climate-vulnerable countries (CIF 2011), and the examined agriculture sector is the mainstay of the Nepalese economy, accounting for about 36 % of its GDP and employing 74 % of the labor force (World Bank 2012).

The case study investigated the following key research questions:

- How can international public climate finance be deployed to encourage domestic private actors to take action in building climate resilience?
- Which drivers and incentives can the public sector lever to create a "business case" for long-term private involvement in resilience?

[1] The PPCR is a target program of the multi-donor Climate Investment Funds (CIF) aimed at supporting highly vulnerable countries to integrate climate risk and resilience into development planning and implementation (CIF 2009).

This chapter, which summarizes the full case study Trabacchi and Stadelmann (2013), is structured as follows: the next section introduces the methodology and the project's approach and goals. Section three presents the results of the analysis, including the process used to develop the project and obtain private actors' interest, a cost-benefit assessment of the project for key stakeholders, and the assessment of the risk allocation among the main project actors. The final section concludes. This chapter reflects atuhors' knowledge at the time of undertaking the analysis, November 2012–September 2013.

2 Methodology

This research adopts a case study approach based on desk-based analysis and semi-structured interviews with project's stakeholders, which were undertaken between November 2012 and September 2013. The group of stakeholders consulted included representatives of the Climate Investment Funds Administrative Unit, Multilateral Development Banks (MDBs), the Government of Nepal and local businesses involved in the project.

The research applies the analytical framework developed under the San Giorgio Group, a public-private platform for exchange and research on green finance (see e.g. CPI 2013). This framework comprises the following main pillars:

- Financial cash flow analysis to estimate the effect of international public finance on potential investment returns for the main private sector actors involved (agribusinesses);
- Cost-benefit analysis of the project for key stakeholders to identify the drivers and incentives the public sector can use to create a more attractive business case;
- Risk analysis to identify and assess the main risks involved in the project and the associated risk management practices, including the risk mitigation impact of international public finance.

The financial cash flow analysis and the cost-benefit analysis use a sugar processing company as an example because of the better availability and quality of data and ex-post assessments from previous similar projects. The annex and the full case study (Trabacchi and Stadelmann 2013) provide more details on the data and assumptions used for the financial model.

For the purpose of this analysis, the private sector is defined as privately owned or controlled companies and individuals. Emphasis is placed on agribusinesses and banks as enablers of farmers' climate resilience.

3 Project and its Institutional Setting

3.1 Sub-project in Strategic Program on Climate Resilience

The studied project "*Promoting Climate-Resilient Agriculture*" is part of a broader set of interventions contained in the so-called Strategic Program for Climate Resilience (SPCR), an investment plan developed by the Government of Nepal with the support of IFC and other MDBs operating under the PPCR in the country and proposed for PPCR support.[2]

The PPCR provided about USD 2.1 million in grants and USD 3.6 million in concessional loans to this project which focuses on three crops cultivated in five selected districts of the major food grain production area of Nepal (Terai),[3] namely: rice, maize, and sugarcane (IFC 2012a). A diagnostic study commissioned by the IFC in 2012 identified these crops and this area as being among the most vulnerable to changing climate conditions.

3.2 Private Sector Involvement in the Project

The following paragraphs describe the key private actors that the project aimed at engaging in order to strengthen the resilience of farmers to climate-induced risks.[4] Figure 1 provides an overview of the project's approach and the roles of its main stakeholders.

- Agribusinesses processing sugarcane, rice, and maize. By building the skills and knowledge of these companies' technical teams and extension officers, the project plans to enable the participating agribusinesses to train a total target of 15,000 farmers in their supply chains over a 4-year time frame, and facilitate their access to stress-resilient seeds, irrigation technologies, and fertilizers.[5] At the time of writing of the case study (November 2012-September 2013), IFC was in the

[2] See Trabacchi and Stadelmann (2013) or the CIF website (http://www-cif.climateinvestmentfunds.org/country/nepal/nepals-ppcr-programming) for details on the Nepalese SPCR.

[3] Terai contributes about 56 % of the country's annual cereal production according to Regmi (2007).

[4] The project also envisaged the involvement of an Indian irrigation technology provider with the aim of promoting access to, and adoption of, water-efficient irrigation technologies in the Nepalese agricultural sector. However, we did not analyze its role in detail, as at the time of writing its involvement was not yet secured.

[5] The project adopts a so called 'trainers of trainers' approach i.e. 15 experts will strengthen companies' technical teams as well as dealers and vets skills, and/or embed new expertise in these companies. In turn, trained staff will then provide training to lead farmers (from individuals to farmers' cooperatives), who are then expected to train other farmers through demonstration and replication of practices. The set-up of demonstration plots within these companies will serve as learning-by-doing grounds for the farmers trained under the project, but also for others in surrounding areas.

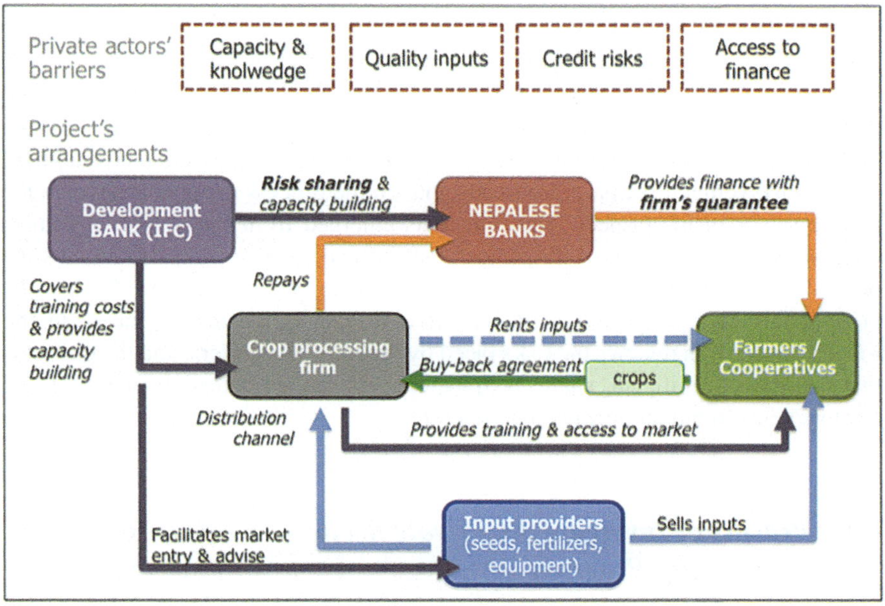

Fig. 1 Project's approach to addressing key barriers to climate-resilient agriculture. (Sources: Authors' elaboration based on CIF (2011, 2012a), IFC (2013b, c). Note: the processing firm could also act as intermediary of finance

process of finalizing the cooperation agreements that define the detailed terms, conditions, and specific responsibilities of these companies in the project.

- Local commercial banks. By establishing risk-sharing mechanisms, providing advice on how to improve internal risk management systems, and supporting the design of appropriate financial products (e.g. working capital, input finance, and value chain finance), the project aims to address the risks and lack of capacity that prevent these banks from lending to the agricultural sector. This, in turn, is expected to enhance farmers and other agricultural supply chain members' access to finance and financial products tailored to their needs.
- Farmers cultivating sugarcane, rice, and maize. By participating in the training activities provided by the agribusiness company to which they are linked in the value chain, and contracting loans, the aim is to empower and incentivize farmers to invest in climate-resilient measures.

This project is innovative in the context of adaptation given that it develops a market-driven model for financing climate resilience that has not yet been used.[6] The long-term goal of the project is to build a sustainable business case for the involved agribusiness companies to train farmers on climate-resilient agricultural practices beyond the project's life (IFC 2013b).

[6] The project developer and implementer IFC has already applied this approach in the context of 'business-as-usual' agricultural development (see e.g. IFC 2012a, b, 2013a).

4 Results

This section describes the main analyses undertaken to understand how international public climate finance could be deployed to encourage domestic private actors to take action in building climate resilience and which drivers motivate these actors' involvement. It first presents insights on the process adopted to involve the private sector in the project. These insights emerged from the review of project documents and interviews with project stakeholders. The section then shows the results of the cost-benefit assessments for the main stakeholders involved in the project, particularly focusing on the incentives driving agribusinesses' involvement. Finally, the section reviews the risks faced by project stakeholders, and the elements that can help to mitigate, share, or transfer them among actors, focusing on the arrangements to unlock local banks' lending.

4.1 Process of Involving the Private Sector in Developing the Country Program

The private sector was engaged early in the process leading to the development of the country's strategic investment plan for climate resilience and the related projects, including the assessed "*Promoting Climate-Resilient Agriculture*". It was engaged through an extensive consultation process supported by diagnostic analyses and the establishment of a 15-member Technical Private Sector Working Group representing all the major sectors where contributions from the private sector had been foreseen. These measures, which were carried out to understand local issues associated with changing climate conditions, to identify possible impacts, and to develop countermeasures to mitigate risks were central to:

- Overcome the initial reluctance of the government to involve the private sector in contributing to the country's adaptation efforts under the PPCR by improving its understanding about the current and potential role of private actors in climate change adaptation (ADB 2013; CIF 2012b; IFC 2013a). This was a prerequisite to its decision to allocate PPCR funds to private sector activities in the country's strategic program (ADB 2013) given that this is a responsibility of the recipient government.
- Overcome the reluctance of private actors to participate in the project by building their awareness of their vulnerability to climate-induced risks and/or the opportunities that climate change could present.
- Identify possible private sector partners for implementation (ADB 2013; Poshan 2010), as private sector representation in the Working Group itself does not guarantee any private sector involvement in the actual implementation where private sector players have to invest substantially more time and finance when compared to the consultation process.

4.2 Assessment of Costs and Benefits per Main Stakeholder

This section first presents the costs and benefits to agribusinesses and banks then describes the potential benefits associated with the engagement of these actors for farmers and the Government of Nepal.

4.2.1 Agribusinesses

Agribusinesses have a strong incentive to engage in the project because poor farming practices and vulnerabilities to climate-induced risks directly impact the profitability and returns of their businesses. Insufficient supply of consistent quantity and quality of local rice, sugarcane, and maize crops, has already affected their businesses: some companies are running their plants below capacity (PwC 2012; Golchha 2013), while others have to procure supplies abroad at higher prices (Nimbus 2013a). The sugar processing company taken as examplary for the analysis, for instance, is running its plants at 75–80 % of its capacity due to insufficient supply of sugarcane, but also delays in national price fixing (Bhaghat 2010; Golchha 2013), resulting in foregone revenues of about USD 0.4–0.5 million per year (authors' elaborations).[7] Some Nepalese sugar mills even had to temporarily close down operations due to insufficient sugarcane supply (see e.g. Sugaronline 2013).

Changing climate conditions could result in even lower yields, causing plants to run further below their potential capacity, and shutdowns to become more common. Current climate risks such as floods and droughts are projected to intensify and become more frequent. In the Terai region, where the project is located, productivity is expected to fall by 4–8 % for sugarcane, 5–6 % for rice, and 15–16 % for maize by 2030 (PwC 2012).

The assessment of the "business case" reveals that agribusinesses stand to benefit both financially and strategically from their investment in climate-resilient activities.

Financial benefits are expected to mainly stem from enhanced and more regular supply of crops and improvement in the quality of crops, as farmers' training is expected to lead to the adoption of improved farming practices. Assuming that farmers increase their productivity by 20 % (standard scenario, see Appendix B: "Outcome of Training Measures: Literature Review" for plausibility check), the sugarcane processing company for which we elaborated a financial model[8] could

[7] These calculation are based on the following assumptions: sugarcane crushing capacity of 3,000 tonnes per day, plant used at about 75–80 % of their capacity (Golchha 2013), and running 150 days a year (Sugaronline 2013); sugarcane purchases being responsible for 75 % of overall sugar production costs; a net profit margin of 6 % (see Appendix "Model Inputs/Assumptions for Sugarcane Farmers Training"), and prices of USD 54.4 per tonne of sugarcane (Ekantipur 2013, exchange rate from Oanda 2013).

[8] The sugar processing company was chosen for the financial model because of two main reasons: it reaches most farmers out of the selected companies, and it has direct relationships to farmers, so

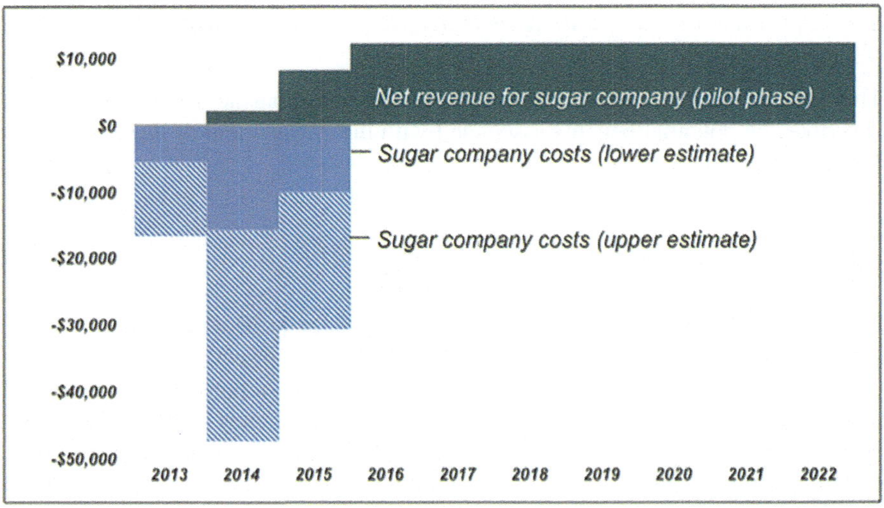

Fig. 2 Cost and revenues of a sugar processing company under the pilot project (standard scenario with 20 % production increase, in current USD). Source: Authors' elaboration. Note: costs and revenues calculations assume project-related training to 567 farmers over 2 years and a 20 % production increase; 340 trained sugarcane farmers adopt improved practices 1 year after the training, and that neither decay nor diffusion of farmers' knowledge will happen after the project. If farmers' knowledge were to decay, the revenues of the sugar company would decrease over time. If knowledge were to diffuse to other farmers, the revenues of the sugar company could increase further (See Appendix A: "Model Inputs/Assumptions for Sugarcane Farmers Training" and B: "Outcome of Training Measures: Literature Review" for further assumptions and sources for data)

expect to increase net revenues by at least USD 12,000 a year once all targeted farmers are trained and take up improved practices. These benefit estimates are to be compared with costs in the order of USD 32,000 (lower estimate) to USD 95,000 (higher estimate) over 3 years for staff time, facilities, and demonstration plots (see Fig. 2).

Companies' investment of up to USD 95,000 each in the project (made up of both cash and in-kind contributions of staff time, facilities, and demonstration plots), is projected to be recouped around 5 years after all targeted farmers (about 5000 per company) are trained, as improved supply of agricultural products should lead to higher turnover and profits. The use of USD 1 million in PPCR grants to cover the incremental costs associated with training farmers reduces the payback time by several years.

Estimates suggest that the sugar company's investment pays for itself in 3–8 years under a productivity increase scenario of 20 %. Without PPCR coverage of some costs, the pay-back time would be at least 11 years and, therefore, unlikely to be attractive for sugar processing company (see Table 1). Without IFC's know-how

there is less uncertainty in the calculations. The results for other agribusiness companies (rice, maize) may be different, there are both theoretical arguments for lower performance (less direct channel for training) and or higher performance (higher motivation to improve relationship), see Box 3 Trabacchi and Stadelmann (2013) for details.

Table 1 Impact of PPCR grants on the pay-back time of the sugar company's contribution to the project

Productivity increase scenarios	Pay-back time with PPCR cost-coverage		Effect of PPCR cost-coverage on pay-back time*
	Upper estimate for contribution (*USD 0.1 million*)	Lower estimate contribution (*USD 0.03 million*)	All estimates of sugar company contribution
Low 10 %	15.7 years	5.3 years	−27.6 years
Standard 20 %	**7.9 years**	**2.6 years**	**−13.8 years**
High 30 %	5.3 years	1.8 years	−9.2 years

Source: Authors' elaboration. Note: Standard scenario in bold
* Assumes PPCR cost coverage of about USD 700,000 in the two-year pilot phase. This is 70 % of the overall PPCR cost coverage over the four years and assumes that costs in the first 2 years of the project will be higher because of the development of training material and training experts in agribusiness

and about USD 1 million in grant funding from the PPCR to cover start-up costs, agribusinesses would not have engaged in training farmers in practices leading to enhanced climate resilience.

Agribusinesses' decision to take part in the project goes beyond pure financial metrics. Interviews with one of the involved agribusinesses (Probiotech) and IFC indicate that there are strategic benefits not immediately linked to the direct financial outcomes of the project that companies seem to value even more (Nimbus 2013a; IFC 2013a, d).[9] These benefits include:

- Increased know-how of training and climate-adaptive practices.
- Monitoring and evaluation of the outcome of the training activities through IFC (independent evaluator).
- Improved relations with farmers. Improved relationships, in fact, may lead to reduction in margins paid to intermediaries. Moreover, farmers can also represent a potential customer base for agribusinesses selling seeds and fertilizers.

If the project performs as planned, it has the potential to incentivize agribusinesses to sustain the training activities beyond the project's life, thereby ensuring the sustainability of the intervention once public support is phased-out.

At the end of the project's pilot phase, in fact, agribusinesses may see the benefits of continuing to train farmers beyond the project's life as they may see more benefits than costs in training additional farmers. This is because e.g.: (i) start-up costs should not occur after the pilot period; (ii) learning-effects during the project could help to improve the effectiveness of training farmers and generate efficiency gains; (iii) agribusinesses' increased awareness of the benefits of training programs.

To illustrate the potential long-term business case, Figure 3 considers the case of the sugarcane-processing company to show the costs and benefits of extending training to 4400 additional farmers in the 2 years after the pilot phase. In a standard

[9] These benefits also accrued to a sugar company (DSCL) who was involved in a similar IFC project in India (but not this project in Nepal), as highlighted during an interview with the company's management (DSCL 2013).

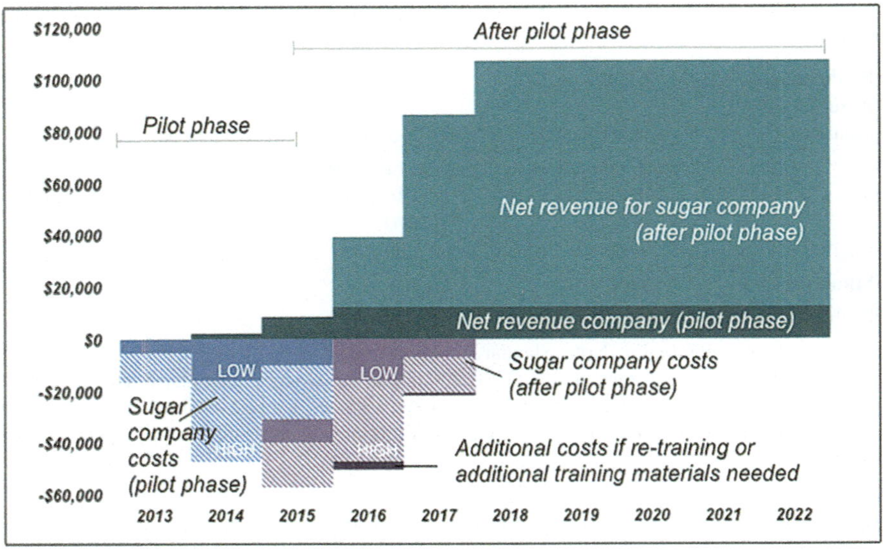

Fig. 3 Sugar company's costs and revenues (in current USD) – pilot phase and beyond. Source: Authors' elaboration. Notes: Projections assume that 4,400 additional farmers are trained after the pilot phase, of which 2700 are assumed to adopt improved practices and increase their productivity by 20 % (standard scenario). The costs of training per farmer are much lower after the pilot phase as training materials, including demonstration plots, are already available and training experts are already available

scenario, training of additional farmers will pay back in only 2–7 months, as more farmers will be trained per year.

Assuming the project generate learning and result in efficiency gains, the benefits to companies may be even higher, and the payback period for their investment shorter. Additional gains can be generated by the demonstration effect of the project; In fact, farmers not involved in the training may replicate practices adopted successfully by "lead farmers".

Public financial support will likely still be needed to continue farmers' training after the pilot phase, but can be gradually phased out. PPCR funds and IFC support may be required for delivering additional training to reach an increasing numbers of farmers or to invest in content or tools, or because productivity increases are lower than expected.

In other IFC agricultural projects (not targeting climate resilience), IFC was able to reduce or even phase out the financial support provided, and agribusinesses remained engaged beyond the initial intervention (IFC 2013a).

4.2.2 Local Commercial Banks

The analysis undertaken revealed that there are four main incentives that can motivate commercial banks to participate in the PPCR project:

- Market potential of the agricultural sector. It is estimated to be around USD 1.2 billion, with an unmet credit demand of 36 % (PwC 2012)
- Compliance with regulations on agricultural lending. The Nepalese Central Bank mandated commercial banks to enhance lending to productive sectors like agriculture to at least 10 % of their total loan portfolio by mid-July 2014. At the time of writing (November 2012-September 2013), agricultural lending represented only about 3 % of Nepalese commercial banks' total loan portfolios;
- Addressing regulatory constraints on the Capital Adequacy Ratio (CAR). Agricultural loans have implications for banks' CAR as agriculture is ranked among the most risky activities in banking regulation. Through the project's risk-sharing mechanism, these banks could increase their exposure to the agricultural sector without having to recapitalize.
- Enhanced capacity to evaluate and manage the risks specific to agricultural lending and to develop appropriate financial products.

Ultimately, profitability will depend on farmers' demand for finance, borrowers' risk profiles, the interest rate applied, and the fee associated with the risk-sharing facility.

On the risk/cost side, banks have to bear additional risks associated with the sector, lending to the non-traditional agricultural practices promoted within the project, and opportunity costs, such as lower lending to relatively more profitable and less risky sectors such as manufacturing (NRB 2012b).

4.2.3 Farmers

The project set a target of increasing trained farmers' productivity and incomes by 20 % through training tools and technologies helping them adapt to climate change risks (IFC 2013a, c). Similar projects suggest that this goal is achievable (see e.g. Nimbus 2013b; CIMMYT 2010, and Appendix B: "Outcome of Training Measures: Literature Review" for a literature review).

The engagement of agribusiness firms as training providers has the potential to generate benefits to farmers beyond improving their knowledge of agricultural practices. As agribusinesses and farmers are linked through the supply chain, farmers may benefit from:

- More secure markets for their supplies if companies promote contract-farming arrangements, or offer purchase guarantees Improved seed varieties, fertilizers and/or technologies otherwise not easily accessible.
- Finance, thanks to agribusinesses intermediation of loans and/or purchase guarantees.

The involvement of commercial banks through a risk-sharing facility is expected to:

- Lead to more affordable terms and conditions for loans, at rates lower than those prevailing on the market, as suggested by the rates offered by banks with a guarantee from a rated financial institution (see Figure 4).
- Enable access to a higher volume of credits and services.
- Promote farm diversification, which can help to increase farmers' adaptive capacity.

Under the project, farmers will not be asked to pay for the training activities but they will have to pay the cost of investing in improved and climate-resilient farming practices. It can be expected that they will only invest in measures where benefits should more than repay investment costs.

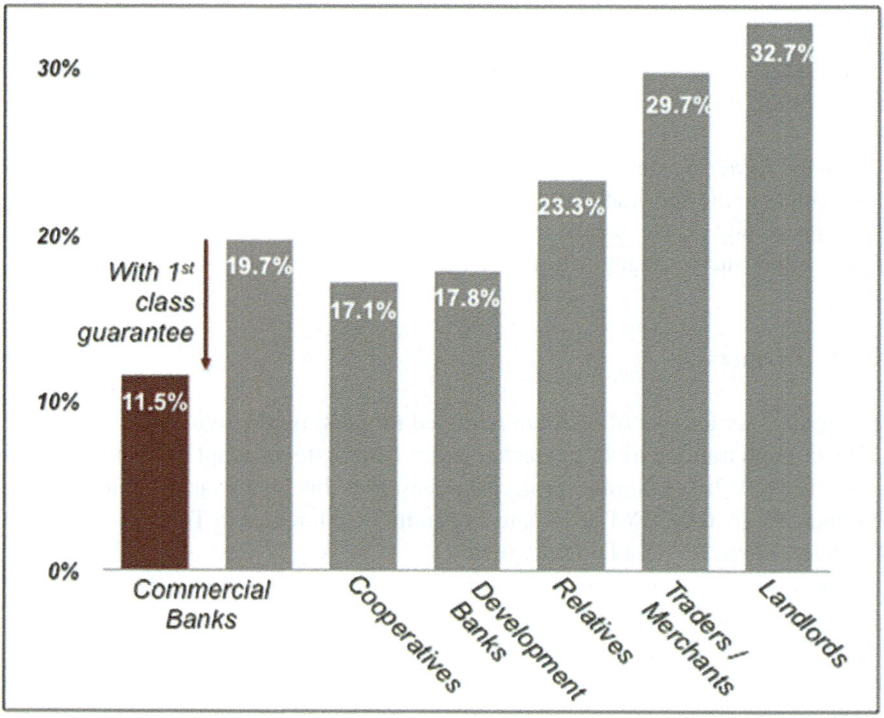

Fig. 4 Level of interest rates demanded by various lenders in the agricultural sectors. Notes: Due to data availability issues, the rate applied for loans with a first class bank guarantee from a rated bank are not specific to agricultural lending, and refer to the average made available by the institution that showed most interest to partner with IFC in the project (Sources: authors' elaboration based on PwC (2012); Banks' web sites; NRB 2012a; IFC 2013a)

4.2.4 Overview of Benefits for the Private Sector

Table 2 shows project inputs, expected direct outputs, interim outcomes and final benefits for different private sector stakeholders.

4.3 Project's Risks and Management

Our assessment and the interviews with key stakeholders (IFC 2013a) clearly show that private companies in Nepal are unlikely to invest in or borrow for improved agricultural practices unless the public sector reduces the risks associated with these non conventional investments.

The project's knowledge building, financial assistance, and risk-sharing arrangements can mitigate many of the risks, but several outcome risks still remain, particularly, uncertainties about farmers' investment in and adoption of improved agricultural practices.

Table 3 shows the inherent risks faced by key project stakeholders, the effect the most important risks have in case of occurrence, and the strategies used to mitigate those risks.

Among the risks identified, Table 3 focuses on those with high to very high impact, whatever the probability of occurrence.

The dynamic risk matrix (Figure 5) illustrates two aspects: risk allocation — where identified risk events originate and sit at project initiation; and risk response — how risk are managed and/or shifted/would shift among project's stakeholders through the use of risk transfer mechanisms (see arrows). Risks are categorized according to the three main phases of the project:

- Development risks i.e. risks incurred before the project begins implementation
- Operation risks i.e. risks incurred when implementing and running the project
- Outcome risks cover the risks of not achieving the public objectives.

The most interesting risk management measures in this project are the risk-sharing facility and the involvement of value chain actors in lending activities. Both measures have the aim to attract the interest of a local bank to lend for climate-resilient agriculture, see Box 1.

5 Conclusions

Private sector engagement is critical to "climate-proof" key sectors of countries' economies and thereby secure climate-resilient development.

The assessed IFC-PPCR project in the Nepalese agricultural sector shows that private actors have economic interests in embedding adaptation to climate change in their business practices. For agribusinesses, farmers' exposure to weather events

Table 2 From projects inputs to final benefits for private sector players

Input	Output	Interim outcome	Expected final benefits
Public capital:			**Agribusinesses**:
≈USD 2.1 million in PPCR grant resources for capacity building activities	Development of crop-specific training modules and methods to deliver climate resilience	Enhancement of at least 1700 farmers' knowledge of climate-resilient, higher yielding farming practices	Enhanced capacity and expertise on how to increase crops productivity and train farmers
≈USD 10 million in guarantees for a risk-sharing facility with at least one bank	15 trainers instructed on climate resilience, embedded in companies' technical teams	Enhanced availability of improved stress-resilient and high-yielding seed varieties and of improved technologies	Improved value chain relationships
			More secure supply of rice, maize, and sugarcane, of higher quality; higher operational efficiency and profitability
Private resources:			
≈USD 0.3 million in cash and in-kind contributions from agribusinesses	Development of financing products and risk management training to the staff of the partnering bank(s)	Banks sustain the adoption of improved practices through the promotion of tailored financial products	
			Banks:
≈USD 10 million from the target bank for the risk-sharing facility. Additional resources will be invested to cover the cost of the facility and to lend to projects			Improved risk-management practices and ability to satisfy agricultural market needs
			Farmers:
			Improved ability to cope with climate-related risks
			Improved value chains relationships
			20 % higher yields and income

Table 3 Main risks identified, their effect on project's stakeholders and risk management measures

Risk →	Effect →	Risk management measures
Failure to engage suitable private actors in the country's climate-resilient program due to, inter alia,	IFC and the PPCR bear this reputational and financial risk that could negatively impact on their credibility and result in losses of the public resources invested to develop the project	Intense consultation phase supported by the establishment of a Technical Private Sector Working Group and analytical studies (see e.g. IFC 2013a; ADB 2012a)
Private actors' limited understanding of climate-related risks and opportunities	The most vulnerable group of actors, farmers, will remain exposed to climate impacts	IFC's previous experience in projects applying similar approaches
		IFC's existing relationships with some of the private partners involved (IFC 2012a)
Government's limited understanding of the private sector's role in building climate resilience		IFC's screening, appraisal tools, and safeguard standards
Failure to remove barriers to finance due to:	IFC and the PPCR bear this reputational and financial risk that could negatively impact on their credibility and result in losses of the public resources invested to develop the project	IFC expertise in operating with financial institutions in developing countries, and in structuring risk-sharing facilities in the agricultural sector (e.g. see IFC 2010)
The inability to structure risk-sharing mechanisms (i.e. risk-sharing facility and involvement of agribusinesses in lending activities) sufficiently attractive to local banks to effectively unlock their resources for lending to climate-relevant measures	The most vulnerable group of actors, farmers, will remain exposed to climate impacts	Existing relationships with some local banks (see Ekantipur 2013; IFC 2013a)
		PPCR resources to provide the first loss coverage in the risk-sharing facility under discussion
Inability of the planned training measures to stimulate demand for finance, thereby generating a deal flow		Technical assistance at the farm level
		Involvement of agribusinesses as loan intermediaries and/or guarantors

(continued)

Table 3 (continued)

Risk →	Effect →	Risk management measures
Failure to deliver effective training measures due to:	IFC, PPCR, agribusinesses and farmers share this risk that could have negative effects on IFC/PPCR/agribusiness reputation, the financial resources invested and in case of farmers and agribusiness on their ability to make their production more climate-resilient	IFC Previous experience in the development and provision of training to and through value chain actors, and in sourcing experts
The lack of skilled human resources to develop training activities relevant for coping with climate variability and change over time, or		IFC knowledge of crop-specific vulnerabilities and barriers thanks to in-depth studies
		Adoption of a phased training approach, including follow-up training activities and monitoring (IFC 2013c)
		Technical backstopping
The inadequacy of the delivery model to farmers' literacy level and culture		Agribusiness provision of services to farmers (e.g. financing, training, inputs etc.) to manage the risk of side-selling or non-adoption of improved agricultural practices

and their limited productivity can both affect their profitability, thereby incentivizing them to strengthen farmers' capabilities and resilience. Farmers also have strong incentives to participate in the IFC-PPCR project in order to learn how to avoid climate-induced losses, and how to increase their income through improved yields, better quality production and stronger ties with output markets. However, even though self-interest should incentivize them to undertake measures that reduce their climate vulnerabilities or to harness potential opportunities that may arise from a changing climate, knowledge, capacity – both financial and technical – and risk coverage gaps can diminish private actors' ability and reduce their incentives to invest in climate resilience.

The early lessons from the IFC-PPCR project in Nepal provide evidence of how international public climate finance can be deployed to address the knowledge, capacity, and risk gaps hindering private action on climate resilience. Specifically, the study identified the following three key interventions:

- Consultation and involvement of local actors early in the process, with the backing of evidence-based analyses. This process is essential to increase national governments' awareness of the potential for private actors to participate in building resilience, to educate private actors about climate-related risks and pos-

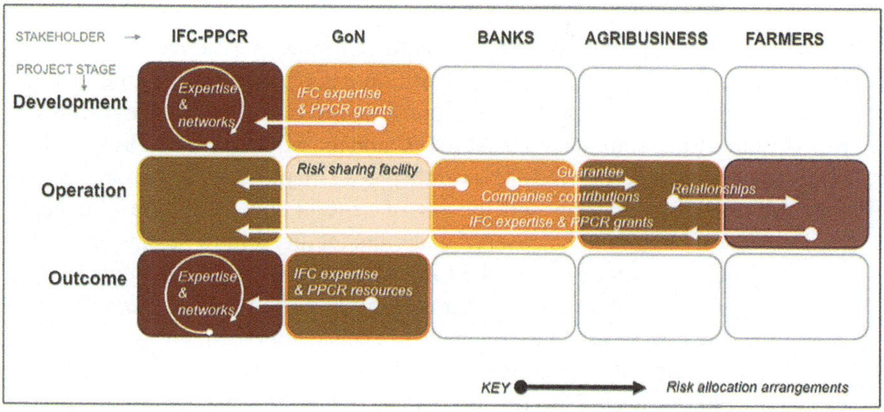

Fig. 5 IFC-PPCR project dynamic risk matrix. Source: Authors' elaboration. Note: Colors refer to the relative risks taken by stakeholders: the darker the color, the higher the risk. Risks are categorized according to their potential 'magnitude' multiplied by the 'likelihood of risk': from 'very high' in *dark red*, to 'high' in *orange*, 'moderate' in *light orange* and 'low' in *yellow*. Given the early stage of this project, we acknowledge the subjectivity of the weighting system

sible opportunities, and to identify business models enabling private sector participation.

- Tailored knowledge and capacity building measures, and innovative financing mechanisms to create private delivery models for adaptation interventions. This lays the foundation for long-lasting results, maximizing the potential outcomes from each dollar of public finance invested. The project's provision of know-how to agribusinesses is important to enable them to train farmers on improved agronomic practices. Capacity building measures at the farm level are then essential to create demand for investment and generate a deal flow.
- A supply-chain approach leveraging existing relationships and alignments of interest between buyers and suppliers of agricultural products to build the "business case" beyond the project's life.

The Nepalese experience also highlights that the level of public support in private-sector oriented projects has to be limited to avoid market distortions and to ensure that public resources are spent effectively. In the Nepalese case, agribusinesses and banks are asked to make increasing contributions to the project as they get results and gain experience in order to limit the risk of moral hazard behavior. Still, grant funding is an essential component to build their knowledge and capacity and incentivize them to engage in ventures with uncertain outcomes.

If proved successful, the project's model could be scaled-up to reach out to more farmers, both through the agribusinesses involved in the project – which have around 22,000–36,000 farmers in total in their supply chains – and by engaging more agribusinesses processing the same crops or others in the country. Context and crops-specific conditions, however, could represent new challenges and would need to be taken in duly account in project development.

Box 1: Addressing the Risk of Lending for Climate Resilience

IFC is working to structure a USD 20 million risk-sharing facility to transfer 50% of the risks associated with a portfolio of eligible 'climate-resilient loans' from a local commercial bank to IFC and the PPCR (IFC 2013a). At the time of writing the case study (November 2012-September 2013), the deal was still under negotiation and the set of criteria specifying the climate-resilient assets eligible to be covered under the facility still to be agreed.

The facility is planned to take the form of an unfunded partial credit guarantee, shared *pari passu* with the partnering bank. Its structure, terms, and conditions, are critical to induce the bank to expand lending to the agricultural sector and experiment climate-relevant lending while avoiding moral hazard behavior. The pricing of the facility, which is linked to the profitability of lending, is another critical element as it can tip the balance in terms of demand and utilization of the facility itself (IEG 2009; Mignucci et al. 2013). Narrow margins, in fact, would reduce the motivation of the bank to pay for the facility.

There are various options for allocating first and principal loss, depending on the specific needs of the partnering institutions and the nature of the assets to be covered by the facility. As illustrated in Fig. 6, each approach implies different risk-sharing/pricing trade-offs.

The use of PPCR's funds to cover part of the first loss tranche has the following effects:

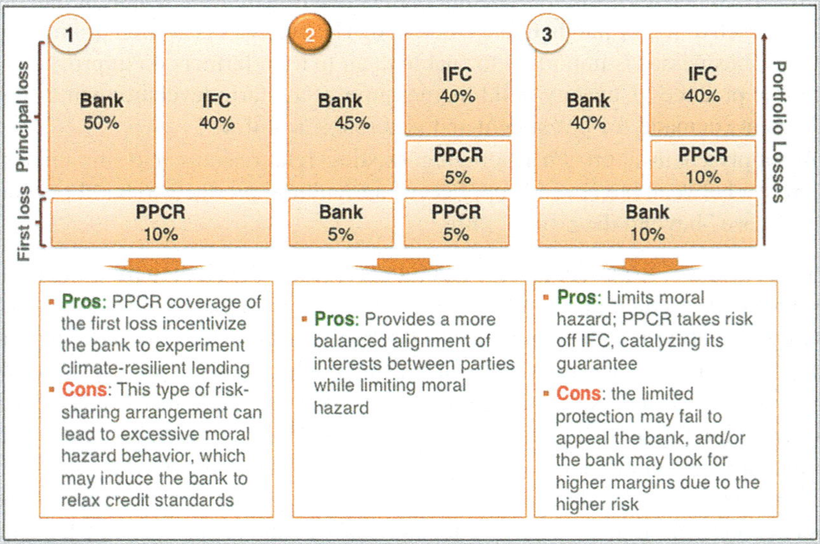

Fig. 6 Three different and possible options for the risk-sharing facility (Source: author's elaboration based on IFC (2013a) and IFC (2009))

(continued)

Box 1: (continued)

- Removing some of the risks from IFC, which might have not been willing to engage in this venture without third party backup;
- Encouraging the bank to kick-start climate-relevant agricultural lending;
- Lowering the price of the facility.

IFC's direct involvement and partial cover of the principal loss – to be used after the exhaustion of the first loss – enables the local commercial bank to share the risks with an international financial entity relatively more suited to manage them. This could lead the partnering bank to reduce the interest rates applied for loans, to the advantage of farmers (see Fig. 5).

Involving Value Chain Actors to Reduce Risks for Banks

The involvement of agribusinesses as vehicles to disseminate and collect back the loans through value chain financing can further buy down credit risks by enhancing farmers' creditworthiness via purchase guarantees on the produced crops. It can also help to reduce the transaction costs of lending to farmers, which typically prevent banks to reach out to small holder farmers.

Agribusiness involvement can differ according to the characteristics of the supply chains. Figure 7 shows in A the option where relationship between companies and farmers are looser due to the number of intermediaries between the two (e.g. rice and maize), in B the option where relationship are closer i.e. farmers and the processing company have direct interactions (sugarcane). In A, banks could lend to farmers' cooperatives on the basis of farmers group guarantees and agribusiness buy-back guarantee on farmers' products.

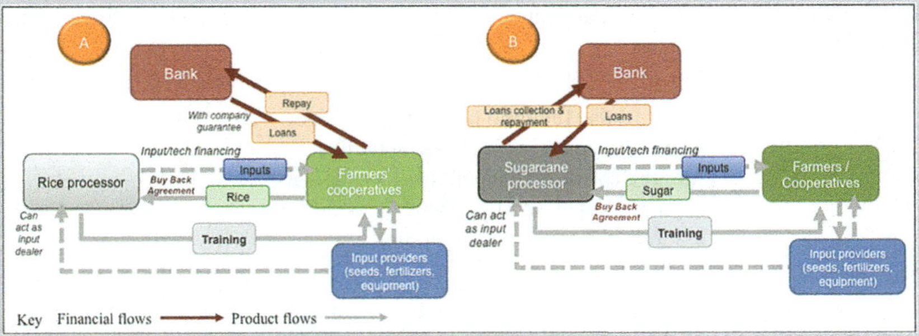

Fig. 7 Possible financing arrangements among supply chain stakeholders (Source: Author's elaboration based on IFC (2013b))

At the time of writing the case study (November 2012–September 2013), the project was unfolding and, therefore, it was too early to fully assess the effectiveness of the IFC-PPCR model in involving private actors in building countries' resilience. Close monitoring and post-project evaluation will be essential to understanding what worked and what did not and thereby contribute to the creation of the knowledge base needed for shaping future interventions, including their scale-up and replication in other contexts.

Acknowledgments The authors wish to thank the following experts for their cooperation and valued contributions including, in alphabetical order, Fisseha Abissa, Andrea Bacher, Sofia Bettencourt, Patricia Bliss-Guest, Stephanie Borsboom, Nella Canales Trujillo, Jan Corfee-Morlot, Craig Davies, Rowena Dela Cruz, Jane Olga Ebinger, Dieter Fischer, Dinesh Gautam, Saleemul Huq, Shaanti Kapila, Patrick Karani, Bhuban Karki, Krishna Kumar Shrestha, Andrea Kutter, Nicolina Lamhauge, Rita Lohani, Pradip Maharjan, Cindy Malvicini, Anthony Mills, Joyita Mukherjee, Smitha Nakooda, Haruhisa Ohtsuka, Prabhakar Pathak, Khadga Bhakta Paudel, Narayan Prasad Paudel, Doug Proctor, Sunil Radhakrishna, Neha Rai, Bradford Roberts, Ahmad Slaibi, Martha Joan Stein-Sochas, Vladimir Stenek, Ashish Jung Thapa, Nara Bahadur Thapa, Harsh Vivek. A special thanks to Anupa Aryal Pant, Kanta Kumari Rigaud, and Nancy Chaarani-Meza, who advised us during the development of the report.

Finally, the authors would like to acknowledge the suggestions provided by staff at Climate Policy Initiative.

Appendixes

Appendix A: Model Inputs/Assumptions for Sugarcane Farmers Training

Input	Description	Source
Sugar mill crushing capacity	3000 tonnes of sugarcane per day, for each of two mills	Golchha (2013)
Operating time of sugar mills	150 days per year	Derived from Sugaronline (2013)
Capacity factor of sugar mills	75–80 % (Golchha 2013) (80 % is also the ratio of the 120 days per year sugar mills in Nepal are actually running (MoAD 2013) divided by the 150 days the Nepalese Sugar Mill Association assumes the sugar mills can run)	Golchha (2013), MoAD (2013)
Procurement of sugarcane as a share of total operating costs	75 % This is the average observed in a number of Pakistan sugar mills, and here used as a proxy	Fatima (2011), Mirpurkhas (2012), Mehran (2012)

(continued)

Input	Description	Source
Net profit margin of sugar mills/ producers	6 %, median derived from analyzing yearly net profit rates for the years 2008–2012 for three sugar mills in Pakistan and eight sugarcane producers in India	Fatima (2011), Mirpurkhas (2012), Mehran (2012), Moneycontrol.com (2013)
Sugarcane price 2013	54.4 USD per tonne, calculated by multiplying 481 Nepalese rupees per quintal, including VAT (Ekantipur 2013) with 100 and an exchange rate of 0.0113 USD per Nepalese rupee, average of 1st June 2012 – 1st June 2013 (Oanda 2013)	Ekantipur (2013), Oanda (2013)
Productivity increase	Standard scenario: 20 % improvement, project goal	IFC (2013c). See Appendix B: "Outcome of Training Measures: Literature Review" for past achievements
	Lower scenario: 10 % improvement, lower-end of achievements as reported by the literature and implementing organizations (see Appendix B: "Outcome of Training Measures: Literature Review")	Based on IFC (2013e)
	Upper scenario: 30 % improvement, optimistic scenario, considering that the 52–56 % achievement in a similar sugarcane project in India (DSCL) may be too optimistic in this case as the project does not only target productivity but also climate resilience	
PPRC grant for farmer's training	USD 1 million over 4 years. We assume that 70 % will be spent in the pilot phase (first 2 years), as start-up costs for e.g. hiring experts and developing training tools will mainly occur in the first years	IFC (2013a, b)
Net revenue	Assumed to be additional revenues from sugar production due to farmer's training multiplied by the net profit margin. This assumes that the net profit margin overall does not change when the company produces additional sugar, which can be justified as, in general, sugar mills spend only around 2.5 % of their income on fixed plants costs (incl. operation and maintenance), while the rest (particularly 70–80 % sugarcane procurement costs) can be assumed to be variable	Fatima (2011), Mirpurkhas (2012), Mehran (2012) for the cost break-down of sugar mills operating in Pakistan used as a proxy
Farmers	Trained in first 4 years: 15,000, equally split between maize, rice and sugarcane	IFC (2013a, c, f)
	"Adopters" increasing their production: 9000 equally split between the three crops	

(continued)

Input	Description	Source
Farmers trained per year	Targeted farmers trained per year, project goal	IFC (2013f)
	Year 1 (pilot phase): 600 farmers (4 %)	
	Year 2 (pilot phase) 1100 farmers (7 %).	
	Year 3 (2nd phase) 8100 farmers (50 %)	
	Year 4 (2nd phase) 4800 farmers (39 %)	
Time lag between training and effects on productivity	1 year, resulting from original goal to reach all farmers by 2016, and productivity goal in 2017	Based on IFC (2013f) goals
Baseline production of sugarcane	46.5 t sugarcane per ha, average over the two targeted districts (Sarlahi, Morang) close to planned sugar mills, from years 2006/2007 to 2011/2012	MoAD (2012), CEAPRED (2013)
Average farm size	0.83 ha per farmer	NPC-WFP-NDRI (2010)
Future production changes in the baseline	No change. This assumes that future productivity increases are approximately balanced out by losses due to climate change, which are projected to be 4–8 % in case of sugarcane in Nepal by 2030	World Bank (2013a), PwC (2012) for changes in crops productivity
% of additional production sup- plied to training sugar company	100 %. This is likely because there is only one mill per district, and transporting sugarcane is very expensive, given the substantial weight of sugarcane (sugar only weighs around 10 % of the raw material, see Fatima 2011; Mirpurkhas 2012; Mehran 2012) and the low quality of roads in rural Nepal (World Bank 2009). No sugarcane has been exported from Nepal to India in the last 10 years (FAO 2013), even when the border to India is very close for most sugarcane production areas	Based on, Fatima (2011), Mirpurkhas (2012), Mehran (2012), World Bank (2009), FAO (2013) IFC (2013a), PwC (2012), Nimbus (2013a)
In-kind and cash contribution of/costs for agribusiness companies	Lower estimate: USD 32,000, assuming, two additional employees over 2 years with a wage of USD 8000 per year. Does not include costs related to facilities and demonstration plots. Trainers employed are assumed to be paid by IFC. Higher estimate: USD 95,000, recent IFC (2013b) estimation for 2-year pilot period	Wage and number of employees: Nimbus (2013b); Duration of the project: IFC (2012a, 2013c)
Annual inflation (USD)	All calculations were conducted with constant 2013 USD. For transformation to 2013 USD, annual inflation of 2.3 % assumed (average of years 2003–2012), using data from OECD (2013)	OECD (2013)

Appendix B: Outcome of Training Measures: Literature Review

Country	Project developer	Product	Measure for outcome	Observed increase	Source
Nepal	ADB	Maize/rice (irrigation)	Agricultural production	12–13 % overall	ADB (2012b)
Nepal	Various	Crops in general	Households income	16 % overall	Dillon et al. (2011)
Uganda	Various	Crops in general	Agricultural production	13–19 % overall	Pender et al. (2004)
Nepal	IFC	Poultry	Feed-conversion rate	20 % overall	Nimbus (2013a)
India	IFC	Sugarcane	Agricultural production	52–56 % overall (compared to control group)	Derived from IFC (2013e)
Kenya	World Bank	Crops in general	Agricultural production	3–7 % per year (target)	World Bank (2013b)

References

Asian Development Bank (ADB). (2012a). *Stakeholder engagement in preparing investment plans for the climate investment funds. Case Studies from Asia.* Manila: Asian Development Bank and Climate Investment Funds.

Asian Development Bank (ADB). (2012b). *Shallow tube- well irrigation in Nepal: Impacts of the community groundwater irrigation sector project impact evaluation study December 2012.* Manila: Asian Development Bank.

Asian Development Bank (ADB). (2013). *Stakeholder engagement in preparing investment plans for the climate investment funds. Case Studies from Asia* (2nd ed.). Manila: Asian Development Bank and Climate Investment Funds.

Bhagat, S. (2010). Capacity utilization in Nepalese sugar industry. In Tribhuvan University Journal, 27 (1–2). At: http://tujournal.edu.np/index.php/TUJ/article/view/260.

Biagini, B., & Miller, A. (2013). Engaging the private sector in adaptation to climate change in developing countries: Importance, status, and challenges. Review Article. *Climate and Development, 5*(3), 242–252.

Center for Environmental and Agricultural Policy Research, Extension and Development (CEAPRED). (2013a). Datasheet received per email, 27th May 2013, compiling information from the yearbooks Statistical Information on Nepalese Agriculture 2006/2007, 2007/2008, 2008/2009, 2009/2010, 2010/2011, issued by the Ministry of Agricultural Development, Government of Nepal, Khatmandu. CEAPRED. Khatmandu, Nepal.

CIMMYT [International Maize and Wheat Improvement Center]. (2010). Nepal and CIMMYT: 25 years of fruitful partnership. At: http://www.cimmyt.org/en/.

Climate Investment Funds (CIF). (2009). Programming and financing modalities for the SCF targeted program, The Pilot Program for Climate Resilience (PPCR), July 2009. At: https://www.climateinvestmentfunds.org/cif/sites/climateinvestmentfunds.org/files/PPCR_Programming_and_Financing_Modalities.pdf.

Climate Investment Funds (CIF). (2011). Strategic Program for Climate Resilience Nepal. PPCR/SC.8/7, Cape Town, South Africa, June 2011. At: http://www.climateinvestmentfunds.org/cif/sites/climateinvestmentfunds.org/files/PPCR%207%20SPCR%20Nepal.pdf.

Climate Investment Funds (CIF). (2012a). Nepal: Building climate resilient communities through private sector participation. At: https://www.climateinvestmentfunds.org/cifnet/?q=country/nepal.

Climate Investment Funds (CIF). (2012b). Proposal for additional tools and instruments to enhance private sector investments in the CIF, April 2012. At: http://www.climateinvestmentfunds.org/cif/sites/climateinvestmentfunds.org/files/CTF_ SCF_8_Proposal_for_Additional_Tools_and_instruments_for_private_sector.pdf.

CIDCM Shriram Consolidated Ltd. (2013). [Cited as DSCL. 2013]. Discussion with Sunil Radhakrishna, Senior Executive Director, conducted 12th June 2013.

Climate Policy Initiative (CPI). (2013). San Giorgio Group case studies, Venice, Italy. At: http://climatepolicyinitiative.org/publication/san-giorgio-group-case-studies.

Dillon, A., Sharma, M., & Zhang, X. (2011). *Estimating the impact of access to infrastructure and extension services in Rural Nepal* (Research monograph). Washington, DC: International Food Policy Research Institute.

Ekantipur. (2013). *Bank loans to farm sector jump 65pc*. 3 October 2013. Accessed in May 2013. At: http://www.ekantipur.com/2012/10/03/business/bank-loans-to-farm-sector-jump-65pc/361083.html.

Fatima. (2011). *Annual report 2011*. Multan: Fatima Sugar Mills Ltd, Fatima Group. At: http://fatima-group.com/relianceweaving/pdf/RelianceReport- June2011.pdf.

Food and Agriculture Organization of the United Nations (FAO). (2013). FAOstat database. Data retrieved in 9th May, 2013, at: http://faostat.fao.org/site/535/DesktopDefault.aspx?PageID=535.

Golchha Group. (2013). [Cited as Golcha 2013]. Interview with Dunakar Golchha, Executive Director, conducted on 4th July 2013.

Hervé-Mignucci, M., Frisari, G., Micale, M., & Mazza, F. (2013). *Risk gaps: First-loss protection mechanisms*. A CPI Report, January 2013. Venice, Italy. At: http://climatepolicyinitiative.org/wp-content/uploads/2013/01/Risk-Gaps-First-Loss-Protection-Mechanisms.pdf.

Independent Evaluation Group of the World Bank Group (IEG). (2009). The World Bank Group Guarantee Instruments 1990–2007. An Independent Evaluation. Washington, DC.

International Finance Corporation (IFC). (2009). Structured and Securitized Products. Risk Sharing Facility Product Description. At: http://www.ifc.org/wps/wcm/connect/1d022f00487c8d409ca4bd84d70e82a9/Risk+Sharing+Facilities.pdf?MOD=AJPERES.

International Finance Corporation (IFC). (2010). Ethiopian coffee. Summary of proposed investment. http://www.ifc.org/ifcext/spiwebsite1.nsf/ProjectDisplay/SPI_DP29228.

International Finance Corporation (IFC). (2012a). Strengthening critical segments of the poultry supply chain sustainable business advisory – IFC advisory services in South Asia: http://www.commdev.org/userfiles/SBA_PC_Nepal%20Poultry_Sep2012_EN.pdf.

International Finance Corporation (IFC). (2012b). Sugar: A sweet spot for IFC advisory services. At: http://www.commdev.org/userfiles/SBA_PC%20-%20DSCL%20Sugar%20-%20Sept%202012.pdf.

International Finance Corporation (IFC). (2013a). Discussions with IFC staff (Anupa Aryal Pant, Ahmad Slaibi, Dieter Fischer, Haruisha Ohtsuka, Bradford L. Roberts, Vladimir Stenek, Harsh Vivek conducted between January and September 2013.

International Finance Corporation (IFC). (2013b). Project Database [Last accessed October 2013]. At: https://ifcndd.ifc.org/ifcext/spiwebsite1.nsf/d011bd56046289dc85257b6000260169/6fa6c4d40a8956ba85257b63007089e1?opendocument.

International Finance Corporation (IFC). (2013c). Logic Flow for Nepal PPCR – Resilient Agribusiness. Nepal.

International Finance Corporation (IFC). (2013d). DSCL Sugarcane: Improving Farmer Productivity in India. Washington, DC: International Finance Corporation. 1st June 2013. At: http://www.ifc.org/wps/wcm/connect/e19e34804e65cf28b4b7bcfce4951bf6/SBA+Project+Examples+-+DSCL+Sugar.pdf?MOD=AJPERES.

International Finance Corporation (IFC). (2013e). Working with Smallholders. A Handbook for Firms Building Sustainable Supply Chains. Washington, DC.

International Finance Corporation (IFC). (2013f). Terms of Reference for Project Implementation Partner Promoting Climate Resilient Agriculture Pilot Program for Climate Resilience (PPCR) – Nepal.

Mehran. (2012). Annual Report 2012. Karachi: Mehran Sugar Mills Limited. At: http://www.mehransugar.com/annual.php.

Ministry of Agricultural Development (MoAD). (2012). Statistical Information on Nepalese Agriculture 2011/2012. Ministry of Agricultural Development, Government of Nepal, Khatmandu.

Ministry of Agricultural Development (MoAD). (2013). Email by Prabhakar Pathak, Joint Secretary, Gender Equity and Environment Division, 8th July 2013. Ministry of Agricultural Development, Govern- ment of Nepal, Khatmandu.

Mirpurkhas. (2012). *Annual report 2012*. Karachi: Mirpur- khas Sugar Mills, Ghulam Faruque Group.

Nepal Rastra Bank (NRB). (2012a). Quarterly Economic Bulletin. Volume 46, Number 4. Mid-July 2012, Research Department, Kathmandu, Nepal. Available at: http://red.nrb.org.np/publica.php?tp=economic_bulletin&&vw=5.

Nepal Rastra Bank (NRB). (2012b). Banking and Financial Statistics. No. 58, Mid-July 2012. Bank and Financial Institutions Regulation Department, Statistics Division, Kathmandu, Nepal. Available at: http://bfr.nrb.org.np/bfrstatistics.php?tp=bank_fina_statistics&&vw=15.

Nimbus. (2013a). Interview with Dinesh Kumar Goutam – Manager, Probiotech Industries, Nimbus Holdings, conducted 15th May 2013.

Nimbus. (2013b). Web Site [Last Accessed August 2013]. At: http://www.nimbusnepal.com/sister-compa-nies.html#sis26.

NPC-WFP-NDRI. (2010). Nepal – Food Security Atlas, July 2010. National Planning Commission–World Food Programme– Nepal Development Research Institute, Khatmandu.

Oanda. (2013). Historical Exchange Rates. Nepalese Rupee to US Dollars. Average 1st June 2012-1st June 2013. Retrieved May 10th, 2013, from http://www.oanda.com/currency/historical-rates/.

Organisation for Economic Co-operation and Develop- ment (OECD). (2013). Deflators for Resource Flows from DAC Donors. Paris: Organisation for Economic Co-operation and Development. Retried 3rd June 2013 from http://www.oecd.org/dac/stats/Deflators%20base%202011.xls.

Pauw, P., & Pegels, A. (2013). Private sector engagement in climate change adaptation in least developed countries: An exploration. *Climate and Development, 5*(4), 257–267.

Pender, J., Nkonya, E., Jagger, P., Sserunkuuma, D., & Ssali, H. (2004). Strategies to increase agricultural pro- ductivity and reduce land degradation: Evidence from Uganda. *Agricultural Economics, 31*(2–3), 181–195.

Poshan, B. KC. (2010). Engaging the private sector in the Strategic Program for Building Climate Resilience (SPCR) in Nepal. Prepared for the International Finance Corporation Nepal. Nepal.

PriceWaterhouseCoopers (PwC) and Center for Environmental and Agricultural Policy Research, Extension and Development (CEAPRED). (2012). [Cited PwC 2012]. Scoping Study on Climate Resilient Agriculture and Food Security – PPCR-Nepal. May 2012. Price WaterhouseCoopers.

Regmi, H. R. (2007). Effect of unusual weather on cereal crops production and household food security. The Journal of Agriculture and Environment, MoAC, GoN, June2007, Kathmandu, Nepal.

Sugaronline. (2013). NEPAL: Shree Ram Sugar Mills shuts down on cane shortage. 8th April 2013. Retrieved from http://www.sugaronline.com/website_contents/view/1211495.

Trabacchi, C., & Stadelmann, M. (2013). Making Climate Resilience a Private Sector Business: Insights from the Pilot Program for Climate Resilience in Nepal. A CPI Report, December 2013. Venice, Italy. At: http://climatepolicyinitiative.org/publication/making-adaptation-a-private-sector-business-insights-from-nepal-pilot-program-for-climate-resilience/.

World Bank. (2009). *Connecting Nepal's Rural Poor to Markets*. Washington, DC: World Bank. At: http://web.worldbank.org/WBSITE/EXTERNAL/PROJECTS/0,,contentMDK:22421682 ~me- nuPK:64282138~pagePK:41367~piP- K:279616~theSitePK:40941,00.html.

World Bank. (2012). Nepal's Investment Climate. Leveraging the Private Sector for Job Creation and Growth. Afram G.G and Salvi Del Pero A., International Bank for Reconstruction and Development/International Development, Washington, DC 20433.

World Bank. (2013a). *World Bank Data. Indicators*. Washington, DC: World Bank. Accessed on 3 June 2013. At: http://data.worldbank.org/indicator.

World Bank. (2013b). Kenya Agricultural Productivity and Agribusiness Project. Results. Washington, DC: World Bank. Retrieved May 14th, 2013. At: http://www.worldbank.org/projects/P109683/kenya-agricultural-productivity-agribusiness-pro- ject?lang=en.

Part C

Shaping Climate Resilient Development: Economics of Climate Adaptation

David N. Bresch

Abstract Climate adaptation is an urgent priority for the custodians of national and local economies, such as finance ministers and mayors – as well as to leaders in the private sector. Adaptation measures are available to make societies, communities and companies more resilient to the impacts of climate change. But decision makers need the facts to identify the most cost-effective investments, they need to know the potential climate-related damages over the coming decades, to identify measures to mitigate these risks – and to figure whether the benefits will outweigh the costs. The Economics of Climate Adaptation (ECA) methodology provides decision makers with a fact base to answer these questions in a systematic way. It enables them to understand and quantify the impact of climate change and to identify actions to minimize that impact at the lowest cost. It therefore allows decision makers to integrate adaptation with economic development and sustainable growth (Adger et al, Nat Clim Change 3(2):112–117, 2013). Case studies in more than 20 different regions around the globe, ranging from Maharashtra in India to the US Gulf coast, showed that a significant portion of expected damage from climate change can be averted using cost-effective adaptation measures – a strong case for preventive action. Since many mountain regions are especially sensitive to natural hazards yet in many instances lack precise and long historic records of pertinent data, the ECA methodology provides a tool to overcome quite some of these limitations by making extensive use of probabilistic risk assessment.

Beyond cost-effective protection, risk transfer is a powerful tool to manage low frequency/high severity impacts; it can protect assets and livelihoods from catastrophic events. Risk transfer puts a price tag on risk and thereby incentivizes investments in prevention measures – thus helps twofold to strengthen climate resilience.

D.N. Bresch (✉)
Swiss Federal Institute of Technology, ETH, Zürich, Switzerland
e-mail: dbresch@ethz.ch

N. Salzmann et al. (eds.), *Climate Change Adaptation Strategies – An Upstream-downstream Perspective*, DOI 10.1007/978-3-319-40773-9_13

1 Introduction

Even if all emissions could be stopped today, the climate will continue to alter in the coming decades. This means we need to reduce emissions as quickly as possible (mitigation, e.g. Stern 2006) and deal with the impact of climate change by making our societies more resilient (adaptation, e.g. Tol et al. 1998).

Improving the resilience (Bresch et al. 2014) of our societies in the face of climate change will increase in importance. A key step in climate adaptation is to create an economic framework for governments to use in developing adaptation strategies on a country and regional level to understand the underlying climate risks and the costs of adaptation. Here, we present a methodology that quantifies local climate risks and provides decision-makers with the facts to design a cost-effective climate adaptation strategy. Adaptation measures include, for example, building defences, improved spatial planning, ecosystem-based approaches, building regulations and risk transfer and insurance against some of the more extreme weather events.

This chapter examines how decision makers can assess potential damages due to climate change and which measures will be most cost effective in averting them, with a focus on one particular methodology developed by the Economics of Climate Adaptation (ECA) working group (2009). Today, annualised damages of 1–12 % of GDP result from existing climate risks and they are – based on the projection of future economic development and analysis of three climate scenarios – likely to rise to up to 19 % of GDP by 2030. Depending on the region, up to 80 % of this increase is driven by economic development in hazard-prone areas, such as (mega) cities in coastal regions. The good news is that the ECA methodology identifies cost-effective adaptation measures with significant potential: In general, between 40 and 65 % of the projected increases in damages can be averted cost-effectively – a strong case for preventive action (ECA working group 2009). The methodology provides decision makers and their stakeholders with the facts to design a climate adaptation strategy, combining risk avoidance, damage reduction, including ecosystem-based approaches, and risk transfer measures – following a pre-emptive approach to manage total climate risk.

A novel methodology to **quantify local total climate risk** is proposed, looking at the combination of today's climate risk, the economic development paths that might put greater population and value at risk and the additional risks presented by climate change. Starting from a comprehensive mapping of hazards and exposed assets, using state-of-the-art probabilistic risk modeling technique, combined with a cost/benefit approach to assess a portfolio of adaptation measures, the methodology provides an economic framework to fully integrate risk and reward perspectives.

Adaptation measures are available to make societies more resilient to the impacts of climate change. But decision makers need the facts to identify the most cost-effective investments. Climate adaptation is an urgent priority for the custodians of national and local economies, such as finance ministers and mayors – as well as to leaders in the private sector. Such decision makers ask:

- What is the potential climate-related damage to our economies and societies over the coming decades?
- How much of that damage can we avert, with what measures?
- What investment will be required to fund those measures – and will the benefits of that investment outweigh the costs?

The Economics of Climate Adaptation (ECA) methodology provides decision makers with a fact base to answer these questions in a systematic way. The methodology serves as a kind of reference, having been applied in 20 case studies across the globe from Maharashtra in India to the US Gulf coast, covering different types of hazards, regions and economic sectors. It enables decision makers to understand the impact of climate change on their economies – and identify actions to minimise that impact at the lowest cost to society. It therefore allows decision makers to integrate adaptation with economic development and sustainable growth.

2 Economics of Climate Adaptation (ECA): Aims and Principles

The methodology has sought to address the following requirements:

1. *Provide holistic analyses linking climate hazards to adaptation measures*: This entailed bringing together a sequence of analyses to quantify the risk from climate hazards based on climate change scenarios, assessing the costs and benefits of adaptation measures (see e.g. Della-Marta et al. 2010; Corti et al. 2011; Schwierz et al. 2010), and considering qualitatively the non-economic benefits of such measures. The *climada* opens-source tool[1] does implement such a holistic approach (Bresch 2014).
2. *Perform a consistent comparison of adaptation measures*: By applying a comparable methodology applicable to all hazards and across all sectors, decision-makers are informed about adaptation trade-offs between economic sectors (Fankhauser et al. 1998).
3. *Be applicable to both the developed and the developing world*: Portions of the analyses required already exist in the developed world, while in the developing world key data sets need to be created, for example, physical hazard models connected to IPCC projections, asset and income census data, and vulnerability of infrastructure.

[1] *climada* – the open-source Economics of Climate Adaptation (ECA) tool – consists of the core module, providing the user with the key functionality to perform an economics of climate adaptation (ECA) assessment. Additional modules implement global coverage (automatic asset generation), a series of hazards (tropical cyclone, storm surge, rain, etc.) and further functionality, such as Google Earth access: https://github.com/davidnbresch/climada (retrieved 31.3.2015).

4. *Serve stakeholder needs*: Weaving these components into a clear and relevant tool for decision-makers in their own countries, regions and cities.

In line with these objectives, the methodology follows a set of guiding principles that are linked to the tangible outputs of the analyses:

- *Assess "total climate risk"*: This consists of current and future risk from climate hazards – that is, not only the expected additional risk from climate change but also risks due to current climate risks – and developed damage models with multiple climate change scenarios to reflect uncertainty. Decision makers must respond to the total risk facing society and not only to the incremental risk. Total climate risk is therefore the sum of risk today, the additional risk due to economic development and the aggravation of the situation due to climate change. Different risk measures can be employed, but the most common is annual expected damage, i.e. the integral measure of all possible outcomes (severity) times their respective probability (frequency).
- *Be transparent*: Prepare to share the underlying steps, assumptions and tools with local decision-makers and a global audience of stakeholders; Gathering as much as possible existing knowledge and shape the focus and scope in an inception workshop does not only help well situate any study, it also lays a solid basis for further interaction with stakeholders, not least needed when specific adaptation measures need to be checked for local applicability and feasibility. And such an approach obviously leads not only to higher acceptance of the results of the study, it also prepares for successful implementation of the recommendations.
- *Build modular tools*: Ensure that the methodology – the models for both risk assessment and cost-benefit evaluation of adaptation measures – allows for modification and refinement (see e.g. Knutti and Bresch 2015; Bresch 2014) based on future findings from researchers (for example, new insights into how climate change affects local hazard patterns);
- *Apply the analysis across sectors*: Quantify economic damage from the "bottom-up" by including detailed risk assessments of physical assets and incomes across different sectors of the economy.

3 Consistent Application

Replicable analytical approaches will ensure consistency, but require streamlining assumptions including:

- Scenario planning to address uncertainty
- Assumptions used to forecast economic and population growth
- Adaptation measures assessed using a cost-benefit analysis

(a) Scenario planning to address uncertainty

Future climate uncertainty needs to be addressed by developing discrete scenarios based on publicly available scientific research. Note that integrated advanced approaches such as decision trees or chaos theory could be applied to more accurately assess the full range of uncertainties. However, in light of the pressing need for rapid decisions and actions in adapting to climate change, these sophisticated models are subject to the law of diminishing returns – they may provide only a slightly more precise answer for significantly more effort invested. In addition, these complex models risk decreasing the replicability of analyses and, more importantly, may become less transparent and traceable to decision-makers who are not climate experts. Providing the required level of detail is key – yet not to disguise uncertainties (Stirling 2010).

(b) Assumptions used to forecast economic and population growth

Use simple assumptions on economic and population growth to increase transparency of the model, rather than leverage general equilibrium methodology concepts. General equilibrium models incorporate the impact of economic investments – including adaptation measures – on future GDP and population growth. These models try to estimate the feedback loop dynamically in a system. However, while the adaptation measures are likely to feed back into future growth, the ECA chose to make economic and population growth independent of investment choices. The advantage of using such simplifying assumptions is that practical and understandable models are more likely to gain acceptance among non-experts.

(c) Adaptation measures assessment using cost-benefit analysis

A societal cost-benefit analysis methodology is used to assess measures. Cost-benefit ratios may not be perfect indicators of the value of adaptation measures: for example, the inclusion of various costs and benefits in net present value cash flow calculations are subject to debate. Nonetheless, cost-benefit approaches are commonly used in national, regional and local decision-making, and are a recognized form of presenting information to support trade-off decisions.

The end product of this analysis is a cost-benefit curve comparing the selected adaptation measures rather than a recommendation to implement specific measures. It should be emphasized that this methodology is designed to support local decision-making processes rather than to provide a prescriptive answer on which adaptation measures a location should implement. A cost-benefit analysis is only one of several decision-making criteria, including the flexibility of measures, capital expenditure constraints, cultural preferences, and the value placed on ecosystems. The local expertise of decision-makers (Conway and Mustelin 2014) is therefore critical in evaluating which measures are most attractive when taking these factors into account.

4 Quantifying Expected Damages and the Costs of Adaptation

Expected damages and costs of adaptation are two complementary ways of examining the impact of climate change. Expected damage is the amount of damage likely to occur in a defined time period (for example, 1 year). It is calculated as a function of the severity and frequency of the climate hazard, the value of assets (for example, buildings) exposed to the hazard, and the vulnerability of those assets to the hazard. A portion – sometimes nearly all – of the expected damage can be addressed by adaptation measures.

The cost of adaptation is the investment required in adaptation measures aimed at minimizing the damage from future climate hazards. Hence, the total cost of climate change is the sum of the cost of adaptation and any residual expected damages not averted by the adaptation measures. The focus on expected damages and adaptation measures at the local level is guided by the practical assumption that climate change will have significant local impacts requiring the urgent focus of local decision-makers. Despite uncertainties and the overlapping effects of climate change in the economic, environmental and social sectors, these steps and calculations are executable even in settings where data is often sparse.

To illustrate above points, we present two examples. In the series of ECA studies done so far, we studied among other hazards, drought risk in Maharashtra, India, and hurricane risk to the energy system along the US Gulf coast. We will first exemplify the methodology and provide some details using the Maharashtra case study. The second case study of the US Gulf coast will be introduced in the next section, in order to discuss and illustrate risk transfer in more detail.

Maharashtra, a large rural state in the centre of India, suffered 3 years of crippling drought between 2000 and 2004. The on-going drought caused terrible hardship for the two-thirds of inhabitants who depend on agriculture and allied activities for their livelihoods. As a result of the drought, crops failed, quality of harvests declined, livestock died, available employment decreased, and household debt increased. Many families fell below the poverty line, some starved and several farmer suicides were recorded. Sporadic migrations of families and the movement of people to cities to find temporary employment negatively affected social welfare in the state (ECA working group 2009). Although Maharashtra has the largest area of drought-prone agricultural land in India, many other parts of India also face the risk of drought from erratic rainfall patterns. This test case therefore serves as a useful initial basis for gauging how the risk of drought might affect agricultural production and across India – and elsewhere. The test case focuses on drought and its impact on agriculture, as the hazard that poses the greatest potential threat to India's economic value and livelihoods over the next 20 years. While Indian agriculture already faces considerable historical drought risk, climate change could worsen this risk significantly, both by increasing temperatures and reducing rainfall. In Maharashtra, India, today's annual average damage potential of drought-exposed agriculture amounts to USD 240 million. Rural development increases this figure by USD 120 million until 2030 (ECA working group 2009).

The climate change scenarios for the test case found that this risk could be exacerbated significantly, even in the next two decades. By 2030, a scenario of "High Change" could result in an 8 % decrease in annual rainfall across the state.[2] This could result in a several-fold increase in the frequency and severity of droughts. It is possible that droughts that currently occur once every 10 years could be occurring as frequently as every 3 years by 2030. We assessed sensitivity to drought for each crop, i.e. change of crop yield as a function of change in rainfall (see e.g. Aggarwal 2008). For jowar, bajra, wheat, rice, groundnut, turn, gram and sugarcane we used analysis on 30 years of production and rainfall data to determine yield and crop area sensitivity (also referred to as the 'damage function' for each crop variety). We used analogous crops for those where data was not available. Applying a probabilistic loss model combining the hazard profile of losses, climate scenarios, asset value and crop yield sensitivity, we were then in a position to quantify the resulting change in yield by 2030. Drought risk in Maharashtra is further aggravated by the potential impacts of climate change – adding another USD 200 million to the annual loss burden, which leads to a total climate risk estimation of about USD 570 million by 2030. Note again that total climate risk denotes here the annual expected damage due to decrease in yield due to damage to crops.

While the focus in the test case was primarily on economic value and average expected loss, we also estimated the number of human lives impacted by drought, as well as the impact on subsistence farmers, which is likely proportionately much greater than the losses to overall GDP. For example, a specific extreme event (a 1-in-25 year drought) may affect up to 30 million people, or 30 % of Maharashtra's population – including 15 million small and marginal farmers. The same event would reduce 14 % of agricultural output and 30 % of food grain production. The impact is particularly severe for small farmers (with an average annual household income of USD 546) and marginal farmers (USD 440). Without any drought, these individuals face an annual deficit because their consumption is greater than what they produce. A 1-in-25 year drought increases their debt by 26 % and 96 %, respectively. In addition to humanitarian concerns, these small and marginal farmers are important because their land represents 41 % of cultivated land by area and 68 % of the number of farming households (ECA working group 2009).

What, then, can decision-makers in Maharashtra – and in other locations in India and across the globe faced with comparable climate risk – do to address the risk and shape climate-resilient development and regeneration paths? Comparing their costs and benefits, a range of measures was evaluated, including infrastructural measures, such as drip irrigation and sprinkler irrigation; engineering measures, such as crop engineering; behavioural measures such as watershed management and soil techniques, and risk transfer measures, including crop and weather index insurance. Through a series of local workshops and interviews we listed an exhaustive set of

[2] Results for 22 GCMs based on IPCC 2007 A1B scenario. The 90th percentile of the GCM range of results indicates a possible decrease in precipitation.

alternatives in these 4 categories of over 30 measures and filtering out those that are not applicable to Indian agriculture or not feasible or recommended by local experts. Some of these measures, including last-mile irrigation, rehabilitation of irrigation systems, ground water pumping, planned irrigation products, and canal lining are planned government projects that are already "factored into" the baseline loss assessment, and so are not considered as additional measures to protect against drought risk (Fig. 1).

A careful cost/benefit analysis of this basket of adaption measures (ECA working group 2009 and also Bresch and Schraft 2011) shows that almost 50 % of the damage under a high climate change scenario could be averted through measures whose economic benefits exceed or approximate their costs (Fig. 2). For example, drainage systems in rain fed settings, soil techniques, and drainage systems in irrigated settings all have negative cost-benefit ratios. Drip irrigation has the highest absolute level of loss averted with USD 547 million. It is the capital cost of drip irrigation that leads to the positive cost benefit ratio result. In total, the 12 measures evaluated have a capital cost of USD 6.7 billion. For events of a very low frequency, insurance measures – to transfer rather than directly prevent the expected loss – may prove a cost-effective component of the portfolio. These measures include increasing the penetration of crop and index insurance (Agrawala and Carraro 2010). The insurance measure on the cost curve is illustrative, given the assumption that risk transfer benefits calculated from the expected loss model are equal to the societal costs. The actual cost benefit ratio of specific insurance measures depends on the type of insurance and transactional costs. In terms of specific insurance options, our analysis suggests weather based index insurance is the most attractive. Weather based index insurance options cover the economic loss of crops based on weather indices, such as measured rainfall or duration of a drought. Its benefit is that it covers up to 70 % of the economic value lost, and pays out within 30 days of the event. Index insurance is not cost-effective in the strict sense, as total cost of expected damage plus capital cost plus expenses is obviously larger than expected damage alone. Nevertheless, insurance is often more attractive – and more effective – than further prevention and intervention measures, as we discuss in the next section.

5 Risk Transfer

While cost-efficient adaptation/prevention measures are available in different locations, no individual, business and public institution can afford to prevent losses from every conceivable risk event. This is especially true for events that are unlikely to occur or that can only be avoided at an enormous cost, as is the case with natural disasters. In these cases, re/insurance can play an important role in helping individuals, communities and businesses recover from the devastation wreaked by severe weather events (Surminski 2013). Transfer of such risks is an efficient way to obtain additional protection for low-frequency natural catastrophe events. Important, however, is that risk prevention and risk transfer are mutually reinforcing. While

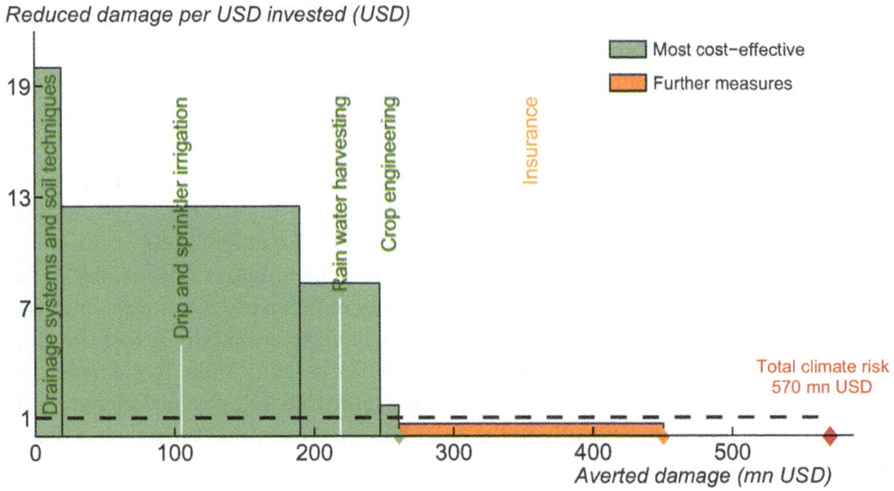

Fig. 1 (Adapted from ECA working group 2009): The adaptation cost curve for drought risk in the state of Maharashtra, India. For each adaptation measure (*rectangle*, *green/orange*), the damage aversion potential (*horizontal axis*) and its benefit/cost ratio (*vertical axis*) is shown (Note that for this particular case, almost 50 % of the damage under a high climate change impact scenario can be cost-effectively averted. See text for details)

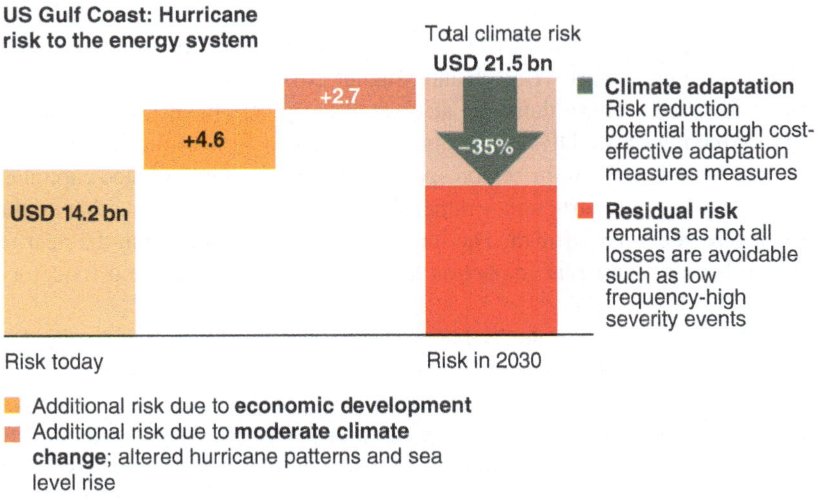

Fig. 2 (Swiss Re 2010, 2014): Risk to the energy system along the US Gulf coast today and by 2030 (Note economic development as key driver of risk (USD 4.6 billion) and climate change aggravating the situation by a further USD 2.7 billion) under a moderate climate impact scenario)

insurance is a useful component in a given adaptation portfolio, keeping insurance prices in check by minimizing residual risks through prevention measures is equally important.

ECA studies show that a balanced portfolio of prevention, intervention and insurance measures is available to pro-actively manage total climate risk and to strengthen a region's resilience. To illustrate this point we provide further evidence taken from a case study in the US Gulf region. The Gulf Coast faces significant risks from hurricanes that extensively damage assets and impact the economy. Hurricanes drive damage through extreme winds and storm surge or flooding. Over the last century, hurricanes have caused significant asset damage (USD 2.7 trillion, 2010 dollars) across Texas, Louisiana, Mississippi and Alabama (Pielke et al. 2008).

In an ECA case study in the US Gulf region, both the impact of hurricanes to the replacement value of physical assets plus the economic value of business interruption has been considered. The team values 23 asset classes spatially across the Gulf Coast (on a zip-code basis) for both property and economic value. Residential and commercial assets are key sources of value, but industrial, oil, gas, and electric utility assets contribute significantly to the asset base in the region. The analysis therefore includes a detailed and granular assessment of oil and gas and electric utility assets. Over 50,000 oil and gas structures (including pipelines, offshore structures and wells), and over 500,000 miles of electric transmission / distribution assets and ~300 generation facilities are modelled across the Gulf Coast. On an average annual basis, current damages to all assets from weather events amount to USD 14.2 billion today. These damages are expected to increase going forward to USD 19 billion (no climate change) and to USD 21.5 billion per year (under a moderate climate change impact scenario) in the 2030 timeframe. Damages may also increase further in the 2050 timeframe, ranging from USD 26 billion to USD 40 billion per year, based on the climate change scenario (Bresch and Mueller 2014a, scenarios informed also by Raible et al. 2011). These damages also represent a significant annual impact of ~2–3 % on the region's GDP. Also, damages amount to ~7 % of the region's capital investment – implying that the region spends about ~7 % of its invested capital each year on rebuilding infrastructure – rather than on capital investments that could be driving future economic growth. The impact of a severe hurricane in the near-term could also have a significant impact on the growth and re-investment trajectory in the region.

In a similar fashion as in the Maharashtra case shown above, we identified over 20 adaption measures for the US Gulf region case study. Key actions considered ranged from infrastructure- or asset-based, technology-based and systemic or behaviour, including ecosystem-based adaptation measures (Reguero et al. 2015). We found that cost-effective adaptation measures could reduce damages by 35 % (Fig. 2). It needs to be noted that implementation is often not straightforward, as actions are dispersed and involve a broad set of structures, activities and stakeholders. For example, some 230 miles of beach nourishment activities, 1000 square miles of wetlands restoration, ~540,000 miles of new, rebuilt or retrofitted distribution lines are required. Actions need to be taken by policy makers (federal, state and local), electric utilities, the oil and gas industry, infrastructure developers and other asset owners, not least of commercial and residential property. There is a strong need for leadership and coordination across stakeholders. However, while significant and broad engagement will be required across the Gulf Coast, these actions are

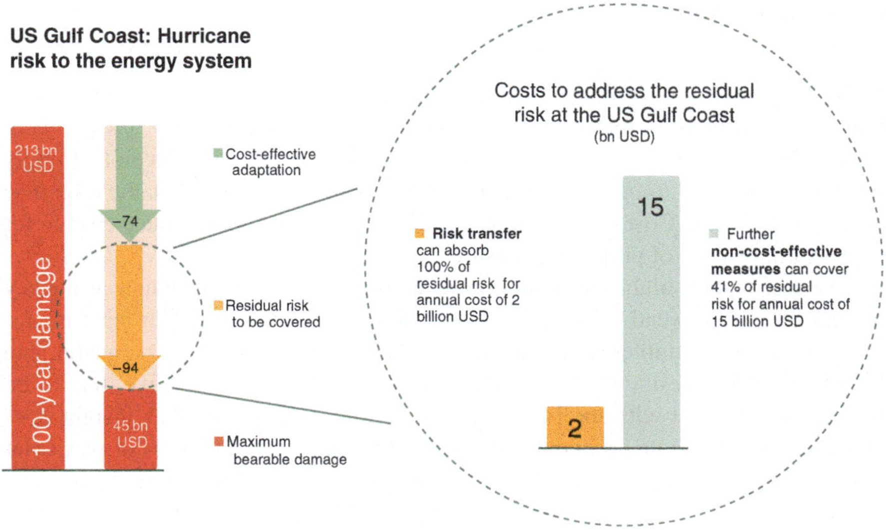

Fig. 3 (Swiss Re 2014): The potential of risk transfer to absorb damages at lower cost than further non cost-effective measures (*orange arrow* and *bar*). Note that the cost for risk transfer is lowered by the fact that cost-effective adaptation measures reduce a substantial amount of risk (*green arrow*). Further non-cost-effective measures (*grey*) would be both less effective (only reducing 41 % of residual risk) and bear much higher cost (USD 15 billion compared to only USD 2 billion for risk transfer)

also essential – in order to place the region on a resilient path going forward. The alternative will be to give in to a spiral of increasing damage and greater economic and human impact.

In addition to risk mitigation or prevention measures, there are also financial measures – including insurance instruments that can be used to transfer risk. These measures are particularly relevant to protect against low frequency, high-impact events. In addition, the price tag conveyed by such risk-transfer measures puts a strong incentive on risk mitigation or prevention measures, as they significantly reduce the price (or premium) for risk transfer.

This can best be illustrated by considering a 100 year damage event, for example, which is estimated to amount to an economic loss of USD 213 billion for the region (Fig. 3). Cost-efficient adaptation measures could lower the damage amount by 35 % or about USD 74 billion for such low-frequency/high-impact events. Meanwhile, the region's public authorities assessed the maximum affordable damage from a single event to be around USD 45 billion. The remaining USD 94 billion in economic damage is residual risk that also needs to be addressed through non-cost-efficient measures or risk transfer. Non cost-effective measures costing USD 4.7 billion per year, such as home elevation and opening protection (for example, shutters) for all existing buildings, could cover only 41 % of the residual risk. Risk transfer, however, presents a more cost-efficient solution by providing more comprehensive coverage for only USD 2 billion a year.

6 Conclusions

In the 20 Economics of Climate Adaptation (ECA) studies carried out so far, we find that the key drivers in many cases are today's weather and climate risk and economic development. The prioritization of adaptation measures is not strongly dependent on the chosen climate change scenario, at least until 2030, the reference time frame chosen. Cost-effectiveness is still valid even without climate change for a substantial subset of proposed measures.

The case studies highlight economic development and climate change as the key drivers for future weather-related losses. The analysis presents a strong case for immediate action: Implementing adaptation measures, including risk transfer, can help build global resilience to climate change. It is far less expensive than to do nothing, i.e. to deal with the rising costs only after they incur. This insight does apply across all regions, economic sectors and hazards studied, but is of particular relevance to mountainous regions. Culture and tradition does often favor such forward-looking management of weather and climate risk – and the proposed methodology supports local communities in identifying the most effective adaptation measures – in essence providing the 'adaptation business case' and therefore not least enables them to seek (financial) assistance where appropriate.

Even so, the gap between economic and insured losses remains large, and natural disasters continue to place a significant burden on the public sector, uninsured individuals and businesses. Risk transfer can protect livelihoods from catastrophic events and increase the willingness of decision-makers to invest in economic development. Additionally, risk transfer puts a price tag on risk and thereby incentivizes investments in prevention measures. In this way, risk transfer provides a powerful tool to strengthen the resilience of local and national economies.

Acknowledgements The author of this chapter co-led the Economics of Climate Adaptation (ECA) report (ECA working group 2009). The present chapter reproduces portions of and is influenced by that report, which builds on and summarizes the first eight case studies researched by Economics of Climate Adaptation Working Group, a partnership between Climate Works Foundation, Global Environment Facility (GEF), European Commission, McKinsey & Company, The Rockefeller Foundation, Standard Chartered Bank and Swiss Re.

References

Adger, W. N., Barnett, J., Brown, K., Marshall, N., & O'Brien, K. (2013). Cultural dimensions of climate change impacts and adaptation. *Nature Climate Change, 3*(2), 112–117.

Aggarwal, P. K. (2008). Global climate change and Indian agriculture: Impacts, adaptation and mitigation. *Indian Journal of Agricultural Sciences, 78*(11), 911.

Agrawala, S., & Carraro, M. (2010). Assessing the role of microfinance in fostering adaptation to climate change.

Bresch, D. N. (2014). Climada – the open-source economics of climate adaptation (ECA) tool. https://github.com/davidnbresch/climada. Retrieved on March 31, 2015.

Bresch, D. N., & Mueller, L. (2014a). Economics of climate adaptation – Shaping climate-resilient development. A global overview of case studies with a focus on infrastructure. http://media.swissre.com/documents/Economics_of_Climate_Adaptation_focus_infrastructure.pdf. Retrieved on March 31, 2015.

Bresch, D. N., & Mueller, L. (2014b), Economics of climate adaptation – Shaping climate-resilient development. A global overview of case studies with a focus on focus on fast-growing coastal communities. http://media.swissre.com/documents/Economics_of_Climate_Adaptation_focus_coastal_communities.pdf. Retrieved on March 31, 2015.

Bresch, D. N., & Schraft, A. (2011). Neue, integrierte Sichtweise zum Umgang mit Klimarisiken und deren Versicherung. *Schweizerische Zeitschrift für das Forstwesen, 162*(12), 464–468.

Bresch, D. N., Egloff, R., Berghuijs, J., & Kupers, R. (2014). A resilience lens for enterprise risk management. In R. Kupers (Ed.), *Turbulence – a corporate perspective on collaborating for resilience* (pp. 49–65). Amsterdam: Amsterdam University Press. www.oapen.org/download?type=document&docid=477310. Retrieved 31.3.2015.

Conway, D., & Mustelin, J. (2014). Strategies for improving adaptation practice in developing countries. *Nature Climate Change, 4*, 339–342.

Corti, T., Wüest, M., Bresch, D. N., & Seneviratne, S. I. (2011). Drought-induced building damages from simulations at regional scale. *Natural Hazards and Earth System Sciences, 11*, 3335–3342.

Della-Marta, P. M., Liniger, M. A., Appenzeller, C., Bresch, D. N., Köllner-Heck, P., & Muccione, V. (2010). Improved estimates of the European winter windstorm climate and the risk of reinsurance loss using climate model data. *Journal of Applied Meteorology and Climatology, 49*(10), 2092–2120.

ECA working group. (2009). Shaping climate-resilient development: A framework for decision-making, a Report of the Economics of Climate Adaptation (ECA) Working Group. http://media.swissre.com/documents/rethinking_shaping_climate_resilent_development_en.pdf. Retrieved on March 31, 2015.

Fankhauser, S., Smith, J. B., & Tol, R. (1998). Weathering climate change: Some simple rules to guide adaptation decisions. *Ecological Economics, 30*(1), 67–78.

Knutti, R., & Bresch, D. N. (2015). Climate change uncertainty and risk: From probabilistic forecasts to economics of climate adaptation. Lecture course at the Swiss Federal Institute of Technology (ETH). www.iac.ethz.ch/edu/courses/master/modules/climate_risk. Retrieved on March 31, 2015.

Pielke, R. A., Jr., Joel Gratz, J., Landsea, C. W., Collins, D., Saunders, M. A., & Musulin, R. (2008). Normalized hurricane damage in the United States: 1900–2005. *Natural Hazards Review, 9*(1), 29–42.

Raible, C. C., Kleppek, S., Wüest, M., Bresch, D. N., Kitoh, A., Murakami, H., & Stocker, T. F. (2011). Atlantic hurricanes and associated insurance loss potentials in future climate scenarios: Limitations of high resolution AGCM simulations. *Tellus A, 2012*(64), 15672.

Reguero, B. G., Bresch, D. N., Beck, M., Calil, J., & Meliane, I. (2015). Coastal risks, nature-based defenses and the economics of adaptation: An application in the Gulf of Mexico, USA. Proceedings of the International Conference on Coastal Engineering (ICCE). DOI: http://dx.doi.org/10.9753/icce.v34.management.25, direct: https://journals.tdl.org/icce/index.php/icce/article/view/7585. Retrieved on March 31, 2015.

Schwierz, C., Köllner-Heck, P., Zenklusen, E., Bresch, D. N., Vidale, P. L., Wild, M., & Schär, C. (2010). Modelling European winter windstorm losses in current and future climate. *Climatic Change, 101*, 485–514.

Stern, N. (2006). The economics of climate change: The stern review, Part V: Policy responses for adaptation. United Kingdom: Cambridge University Press.

Stirling, A. (2010). Keep it complex. *Nature, 468*(7327), 1029–1031.

Surminski, S. (2013). Private-sector adaptation to climate risk. *Nature Climate Change, 3*(11), 943–945.

Swiss Re. (2010). Shoring up the energy coast – building climate-resilient industries along America's Gulf Coast. http://media.swissre.com/documents/Entergy_study_exec_report_20101014.pdf. Retrieved on March 31, 2015.

Swiss Re. (2014). Natural catastrophes and man-made disasters in 2013. Fostering climate resilience. Swiss Re sigma No 1/2014. http://media.swissre.com/documents/sigma1_2014_en.pdf. Retrieved on March 31, 2015.

Tol, R., Fankhauser, S., & Smith, J. B. (1998). The scope for adaptation to climate change: What can we learn from the impact literature. *Global Environmental Change, 8*, 109–123. Retrieved on March 31, 2015.

Building Resilience: World Bank Group Experience in Climate and Disaster Resilient Development

Daniel Kull, Habiba Gitay, Sofia Bettencourt, Robert Reid, Alanna Simpson, and Kevin McCall

Abstract Concurrently addressing disaster risk and the effects of climate change delivers both immediate and longer term development gains, while also reducing fragmentation of the limited human and financial capacity found in many developing countries. Over the last few years, the World Bank Group has been systematically integrating climate and disaster resilience into its support to low and middle income countries. Early lessons indicate the need to pursue the disaster risk management pillars of risk identification, risk reduction, preparedness, financial and social protection, and resilient reconstruction. Institutional arrangements that bring together multiple sectors and stakeholders with support at the highest level of government is needed for sustained climate resilient development effort and outcomes. While investing in climate resilience often requires higher start-up costs, it is cost effective in the long-term. Spatial planning that considers short-to-long-term risks reduces the possibilities of stranded assets, with proactive management of at risk investments needed. Flexible and predictable financing as part of long-term development programmes can address climate and disaster risk, meet the needs of countries, and reduce poverty in the most vulnerable communities and countries.

Keywords Climate resilience • Disaster risk management • Climate resilient development

---

This chapter summarizes contents from *Building Resilience: Integrating Disaster and Climate Risk into Development – The World Bank Group Experience* (World Bank 2013a). The chapter represents the status of experiences and operations as of June 2015.

D. Kull (✉)
Global Facility for Disaster Reduction and Recovery (GFDRR), Geneva

World Bank Group, Chemin Louis-Dunant 3, CH-1202 Geneva, Switzerland
e-mail: dkull@worldbankgroup.org

H. Gitay • K. McCall
Climate Change Policy Team, Climate Change Group, World Bank Group, Washington

S. Bettencourt • R. Reid • A. Simpson
Global Facility for Disaster Reduction and Recovery (GFDRR), Geneva

N. Salzmann et al. (eds.), *Climate Change Adaptation Strategies – An Upstream-downstream Perspective*, DOI 10.1007/978-3-319-40773-9_14

1 Introduction

Weather-related disasters affect both developed and developing countries, with particularly high disaster impacts in rapidly growing middle-income countries, due to growing asset values in at-risk areas. However, low-income and lower middle-income countries have the least capacity to cope and, in general, suffer the highest human toll, accounting for 85 % of all disaster fatalities (Munich Re 2010). Climate-related impacts will continue to increase due to both development and climate drivers (IPCC 2013), and impacts will be felt most acutely by the poor and most marginalized populations, who commonly live in the highest-risk areas. They also have the least ability to recover from recurrent, low-intensity events, which can have crippling and cumulative effects on livelihoods. The impacts of climate change on poverty are expected to be regressive and differential, affecting most significantly the urban poor and highly vulnerable countries in sub-Saharan Africa and South Asia (Shepherd et al. 2013).

Unless measures are taken to reduce risks, climate change is likely to undermine poverty goals and exacerbate inequality for decades to come. Climate and disaster resilient development, therefore, makes sense from both the poverty alleviation and economic growth perspectives. The World Bank Group (WBG) has thus been supporting developing countries to manage these increasing risks through disaster risk management (DRM) focused on weather extremes and climate resilience that addresses the current and likely future changes in climate. It has brought together its extensive work on DRM and more recent experience from climate resilience to support countries on "climate and disaster resilient development." Early lessons learned, tools, instruments and approaches developed for such work are presented in this chapter.

2 World Bank Group Experience

2.1 Overview

Box 1 provides a brief introduction to the WBG. WBG investment in resilient development is measured through the support provided to resilience/adaptation and DRM as part of development assistance. Using this definition, the share of projects with DRM co-benefits, in fiscal years[1] 2013 and 2014 were 11 % and 12 % respectively. This compares to about 9 % in fiscal year 1984. This upward trend is occurring across all regions and country income groups. The WBG has also committed nearly US$13 billion in investments that provide adaptation co-benefits over the past four fiscal years (2011–2014). This represents 8 % of the total lending commitments in

[1] Fiscal years for the WBG are 1 July to 30 June.

fiscal year 2013 and 7 % in fiscal year 2014; with adaptation support to low-income countries proportionally higher at 13 % and 10 % respectively.

In addition, the WBG facilitates access to a menu of climate finance instruments through external resources such as the Pilot Porgram for Climate Resilience (PPCR) within the Climate Investment Funds (CIF), the Global Environment Facility (GEF), and the Global Facility for Disaster Risk Reduction and Recovery (GFDRR). Funding from these sources for adaptation was $279 million in FY14 and close to a total of $850 million over the FY11-14 period. These dedicated climate funds provide technical assistance and capacity support for mainstreaming disaster and climate resilience into country development strategies and investments.

The WBG's private sector investment arm, the International Finance Corporation (IFC), has also been actively engaging with the private sector on climate and disaster resilience. IFC is increasing awareness of climate risks and has begun incorporating climate change into its policies and investments.

Box 1: Introduction to the World Bank Group

Since its inception in 1944, the World Bank mission has evolved from the International Bank for Reconstruction and Development (IBRD) as facilitator of post-war reconstruction and development to the present-day mandate of worldwide poverty alleviation. The WBG is currently composed of five development institutions: the IBRD, International Development Association (IDA), International Finance Corporation (IFC), Multilateral Guarantee Agency (MIGA), and International Centre for the Settlement of Investment Disputes (ICSID).

The WBG is a vital source of financial and technical assistance to developing countries around the world. It is not a bank in the ordinary sense but a unique partnership to reduce poverty and support development, owned by the governments of member nations. The WBG has set two goals for the world to achieve by 2030: end extreme poverty by decreasing the percentage of people living on less than $1.25 a day to no more than 3 %, and promote shared prosperity by fostering the income growth of the bottom 40 % for every country.

The WBG provides investment financing in the form of credits and grants to low-income countries through IDA, and as loans to middle-income countries through IBRD. Some lower-middle income countries qualify for a blend of the two. IFC provides investment, advisory, and asset management services to the private sector.

Climate change is a fundamental threat to sustainable development and the fight against poverty. The WBG is concerned that without bold action now, the warming planet threatens to put prosperity out of reach of millions and roll back decades of development. It is therefore stepping up its mitigation, adaptation, and disaster risk management work, and will increasingly look at all its business through a climate lens.

2.2 *Key Elements of Climate and Disaster Resilient Development*

Over the last decade or so, experience from countries that have integrated risks from climate change into the development planning process exhibits some common elements, as presented in Fig. 1. The process can start through different elements, but most have done so by strengthening institutions, identifying and assessing risks, and enhancing capacity and knowledge.

DRM experience since the 1970s also shows a process with elements overlapping that of climate resilient development, as illustrated in Fig. 2 and summarised in Box 2. The operational DRM framework is organized around five action pillars. Risk identification provides the base for all other actions: to reduce risk (by putting policies and plans in place that will help avoid the creation of new risk or by addressing existing risks); to prepare for the residual risk either physically (preparedness) or financially (financial protection); and to inform improved resilient reconstruction

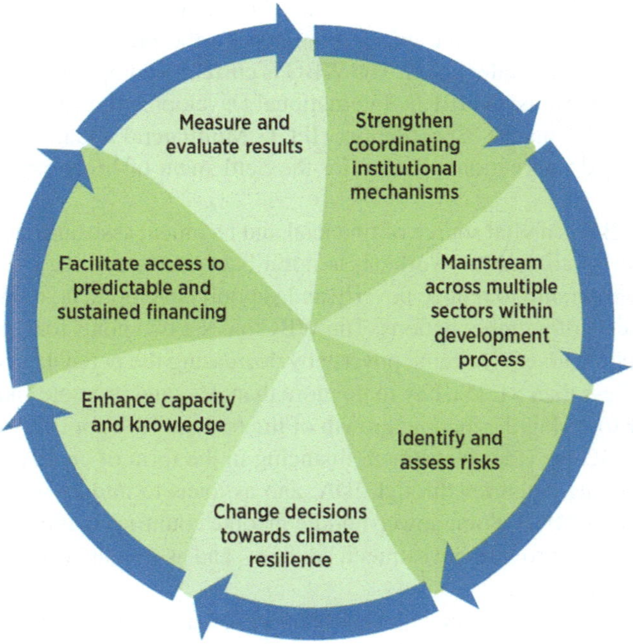

Fig. 1 Process of integrating climate resilience into development (World Bank 2013a)

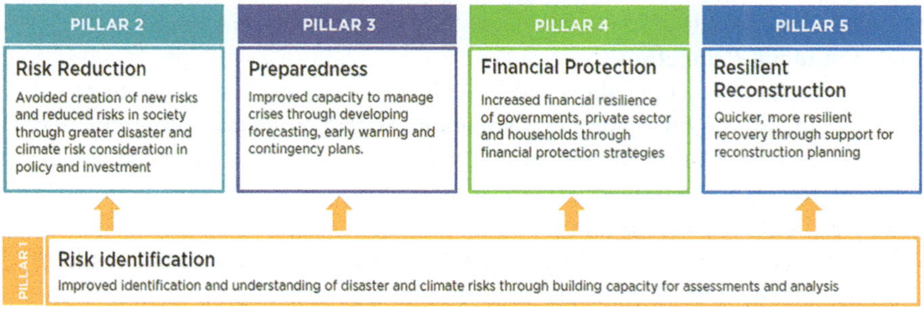

Fig. 2 An operational framework for managing climate and disaster risk (World Bank et al. 2012)

design. The DRM community also recognizes that reconstruction programs provide opportunities to change the status quo and behaviours that contribute to underlying vulnerabilities.

Box 2: Overview of the Pillars of Disaster Risk Management

1. **Risk Identification**
   Risk assessments serve multiple purposes for various stakeholders, ranging from urban risk assessments for disaster preparedness, to multi-country financial risk assessments to support design of financial transfer mechanisms. They can also be used to understand where the greatest benefit to cost ratio investments can be made to reduce risk. Risk assessments are increasingly able to calculate risk under current and future climate and socio-economic scenarios, providing decision makers with additional impetus to act now on the underlying drivers of risk (GFDRR 2014a, b).
2. **Risk Reduction**
   The main driver of growing disaster losses is increasing exposure of people and assets, caused by rapid and unplanned urbanisation. Reducing new risk through anticipatory action is therefore critical, for example through improved territorial planning or building practices. Existing risks can also be addressed for example by retrofitting critical infrastructure or constructing flood protection systems.
3. **Preparedness**
   Considering the context of increasing uncertainty, "planning for the worst" must assume a central role in development. Preparedness forms an integral component of national strategic approaches, helping link disaster response with resilience building. WBG support targets strengthening early warning, national and local coordination, emergency response and civil protection structures, providing real-time impact analysis and enhancing financial preparedness.

(continued)

Box 2: (continued)

4. **Financial Protection**
   Financial protection allows for accelerated resource mobilization in an emergency or pre-emergency situation. Social protection programs and policies help buffer individuals from shocks and equip them to be able to improve their livelihoods.
5. **Resilient Reconstruction**
   Disasters often provide unique opportunities to promote climate resilient development. Politicians and donors alike are attuned to the issue, and the general public may be more amenable to the often-difficult trade-offs necessary for risk reduction. At the same time, accelerated development through multi-sectoral reconstruction investments can produce transformative effects on population and livelihoods dynamics (World Bank 2014a).

Although the approaches used for climate resilience and DRM originated from different disciplines, the two are increasingly converging, partially due to a high proportion of recent disasters being weather related. It is also clear that on the ground for affected areas and communities—particularly the poor—the approaches are indistinguishable; communities and households have to both increasingly consider weather extremes in their decisions and deal with the consequences of the changing climate and the new norms it brings. Details of some key elements are provided below.

2.2.1 Role of Institutions

The role of institutions in climate and disaster resilient development is arguably the single most important—yet the most difficult—part of the process. This is the case for both driving policy change and investment design. As climate change and disasters affect multiple sectors, countries where governance systems are divided across sectoral lines face a particularly complex challenge, since the institutions that have historically driven the climate change and DRM agendas are typically newer and weaker than the more established sectoral ministries, such as agriculture, transport and energy. Often, a leading agency is needed to mobilize and coordinate ministries and development partners, promote information sharing and knowledge management, and influence development planning and the budget in both the short and long terms. Such a lead agency needs to be able to convene decision makers from multiple agencies and levels of government, as well as the private sector and civil society. Emerging experience indicates that in order to have effective convening power, such an agency should be located at the highest possible level of government. While the choice varies, several countries, such as Kiribati, Mexico, Mozambique, Morocco, Samoa and Zambia, have established coordinating agencies under finance and planning ministries, or offices of the President or Prime Minister.

2.2.2 Identifying Risks and Vulnerabilities

The ability of countries to increase their climate and disaster resilience is directly linked to their capacity to generate and analyse data to assess vulnerability (World Bank 2014b) and design appropriate resilience measures. The WBG and GFDRR have been supporting climate and disaster risk assessments through open geospatial data tools, the establishment of the Understanding Risk Community of Practice (now with about 3300 members worldwide), annual Code for Resilience challenges, development of simple climate and disaster risk screening tools, and through technical assistance to over 50 countries. A particular focus has been on promoting open data and information sharing between in-country agencies, the scientific community and decision makers in the field, and in supporting informed decision making for climate and disaster resilient development. As a consequence, access to risk information has improved for an estimated 40 million people in 24 countries that have access to the Internet, and several thousand datasets related to natural hazard risks have been shared.

In an effort to make risk data and analysis available, the Open Data for Resilience Initiative supports governments to develop open systems for disaster risk and climate change information (World Bank 2014c). Complimenting this initiative is the Climate Change Knowledge Platform, an online platform that draws together various international open sources of climate information with links to many of the resources for disaster risk.

Communication and use of risk information is key. As experienced in Tajikistan, direct investment support coupled with facilitation and training helped farmers assume responsibility for sustaining their livelihoods in financially and environmentally sound ways. Participatory planning along with village and household budget limits was an effective mechanism for villagers to prioritize and assess risks of various options, as well as allocate resources (World Bank 2012a). It should also be recognized that political economy realities can sometimes limit the use of strictly science-based approaches to decision support. For example in the Mekong River Basin, the premise that water resource management decisions could be based solely on scientifically derived targets and scenarios proved too constraining. Rather, models have been used to determine the winners and losers of proposed basin development and subsequent negotiations have focused on individual, rather than collective interests (World Bank 2012b).

2.2.3 Risk Reduction and Resilience

The WBG has been supporting a range of risk reduction activities such as improvements in safety standards and building codes, participatory spatial resilient planning and construction of protective and/or resilient infrastructure. In many case, dedicated climate and disaster funds are used for technical assistance to support design and preparation of development projects/programs. Such a process brings in needed and timely technical expertise for risk reduction and detailed resilience measures.

Some examples include a GFDRR-supported assessment of Vietnam's rural roads and national highways that led to climate resilient road designs applied in a nationwide IBRD-funded rural transport project. In Samoa, through PPCR support, the main road is being designed and upgraded to a climate and disaster resilient standard and community-led spatial planning is being implemented to reduce risk and enhance resilience through an integrated planning "ridge-to-reef" approach.

Following tropical storms Ondoy and Pepeng in 2009, the Philippines Department for Public Works and Highways developed the Metro Manila Flood Risk Management Master Plan, which prioritizes policy reform and structural risk reduction investments costing approximately US$8.6 billion. Studies have begun on a plan that proposes alterations for the upstream catchment area and the Laguna Lakeshore, and the government is in discussions with affected communities on new housing and resettlement options. Similarly, the city authorities in the Senegalese capital Dakar, designed a large-scale IDA investment program to protect communities from recurrent floods, improve drainage systems and develop an integrated urban flood risk and storm water management program in flood-prone, peri-urban areas.

For risk to be adequately addressed, stakeholders have to be part of the process and more importantly, own the process and the solutions. This is helping sustain programs as experienced in decentralized watershed management in Uttarkhand, India (World Bank 2014d), building on past experience where it was observed that while fiscal decentralization and community empowerment are necessary, they are not sufficient to promote improved community management of natural resources. It was concluded that more work is needed to strengthen local institutional frameworks and practical mechanisms are needed to tackle externalities arising from insecure property rights (IEG 2011).

Some experiences, for example from the Andes, shows that community-led efforts that have fully engaged the public can help development outcomes, such as improved basic infrastructure, while also contributing to resilience through ensuring functioning ecosystems (World Bank 2014b). Watershed management projects that take a livelihood focused approach perform better than those that do not, with projects combining livelihood interventions with environmental restoration enjoying high success rates, even though effects on downstream communities (such as reduced flooding and improved water availability) and social benefits in both upstream and downstream communities were in the past often not measured (IEG 2010). Such approaches, captured as "ecosystem-based adaptation" are being included in a range of investments such as in Zambia, Samoa, and the Solomon Islands. The sustainability of such efforts can be enhanced by including community driven development as part of national-to-local government development processes.

2.2.4 Early Warning and Preparedness

In many countries, early warning and preparedness are often an entry point for climate and disaster resilient development. Weather, climate and hydrological monitoring and forecasting are essential to inform decision making for climate resilience and provide critical inputs to early warning systems. The WBG's portfolio of projects supporting hydro-meteorological investments have often employed regional approaches to support national capacity by linking with neighbouring regional and global centres of excellence for data, forecast and expertise sharing, for example through a system of "cascading forecasts".

Supply of forecasting services is however not enough. Unlocking strong demand for weather, hydrological, and climate information is necessary in order to sustain the political will to maintain hydrometeorological services (IEG 2012). Building capacity of the agencies involved across the end-to-end service delivery chain improves early warning and preparedness, as well as coordination and information exchange (Rogers and Tsirkunov 2013).

WBG and GFDRR are also supporting a number of countries and communities for enhancing their preparedness for climate and disaster risks. For example, the Senegalese Civil Protection Agency is strengthening its risk management capacity by setting up coordination mechanisms for early warning, preparedness and response. In Burkina Faso, the National Council for Disaster Management and Recovery is developing local contingency and emergency preparedness plans, linking the plans to the existing early warning system, and strengthening community-based preparedness planning, including drills and simulation exercises. In India, the WBG continues to support climate and disaster resilience in Odisha and Andhra Pradesh with the aim of extending early warning systems to the community level, building multi-purpose cyclone shelters and evacuation roads, and strengthening existing coastal embankments. Early indications reveal that project investments are contributing to India's larger efforts to help communities become more resilient to the impacts of natural disasters and the changing climate as shown in the 2013 storms in Odisha.

2.2.5 Financial and Social Protection

The WBG uses a series of instruments (Fig. 3) to support financial protection, which are tailored for national and often regional needs and varying risk profiles. Experience is showing that these need to be part and parcel of climate and disaster resilient development. Much of this work draws on the experience of the DRM community. For example, in 2007, the WBG helped establish the Caribbean Catastrophe Risk Insurance Facility (www.ccrif.org), a Caribbean-owned "parametric"

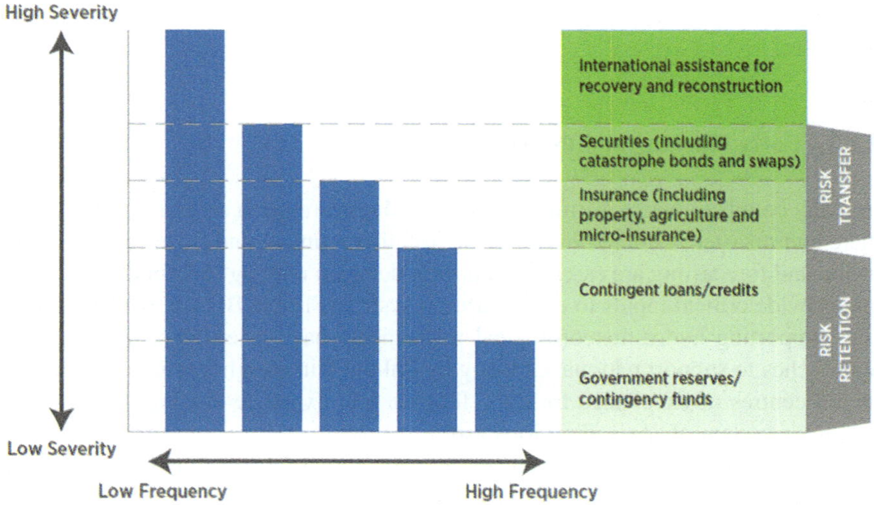

Fig. 3 Financial protection instruments for climate and disaster resilience, developed by the World Bank Disaster Risk Financing and Insurance Team (World Bank and GFDRR 2013), based on Figure 7 in Ghesquiere and Mahul (2010)

insurance pool, which offers fast payout to its 16 member countries upon occurrence of pre-defined hurricane strengths and earthquake magnitudes within defined geographical locations. The Facility offers participating countries an efficient and transparent vehicle to access international reinsurance and capital markets, and is a self-sustaining entity, relying on its own reserves and reinsurance for its financing. The Pacific region has built on this experience in developing the Pacific Catastrophe Risk Assessment and Financing Initiative (PCRAFI) which is helping 15 Pacific Island countries to better understand and address climate-related risk.

The WBG has also expanded the use of its Catastrophe Deferred Drawdown Options, which provide countries with contingent credit lines that can be drawn upon in case of disaster, as the Philippines did in 2011, drawing US$500 million to support response and recovery after tropical storm Washi. Supported by this instrument, Costa Rica has been proactively reviewing the catastrophe risk exposure of public assets and infrastructure, which has helped it develop effective and affordable insurance programs to protect these assets. Results of preliminary work show that a proposed insurance vehicle would improve coverage with a net savings of at least US$50 million over 10 years.

WBG also supports efforts to improve risk transfer for households and individuals, recognizing the limitations to insuring the poorest sustainably. For example, assistance in Mongolia helped to create a livestock insurance pool to protect herders against harsh winters, as well as the Indian government to move towards market-based crop insurance, constituting the largest crop insurance program in the world with more than 25 million farmers insured. It should however be noted that the schemes that have enjoyed significant uptake have mostly required significant subsidies, and that the products generally do not cover landless rural labourers (IEG 2012).

National safety net systems, which in regular times can help minimize the negative impact of economic shocks on individuals and families, can also be designed and funded to scale up in response to a disaster to prevent households from falling into poverty (World Bank 2013b). For example, Ethiopia's Productive Safety Net Program (PSNP)—supported by the WBG in partnership with a number of donors and agencies – was able to scale up its day-to-day activities to disburse $134 million to support 9.6 million food-insecure people during the 2011 Horn of Africa drought. Guidance on how to prepare social protection programs to respond to disasters and climate change can be found in "Building Resilience to Disaster and Climate Change through Social Protection Toolkit", amongst other resources (World Bank and GFDRR 2013).

With increasing climate-related shocks, social protection measures may also need to be complemented by other resilience measures. Community driven development approaches and actions are important elements of an effective poverty reduction and sustainable development strategy, promoting scalable models and approaches to empower poor communities to manage climate and disaster risk and to identify practical ways of getting climate and disaster risk finance directly to the people (World Bank 2013c). Scaling up and sustaining community-based resilience calls for bridging the gap between the local, subnational and national levels, and understanding the complementary roles of formal and informal institutions.

2.2.6 Resilient Reconstruction and Mainstreaming

Given the attention to post-disaster recovery and the increasing climate related disasters, there is increasing attention to resilient recovery. The WBG and GFDRR assist disaster-hit countries through support for Post Disaster Needs Assessments (PDNAs), technical assistance for post-disaster recovery planning and financing and building institutional capacities. PDNAs are country-led and supported by a partnership between the United Nations, the European Union and the WBG, providing a coordinated and credible basis for recovery and reconstruction planning that incorporates current and future climate change risks, risk-reduction measures and financing plans. A lighter version of the PDNA is increasingly used in countries suffering from extensive, recurrent climate-related disasters. In all cases recovery operations with substantial investments included resilient reconstruction or "build back better" principles.

Building back better and integrating resilient approaches into development planning come with upfront costs. However, if the action is well designed and proportionate to the risk, then the outcome will be cost effective and save money in the long run. Experience suggests that "building back better" typically costs between 10 and 50 % more than the cost of simply reconstructing original structures. In the case of transport or irrigation infrastructure that may need to be moved to safer areas, the cost can be several orders of magnitude higher. At the same time, it should be noted that the rushed nature of emergency response projects makes them particularly vulnerable to design and institutional problems, also in terms of securing political buy-in for institutional reforms (IEG 2012).

To avoid creating new risks, a portfolio of measures need to be combined to most efficiently reduce climate and disaster risk. This can include spatial and strategic planning to reduce risks, and changes in standards and norms. These collectively would decrease the probability of having high cost assets (such as ports, dams, and tourism industries) not being able to function in a changed climate and thus being "stranded," and changing incentives and behaviours. This will ensure that the most cost-effective means of building resilient societies into the future will avoid creating new risks.

Elaboration of climate resilient construction codes does not have to be expensive. For example in Madagascar and Mozambique, expenses have ranged from US$160,000 to US$210,000, including for sensitization and training (in Madagascar). Strengthening infrastructure safety standards in Madagascar cost about US$100,000 (for transport) and US$50,000 (for irrigation infrastructure), with an additional US$120,000 envisaged for training. These costs do not include, however, the extensive time required to integrate the new norms into sectoral programs and ensure their effective compliance.

3 Lessons Learned

Some key lessons learned are presented here and are drawn from WBG and its partners' experiences. There is no clear delineation of lessons between those for policy support or for specific investments; much of the choices and trade-offs need to be made in the context of the development planning process. This avoids introduction of inadvertent and new risks, for example by focussing on increased agriculture production without considering the effects of potential run-off on coastal ecosystems and their functions or water availability under a changing climate.

(a) **Provide flexible and predictable financing**

Climate and disaster resilient development requires long-term and flexible programs, based on predictable financing. This allows institutional mechanisms to mature and transcend political cycles, and promotes a learning-by-doing, iterative and flexible approach to identify risks and incorporate resilience into development planning. The latter is particularly important in the face of uncertainties in climate change and development scenarios, which may require frequent adjustments. For this reason, robust monitoring and reporting is of critical importance, to allow programs to scale-up approaches that have been proven to work and to adjust those that have been less successful.

Long-term programs can benefit from an initial phase, focused on planning, institutional coordination and capacity building. Often, this process takes time—typically at least 18–24 months—and entails slow initial disbursements.

However it helps to build consensus and momentum and political will to scale up climate resilient development over the long term.

Finance sources have included grants, credits, loans, and a mixture of national and international funds. Long-term financing is also critical to counteract the perverse incentives that favour short-term disaster financing over long-term risk reduction. At the same time, longer timeframes help optimize opportunities to incorporate climate resilience and improved safety standards immediately after disasters, when public support for risk management is at its highest.

(b) **Foster robust decision-making**
Risk identification needs to be effectively linked to decision making, taking future uncertainties into consideration. By quantifying risks and anticipating the potential negative impacts of climate hazards and disasters, risk assessments can help governments, communities and individuals make better-informed decisions. Systematic screening of risks can also help determine the level of risks to people and assets and guide options for risk management.

Individual investments can themselves actually be less important than their role in catalysing community and national stakeholders and changing behaviours. Currently, the most effective actions appear to be those that combine development benefits in the near term with reductions in vulnerability over the longer term. However, concerted efforts need to be made to ensure that short-term solutions do not increase future risks. This is typically the case with flood protection dykes, which, over the long term, can create a false sense of security and inadvertently expand settlements in high-risk areas. To be robust, decisions should be "stress-tested" across a broad range of climate and socioeconomic conditions.

(c) **Share the responsibility of risk management**
Risk management requires complementary actions at various levels of responsibility—household, community, national and international. Local disaster risks, such as storms or moderate drought, can often be managed by individuals, communities and authorities at the local level, but as risks increase—for example, with major cyclones—national governments and the international community will have to play larger roles. While individuals are able to deal with many risks, they are inherently ill-equipped to manage large or systemic shocks, such as those that arise from climate change, since the past can no longer be considered a reliable predictor of the future (World Bank 2013b). As a result, climate and disaster resilient development needs to occur at different scales—individual, household, community, enterprise, national and international. These different actors have the potential to support climate risk management in different yet complementary ways.

(d) **Institution building and mainstreaming need to take incentives into account**
Capacity building for climate and disaster resilient development needs to be broad based and invest in professionals, especially in early to mid-career, to shield programs from political changes or high staff turnover. In addition, appropriate incentives are required to promote inter-sectoral planning: many

multi-stakeholder committees have failed because line agency participants perceive climate and disaster resilience to be an added responsibility to their already full agenda.

Lack of ownership explains why many stand-alone "adaptation" and DRM projects have not been successful in the past. If, by contrast, they are effectively mainstreamed into line ministries' own programs and budgets, staff are more motivated to perform. For example in Zambia, the Sixth National Development Plan led to the creation of a specific program within the public works sector that considered climate resilience in infrastructure planning, allowing public works staff to participate more actively in the activities of the multi-sectoral Secretariat for Climate Change (under the Ministry of Finance).

In many emerging climate and disaster resilience programs, stakeholder champions frequently emerge to lead and facilitate the process. The result has been the genesis of multi-sectoral and multi-stakeholder processes, which facilitate decisions on incorporating climate risk as part of development planning.

(e) **Stay focused on the poor and the vulnerable**
In the urgency to protect assets, climate and disaster resilient development programs should not lose sight of people. The complexity of most climate and disaster resilient development programs often requires multiple stakeholder meetings and consensus-based decisions, which consume time and resources. By the time decisions are translated into action on the ground, programs may lose sight of their most important objective—to diminish the risk to people and their assets, in particular for the poorest and most vulnerable. Continuously reemphasizing this focus will be critical to achieving climate resilience. Targeted actions will be needed to provide the poor and near poor with the resources, information and knowledge required to become more resilient. Support for community resilience, combined with well-designed social protection mechanisms that can be scaled up in response to disasters, could play a major role in reducing the impacts on the poor and the vulnerable from disasters and climate change.

(f) **Leverage partnerships and share knowledge**
National and regional governments, and international organizations that support them, have accrued a wealth of knowledge on approaches to integrate and mainstream climate and disaster resilience in development planning. International and regional partnerships and South-South knowledge exchange platforms provide opportunities for transboundary learning and cooperation on effective strategies to build long-term resilience. Such platforms are an effective tool to communicate the lessons listed above and share practical approaches for decision-makers at all levels, from national government to community-based organizations. Moreover, these partnerships can accelerate the learning required to invest in the human, institutional, and financial resources that support robust climate and disaster resilient planning.

4 Concluding Remarks

Key drivers—climate change, poorly planned development, poverty and environmental degradation—influence the risk of a climate event becoming a disaster. Thus, these factors need to be managed collectively. In the coming decades, disaster losses are expected to continue to rise due to the increasing exposure of populations and assets, and environmental degradation, compounded by climate change. Therefore, development paths must take the risks of climate change and disasters into account As such, climate and disaster resilience should form an integral part of development planning processes, particularly in the most vulnerable countries.

Given the close interactions between climate change and local/national drivers of vulnerability, it is important to ultimately strengthen all aspects of climate and disaster resilient development, including coordinating institutions, risk identification and reduction, preparedness, financial and social protection, and resilient reconstruction. Getting the institutions and incentives right are the most important issues in climate and disaster resilient development. They can overcome the challenges of limited capacity and reduce the likelihood of introducing new or additional risks. Although an integrated, multi-stakeholder and multi-sectoral approach takes time and may entail slow initial disbursements, it generally results in stronger buy-in from relevant stakeholders and is likely to be more sustainable over the long term. Political cycles favour short-term development decisions, and government employees often have little incentive to participate in inter-sectoral committees to address problems not viewed as part of their mandate. Changing this "culture" is easier when a flexible, learning-by-doing approach is pursued, and the process is relatively independent from political pressures.

References

GFDRR. (2014a). *Understanding risk in an evolving world – Emerging best practices in natural disaster risk assessment*. Washington, DC: Global Facility for Disaster Reduction and Recovery.

GFDRR. (2014b). *Understanding risk in an evolving world – A policy note*. Washington, DC: Global Facility for Disaster Reduction and Recovery.

Ghesquiere, F., & Mahul, O. (2010). *Financial protection of the state against natural disasters: A primer*. World Bank Policy Research Working Paper 5429, Washington, DC.

IEG. (2010). *Water and development: An evaluation of World Bank support 1997–2007* (Vol. 1). Washington, DC: Independant Evaluation Group, World Bank.

IEG. (2011). *Project performance assessment report – India – A cluster assessment of forestry and watershed development activities* (Independant Evaluation Group, report no.: 61065). Washington, DC: World Bank.

IEG. (2012). *Adapting to climate change: Assessing World Bank Group experience – Phase III of the World Bank Group and climate change*. Washington, DC: Independant Evaluation Group, World Bank.

IPCC. (2013). Summary for policymakers. In T. F. Stocker, D. Qin, G.-K. Plattner, M. Tignor, S. K. Allen, J. Boschung, A. Nauels, Y. Xia, V. Bex, & P. M. Midgley (Eds.), *Climate change*

2013: The physical science basis. Contribution of Working Group I to the Fifth Assessment Report of the Intergovernmental Panel on Climate Change. Cambridge, UK: Cambridge University Press.

Munich Re. (n.d.). Münchener Rückversicherungs- Gesellschaft, Geo Risks Research, NatCatSERVICE—as at July 2010. Munich.

Rogers, D., & Tsirkunov, V. (2013). *Weather and climate resilience: Effective preparedness through national meteorological and hydrological services.* Washington, DC: Directions in Development, World Bank.

Shepherd, A., Mitchell, T., Lewis, K., Lenhardt, A., Jones, L., Scott, L., & Muir-Wood, R. (2013). *The geography of poverty, disasters and climate extremes in 2030.* Exeter: ODI, Met Office Hadley Center, RMS Publication.

World Bank. (2012a). *Tajikistan: Community agriculture and watershed management project.* Implementation Completion and Results Report, ICR2093, Washington, DC.

World Bank. (2012b). The Mekong river commission water utilization project. Project Performance Assessment Report, Report No. 70332, Washington, DC.

World Bank. (2013a). *Building resilience: Integrating disaster and climate risk into development – The World Bank Group experience.* Washington, DC: The World Bank.

World Bank. (2013b). *Financial innovations for social and climate resilience: Establishing an evidence base.* Washington, DC: Social Development Department.

World Bank. (2013c). *Climate and disaster resilience: The role for community-driven development.* Washington, DC: Social Development Department.

World Bank. (2014a). *Bolivia: Emergency recovery and disaster management project.* Implementation Completion and Results Report, ICR1384, Washington, DC.

World Bank. (2014b). *Adaptation to the impact of rapid glacier retreat in the tropical Andes project.* Implementation Completion and Results Report, ICR2921, Washington, DC.

World Bank. (2014c). *Open data for resilience field guide.* Washington, DC.

World Bank. (2014d). *Uttarakhand decentralized watershed management project (Gramya I).* Implementation Completion and Results Report, ICR2216, Washington, DC.

World Bank & GFDRR. (2013). *Building resilience to disaster and climate change through social protection. Synthesis note.* Washington, DC: World Bank Group Rapid Social Response and Global Facility for Disaster Reduction and Recovery.

World Bank, GFDRR & Japan. (2012). *The Sendai report: Managing disaster risks for a resilient future.* Washington, DC: World Bank Group, Global Facility for Disaster Reduction and Recovery, and Government of Japan.

The Science-Policy Dialogue for Climate Change Adaptation in Mountain Regions

Thomas Kohler, André Wehrli, Elbegzaya Batjargal, Sam Kanyamibwa, Daniel Maselli, and Urs Wiesmann

Abstract Mountains are among the regions most affected by climate change and they provide some of the most visible evidence of this change such as melting glaciers. While climate change is a global process, adaptation must be based on local contexts, especially in mountain regions with their varied natural and socio-cultural setting and highly differentiated effects of climate change at short distance. The need for mountain-specific adaptation is also given against the background of the key ecosystem goods and services, which they provide to humankind such as freshwater, and which are likely to be affected by climate change. Adaptation has to take place under conditions of uncertainty, but there are options such as Payment for Environmental Services (PES) that allow action under these conditions. Moreover, acting under uncertainty is nothing new for mountain communities. Addressing the science-policy dialogue, we show how this dialogue has been institutionalised in both developed and developing countries, by presenting examples from Switzerland, Kyrgyzstan, and Uganda. We then argue that closing the data gap relating to mountain climates and existing adaptive action could strengthen the science-policy dialogue substantially. To conclude, we advocate the establishment of mountain climate change observatories and the inventorying of promising adaptive action; highlight the need for capacity development and exchange across governmental and

---

T. Kohler (✉)
Centre for Development and Environment (CDE), University of Bern, Bern, Switzerland
e-mail: Thomas.Kohler@cde.unibe.ch

A. Wehrli • D. Maselli
Global Programme Climate Change GPCC, Swiss Agency for Development and Cooperation SDC, Switzerland, Switzerland
e-mail: andre.wehrli@eda.admin.ch; daniel.maselli@eda.admin.ch

E. Batjargal
Central Asia Mountain Hub (CAMH), University of Central Asia, Bishkek, Kyrgyzstan
e-mail: elbegzaya.batjargal@ucentralasia.org

S. Kanyamibwa
Albertine Rift Conservation Society (ARCOS), Cambridge, UK
e-mail: skanyamibwa@arcosnetwork.org

U. Wiesmann
Institute of Geography (GIUB), University of Bern, Bern, Switzerland
e-mail: Urs.Wiesmann@cde.unibe.ch

N. Salzmann et al. (eds.), *Climate Change Adaptation Strategies – An Upstream-downstream Perspective*, DOI 10.1007/978-3-319-40773-9_15

non-governmental institutions including mountain communities; and propose a funding window for climate change adaption in mountains for countries in need.

Keywords Mountains • Climate change • Specific adaptation • Ecosystem services • Science-policy dialogue • Institutionalisation

1 Introduction

Mountains are among the regions most affected by climate change and they provide some of the most visible evidence of this change such as melting glaciers. This present contribution has two major parts: the first part deals with climate change governance with regard to mountains. It shows that while climate change adaptation is an act of global solidarity and national responsibility, locally devised strategies are crucial given the wide range of climates at short distance in mountains. We then highlight the importance of mountains for key ecosystem services which reaches out far beyond mountain regions, and which might be affected by climate change, and illustrate options for action under uncertainty as posed by climate change. The second part turns to science-policy interaction by presenting existing platforms that have institutionalised this interaction. We then posit that the role of science could be much enhance if the data gap on mountain climates would be closed and lessons learnt from proven adaptation action that exist across the mountain world. We conclude by underpinning the importance of capacity development and funding.

2 Climate Change Governance and Mountains

2.1 Global Solidarity, National Responsibility, Local Adaptation

Climate change adaptation has become a key component in the global climate change debate. It was science which put this debate on the development agenda and which continues to play an important role through the IPCC. The scope of the IPCC work is global as there is one atmosphere, where all emitted greenhouse gases are influencing the global climate system. Given this fact, and global disparities in development, climate change mitigation and adaptation have become an act of necessity and global solidarity. IPCC procedures, including its multi-stage review loops, help ensure that global climate policy is rooted in the best available knowledge. This policy is shaped under the UN Framework Convention on Climate Change (UNFCCC). Established as one of the Rio Conventions at the 1992 Earth Summit, the UNFCCC has currently nearly universal membership with a total of 196 countries. Within this process, climate change adaptation is a relatively new component but has gained in importance. The institutional setup on climate change adaptation was broadened, now including a wide array of bodies (e.g. the Adaptation Committee, the Least Developed Countries Expert Group) and instruments (e.g. the "National Adaptation Planning Process" or the Warsaw International Mechanism for Loss and Damage). Moreover, the new legal agreement concluded at the Conference of the Parties COP 21 in Paris in December 2015 does further enhance the emphasis on adaptation.

While mitigation and adaptation are regarded as a matter of global solidarity, implementation is a national responsibility. Every country is called upon to define national climate targets and adaptation strategies and (sectoral) plans and determine the necessary instruments to reach these targets (e.g. legislation, funding), and these elements are laid down in the National Adaptation Plans http://unfccc.int/adaptation/workstreams/national_adaptation_plans/items/6057.php. National policy also defines the specific contributions to be made by sectors of economy and society to reach national climate aims. Each country is thus in the driving seat relating to its climate policy, but faces the delicate task of balancing business as usual against implementing adaptation measures that contribute effectively to reducing the effects of climate change.

This task is compounded by the fact that climate change and its effects are generally **not uniform across a country**. Formulating adaptive measures as a national blueprint is therefore not enough; specific policies and actions are needed to address adaptation problems and opportunities as they exist on the ground, i.e. in their local context. Mountain regions present an exemplary case for this need for local and specific adaptation due to their specific exposure to climate change: Due to the combined effect of topography and orientation (aspect), mountains are characterised by the succession of different eco-climatic zones over very short distance that stretch over hundreds of kilometres in lowland areas. In each of these zones, climate change has specific effects, and the same is true for appropriate adaptation measures. Climate coupled with the effect of relief and gravitation triggers natural hazards that are specific to mountain areas and call for specific adaptation. Moreover, climate change in mountain regions will affect the provision of ecosystem goods and services with effects that reach out far beyond mountains and can have serious impacts for regional and even global development, as will be shown below by taking the example of fresh water.

2.2 *Climate Change in Mountains: A Reality with Far-Reaching Impacts*

Mountains are among the regions most affected by climate change. Some of the most striking evidence such as melting glaciers comes from mountain areas. Many scientists believe that the changes occurring in mountain ecosystems may provide an early glimpse of what could come to pass in lowland environments, and that mountains thus act as early warning systems relating to climate change. Globally, mountain regions have experienced considerable warming over the last 100 years (Fig. 1 top and middle) (Brönnimann et al. 2014/1). In fact, recent research has found that there is evidence for an amplified rate of warming with elevation in many mountain regions of the world (Fig. 2). The picture is complex, as a range of factors contribute to changing temperatures in mountains, including albedo, cloud cover, water vapour, and aerosols. Their interplay is still poorly understood (MRI 2015). While temperatures in mountain areas are expected to continue rising across the globe, and at accelerated rates with increasing altitude in many mountain regions, projections of precipitation display a regionally more differentiated pattern, with some regions expected to receive more rainfall, including the tropical Andes, the Hindu Kush Himalayas, East Asia, East Africa, and the Carpathian region. Regions expected to receive less rainfall include mountains in the Mediterranean, in the Southwest of the USA, in Central America and South Africa, and the Southern Andes (Fig. 1, bottom). Overall, precipitation patterns may change and variability and intensity increase (Brönnimann et al. 2014/1). Shifts in precipitation from snow to rain will reduce the amount of water stored as snow and released during summer, when it is most needed. This may exacerbate seasonal water shortages, especially in river basins which are dominated by snow-melt run off for their water supply. Such basins are home to more than one billion people – or about one sixth of the global population (Adam et al. 2009; Barnett et al. 2005).

As climate change goes on, there is considerable uncertainty as to how supplies of mountain **ecosystem goods and services** will change and whether they will be sustained or not. This uncertainty comes at a moment when the need for these goods and services is higher than ever before across the planet, and growing. Mountain key goods and services include the supply of freshwater to about half of humankind for drinking, irrigation, hydropower generation, and industrial development. Mountains have been called the water towers of the world (Viviroli et al. 2007). Examples of their key role can be found on all continents and in developing and industrialised countries as well as in emerging economies; a non-exhaustive list includes Africa (Egypt, South Africa), the Near East (Basins of Euphrates and Tigris), Central Asia (Amu Darya, Syr Darya), the Hindu Kush-Himalaya (Indus Basin Pakistan), the Pacific coast of South America, the Coastal Range in Brazil, the Western USA including California, and the Mediterranean Basin. Moreover, many of the world's largest cities depend on mountain waters. Demand for water is increasing rapidly due to economic development and population growth, but provision could become even more limited due to the effect of climate change, especially in those regions across the world which largely depend on mountain waters (Fig. 3).

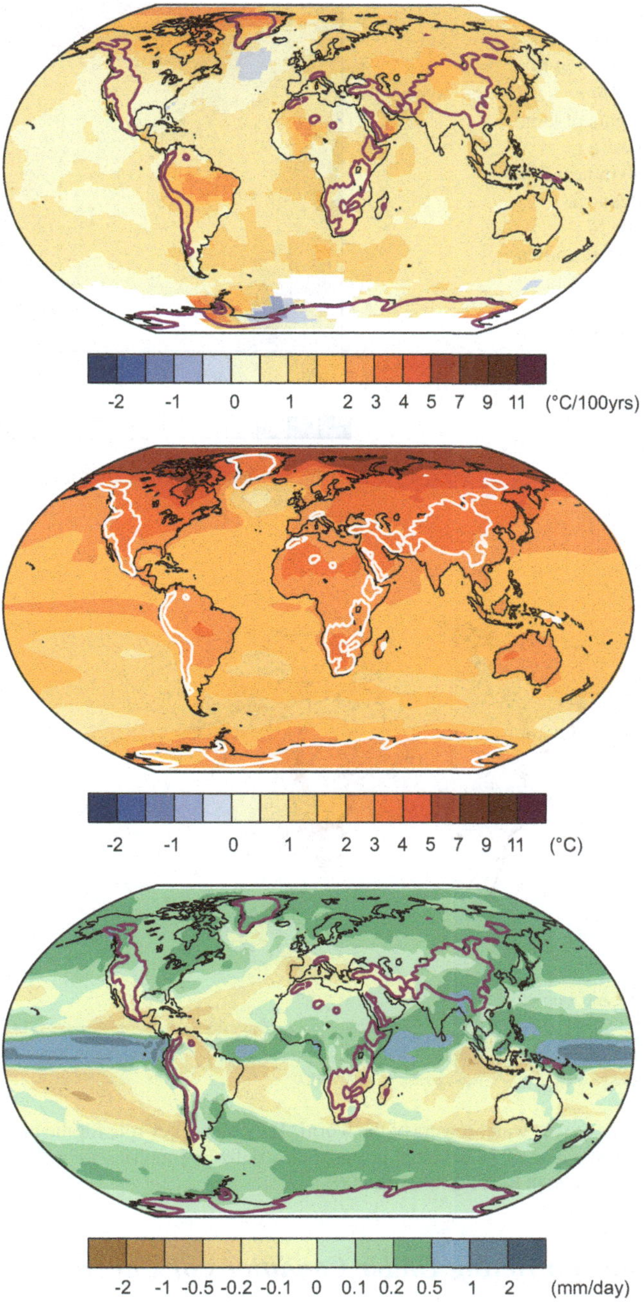

Fig. 1 Linear trend in annual mean surface air temperature (*top*) from 1900 to 2013. Data source: NASA/GISS (Hansen et al. 2010). Modelled changes in temperature (*middle*) and precipitation (*bottom*) from 1985–2005 to 2081–2100 according to a moderate-to-high emissions scenario (RCP6.0, CMIP5 Atlas subset from KNMI Climate explorer; see Collins et al. 2013). *Purple* and *white lines* indicate topography over 1000 m (Figure provided by Brönnimann et al. 2014/1)

Fig. 2 Linear trend in zonal annual mean temperature from 1979 to 2013. The *green line* denotes the heights of large mountain ranges (e.g. Andes ca. 4000 m near 20° S, Tibetan Plateau ca. 5000 m near 30° N); individual peaks are shown as *green triangles* (Data source: ERA-Interim reanalysis Dee et al. 2011)

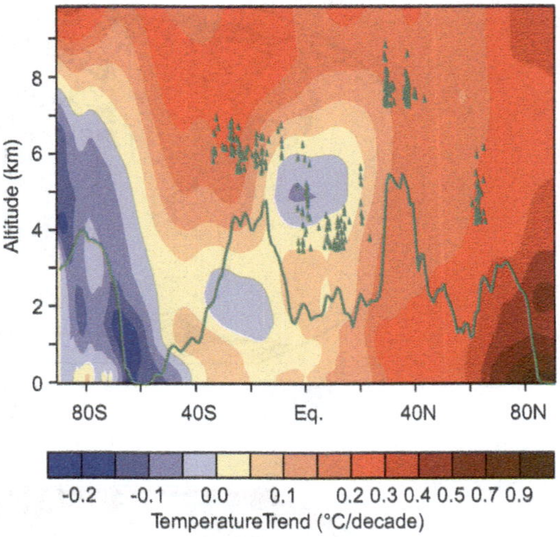

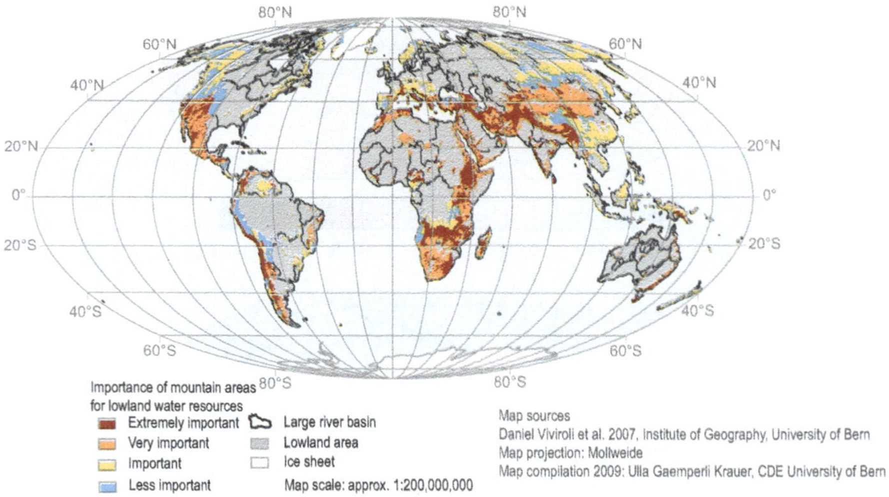

Fig. 3 Mountains are the water towers of the world. They play a key role for the provision of fresh water on all continents and in developing and industrialized countries as well as in emerging economies (Map provided by Viviroli et al. 2007)

2.3 *Options for Acting Under Conditions of Uncertainty*

As the above deliberations show, adaptation for climate change means acting under conditions of uncertainty. Acting under uncertainty is not optimal. However, many decisions cannot be delayed until all uncertainty concerning climate change and its

Box 1: Principles of Decision Making and Planning Under Conditions of Uncertainty

- *Selecting "low-regret" (or "no-regret") options that yield benefits even in absence of climate change and where the costs of the adaptation are relatively low vis-à-vis the benefits of acting;*
- *Selecting "win-win (-win)" options that have the desired result in terms of minimising climate risks or exploiting potential opportunities but also have other social, environmental or economic benefits.*
- *Favouring reversible and flexible options enabling amendments to be made;*
- *Adding "safety margins" to new investments to ensure responses are resilient to a range of future climate impacts;*
- *Promoting soft adaptation strategies, which could include building adaptive capacity to ensure an organisation is better able to cope with a range of climate impacts* (e.g. *through more effective forward planning*);
- *Reducing decision time horizons* (e.g. *the forestry sector may choose to plant tree species with a shorter rotation time*);
- *Delaying action (which should not be confused with 'ignoring the future'). This may be appropriate as part of an active long term adaptation strategy where it has been determined that there is no significant benefit in taking a particular action immediately.* (European Climate Adaptation Platform 2015)

scope and effects will be removed – quite apart from the question whether such knowledge will ever be available. Moreover, there exist a series of principles for moving forward under conditions of uncertainty, and these open up options for adaptation planning and implementation – also in mountain regions (Box 1).

Taking the case of freshwater supplies, arguably the most important service provided by mountains in a global perspective, **Payment for Environmental Services** (PES) is such an option. While the rationale of PES schemes is to help secure flows of critical ecosystem goods and provision of services – in this case water from upstream areas for the benefit of downstream users – they **combine several of the principles for decision making under conditions of uncertainty**: they are no or low regret, because they support the supply of freshwater also in the absence of climate change. They contain a win-win element, as upstream land users are compensated for their efforts to sustainably manage watersheds, while downstream agriculture, settlement and industry benefit from sustained water supply. Numerous PES-schemes for the supply of fresh water are already in operation in many parts of the world. Most of them operate within a watershed and hence at a subnational level (Martin-Ortega et al. 2013; IDB 2012). They thus fulfil the condition of presenting locally adapted action that responds to local needs. Moreover, as PES target watersheds, they might also address other important services such as disaster risk reduction.

There are a number of **challenges associated with PES schemes**, and these also appear in water schemes: First, watersheds are often territorially discordant with political-administrative units such as provinces or districts. Second, the watershed approach means multi-sectorial management given the many demands both in terms of quantity and quality that water satisfies. Both these points necessitate inter-institutional and inter-sectorial coordination, which is a challenge, as authority and action are organised along sectorial and administrative boundaries. Seen from a national perspective, engaging in PES schemes thus requires considerable human, institutional and financial resources, and above all the political will, to engage in the requisite new institutional set ups. Third, and often overlooked: PES schemes are based on a large number of beneficiaries who pay a small number of providers, i.e. small upstream communities are compensated by large downstream populations, or by large downstream private enterprises or public utilities. Costs for individual service beneficiaries are thus low, while the benefits that accrue to providers are substantial enough to warrant sustainable provision of the service. In many tropical regions, however, the situation is reversed. In East Africa, for example, population and economy are concentrated in uplands and highlands, while downstream regions have low population numbers, weak economies, and widespread poverty (Wiesmann et al. 2014). As a result, the payments will be shared by so many that they will be too low to be an incentive. Compensation mechanisms in such situations can still work, but rather *within* mountains and uplands, for example for securing water supplies to cities and town located in these uplands. Nairobi and Addis Ababa present cases in point.

However well PES for water supply will be arranged, they will not suffice to continue business as usual relating to water consumption in many parts of the world, given the increasing demand and the effects of climate change, even if management will increasingly rely on large-scale technological options like large storage facilities and intra- and interbasin transfers. What will be needed to avoid a water crisis is a shift from supply management to demand management, including more effective and efficient use of water (Weingartner 2014).

While PES present a relatively new option for acting under uncertainty, such action is not new, especially so for mountain communities. Across the world, they have always had to face uncertainty including the vagaries of climate. In response they typically have organised activities and assets that provide for their livelihoods in specific ways. For example, farmers and pastoralists have organised their operations across several altitudinal or climatic zones both as part of an adaptation and a risk minimizing strategy. Each of these zones has distinguished climate characteristics relating to temperature regime, amount and distribution of precipitation, and hazards related to extreme events. In order to persist and make a living from their environment, mountain people – not only those dealing with agriculture – always had to adapt to a wide array of climate regimes, an experience that distinguishes them from lowlanders. Adaptation to *different and simultaneous* climate regimes has become deeply rooted in land use practices and livelihood strategies as well as in the culture of many mountain communities. This experience is likely to make them more resistant to the effects of future climate change –

somewhere in their environments, they may have experienced in the past what the future might hold as a novel phenomenon for others (Wymann et al. 2013).

What does this tell us about climate change adaptation in mountain regions? First, adaptation can benefit from the experience of local communities, as these incorporate locally specific traits of climate and locally specific capacities, i.e. response and innovation. We therefore posit that adaptation planning in mountain areas is well advised to take into account ecosystem-based adaptation (EbA – e.g. how much water is available) as well as community based adaptation (CbA – e.g. how to manage and share supplies). Second, in order to tap on local communities' experience, adaptation planning has to engage in a transdisciplinary process which includes mutual learning for adaptation that involves local communities alongside science and policy. Third, promising local activities should be documented – including lessons learnt from PES, taken up in national action plans and be supported in appropriate ways. Fourth, this all calls for substantial investment in knowledge management for adaptation, which includes knowledge generation based on natural and social science approaches, knowledge dissemination, and institutional and human capacity development.

3 The Science-Policy Dialogue: Challenges and Opportunities

3.1 Established Platforms for Science-Policy Interaction

As said earlier, science has been an important partner for policy and decision makers at the global level, mainly through the IPCC and its reports. But also at the national level, many countries have established formal links between science and policy makers as climate change became a topic of political and societal relevance.

In Switzerland, for example, the science-policy dialogue was institutionalised in the 1990s with the formation of the OcCC (Advisory Body on Climate change for the Swiss Government) (Brönnimann et al. 2014/2). Chaired by a member of parliament, OcCC is an interdisciplinary panel of experts under the Federal Department (Ministry) of Environment, Transport, Energy and Communication (OcCC 2015). While its mandate is to advise the government, it also reaches out to economic circles and society at large as well as to the media. The results of its 2007 assessment report (OcCC and ProClim. 2007) led to the inclusion of climate change adaptation as a second pillar alongside mitigation in Swiss climate policy. The Federal Council has prepared a climate change adaptation strategy (2012) that is now being implemented, including a nation-wide analysis of risks and opportunities induced by climate change, main challenges for adaptation, sectorial strategies, and a joint plan of action (2014) (Federal Council 2012). The strategy identifies eight key challenges for adaptation, and mountains are featured prominently: Two of the challenges affect only mountain regions, but with impacts for the whole country, and another four affect the whole country including mountain regions (Table 1).

Table 1 The main challenges in adapting to climate change in Switzerland

Key challenge	Concerns whole/ most of country	Concerns mostly mountain areas	Concerns mostly non-mountain areas
Greater heat stress in agglomerations and cities			X
Increasing summer drought	X		
Greater flood risk	X		
Decreasing slope stability/more frequent mass wasting		X	
Rising snowline		X	
Impaired water, soil, and air quality	X		
Change in habitats, species composition and landscape	X		
Spread of harmful organisms, disease and alien species			X

Source: Federal Council (2012)

In Central Asia, especially in Kyrgyzstan and Tajikistan, with more than 90 % of their territories above 1500 m, climate change in mountains is highly relevant as it increases vulnerability in key sectors of development (Table 2).

In Kyrgyzstan, a multi-stakeholder platform for participatory climate governance was launched for ensuring science and policy interaction. The platform, called Climate Change Dialogue Platform of Kyrgyzstan (referred as CCDP-Kg) was set up by the State Agency on Environmental Protection (SAEPF) and National Centre on Climate Change in 2014 following the Public Panel Discussion organized by the Mountain Hub of the University of Central Asia (UCA) earlier in the same year (http://climate-l.iisd.org/news/kyrgyz-climate-change-platform-launched/). The platform brings together government, academia and science, and also NGOs, international development partners and the private sector for the first time and acts as a working body and forum under the National Coordination Committee on Climate Change (NCCCC) chaired by the First Vice Prime Minister of Kyrgyzstan (http://nature.gov.kg/index.php/en/news-2/839-zasedanie-koordinatsionnoj-komissii-po-problemam-izmeneniya-klimata.htm; and http://ekois.net/ustojchivoe-gornoe-razvitie/).

The SAEPF Director serves as the co-chair for the platform. Operations are run by a working group elected from the participants, representing the different stakeholders on the platform. Scientific inputs are provided by the national hydro-met agency, by CAIAG (Central Asian Institute of Applied Geoscience) and by experts from sectoral research institutes. The aim is to discuss and amend sectoral adaptation plans and development initiatives. In 2015, the platform was engaged in preparing the national positions on iNDC (intended nationally determined contributions) and supported the Kyrgyz delegation at the UNFCCC CoP21 in Paris, including side events and bilateral and international meetings (http://nature.gov.kg/index.php/news/910-zasedanie-kkpik-po-itogam-konferentsii-v-parizhe.html).

Table 2 Economic sectors highly vulnerable to climate change and the relevance of mountains, Kyrgyzstan and Tajikistan

Priority sectors relating to climate change vulnerability	Mountain relevance	Kyrgyzstan	Tajikistan	Observed changes
Water resources	High	X	X	Experiencing too-much, too-little water phenomena; rapid and earlier glacier melting
Agriculture	High	X	X	10 days shifts in ripening season (AGOCA 2015); farm yields are reported to have dropped by 20–30 % across the region since the 1990s (ADB 2014)
Hydropower generation	High	X	X	Frequent power cuts due to lowered water reservoir levels during dry summers (AGOCA 2015)
Human health	High	X	X	Average number of deaths due to climate induced disasters in Kyrgyzstan increased from 61 (2003–2007) to 281 (2008) (MoH Kyrgyzstan 2011)
Natural ecosystems (biodiversity and forest)	High	X	X	Ecosystem shifts upward; species changes in high mountains; apples and apricots now move higher up (AGOCA 2015)
Climate-induced natural disasters/extreme weather events	High	X	X	Between 1990 and 2008, more than 850 incidents of floods and mudslides registered, 92 of which occurred during the first 9 months 2009 (http://sdwebx.worldbank.org/climateportalbhome.cfm)

Source: National Communications to UNFCCC

In Uganda, mountains are important water towers for communities living in and around mountain areas. Mountains are also biodiversity hotspots and key tourism destinations. Examples include the Rwenzori massif which is Africa's third highest peak, and the Greater Virungas that accommodate the famous and endemic Mountain Gorillas which according to the IUCN Red List, are critically endangered. Mountains are also densely populated agricultural areas. But their services are threatened by high population growth, local politics combined with loose governance structures (Myhren 2007), and climate change.

In light of this situation, the Ministry of Water and Environment (MoWE) in collaboration with the Albertine Rift Conservation Society (ARCOS), has set up the Uganda Mountain Stakeholder's Forum (Table 3) to address issues affecting mountain ecosystems and people in the country, including climate change. Uganda's leadership in promoting Sustainable Mountain Development is also marked by the hosting of the World Mountain Forum in October 2016 (http://wmf.mtnforum.org/WMF16/en).

Based on the Forum, the MoWE recently developed the Ugandan National Sustainable Mountain Development Strategy. The initiative was carried out with financial contribution from the Swiss Agency for Development and Cooperation (SDC) through ARCOS. The strategy has the goal to maintain provision of environmental goods and services from the country' mountains to sustain biodiversity as well as socio-economic and livelihood needs. It defines the roles of various institutions for promoting SMD under the coordination of MoWE. These institutions include government ministries, lead agencies such as Uganda Wildlife Authority, National Forestry Authority, National Environment Management Authority, inter-governmental bodies operating in mountains such as the Greater Virunga Transboundary Collaboration (GVTC), as well as non-governmental organizations such as ARCOS.

The strategy also draws much from science. For example, the Makerere University's Mountain Research Centre which will be involved in monitoring and evaluation, to ensure the inclusion of scientific evidence, including climate change, in sustainable mountain development.

3.2 Closing the Mountain Climate Data and Information Gap

The role of science in the policy dialogue on climate change and change adaptation could be enhanced if the data gap on mountain climates could be closed. The paucity of data is particularly marked for mountain areas of the global South and East. Here, climate stations with long term records (ideally more than 30 years) are very few at higher elevations. In the latest (2011) version of the Global Historical Climatology Network (GHCNv3) database which includes over 7000 stations worldwide, only 190 stations (3 %) are above 2000 m (the area above this altitude represents 6.4 % of the global land mass[1]), and as few as 52 (0.7 %) are above 3000 m (the area above

[1] Without Antarctica.

Table 3 Mountain stakeholders mobilized under the Ugandan Mountain Stakeholders' Forum

Category	Stakeholder	Interest	Engagement in SMD
Primary stakeholders (stakeholders with low power, low mandate but with high interest)	Inhabitants or mountain communities and lowland communities	Ecosystem goods and services (water, energy, soil fertility, etc.)	Implementation of options for land management, natural resources development and management, biodiversity conservation, energy and water use efficiency
		Resilience to effects of climate changes and disasters	
		Access	
		Social services	
Primary stakeholders (stakeholders with high power, high mandate and high interest)	Ministries	Integrity of mountain ecosystem	Ecosystem protection
	Protected areas agencies	Protection of fragile components	Sustaining ecosystem goods and services
	Districts	Resilience of mountain ecosystems and people	Mitigation and adaptation to changes in ecosystem
		Disaster management	Resilience of ecosystem and inhabitants
		Knowledge generation	Knowledge generation
Secondary stakeholders (stakeholders with low power, low mandate but with interest)	Development agencies (NGOs)	Integrity of mountain ecosystems	Natural resources development
	Research agencies	Resilience of mountain ecosystems and people	Mitigation and adaptation technologies
		Disaster response and management	Knowledge generation
		Natural resources development	
		Mitigation and adaptation technologies	
		Knowledge generation	
Partner stakeholders (stakeholders with high power, high mandate but with low interest)	Security agencies, infrastructure developers	Infrastructure development	Infrastructure development
	Private sector	Provision of economic and social development services	Provision of economic and social development services
Tertiary stakeholders (stakeholders with low power, low mandate and low interest)	Social service providers	Provision of social development services	Provision of social development services

Fig. 4 View of Muhabura, part of Great Virungas Mountain range, between Rwanda, Uganda and Democratic Republic of Congo (Photo by ARCOS)

this altitude represents 3.1 % of the global land mass) (MRI 2015). Their geographical distribution across the mountain world shows important gaps (Fig. 4). Moreover, many stations are located in valleys, presenting a biased picture of mountain climates especially relating to precipitation. For example, simulations for the Upper Indus Basin based on satellite imagery interpretation of glacier extent suggest that rainfall at high elevations may be over 2.5 times higher than the amounts recorded by the current rainfall stations located in the valley bottoms (Immerzeel et al. 2012). The lack of information stands in sharp contrast to the great variation of climate within mountainous regions at the local scale as compared to lowland areas and the arctic, which as stated earlier is due to the marked elevation range in mountains, as well as to aspect and slope. Site-based records are crucial, as satellite data provide proxy data for temperature, but not for precipitation, are of limited duration and do not provide continuous records; frequent cloud cover in mountain areas also puts a limit to their use (Fig. 5).

In order to devise effective adaptive action, there is thus an urgent need to establish long-term climate observation in mountain regions, a national responsibility. A strategy towards this end must be based on the combined use of monitoring sites, remote sensing, and improved regional climate models that build in the complex mountain topography (Neu 2009). Records should be standardised so as to allow comparison across countries and mountain ranges, and between mountains and other regions. In addition to new observatories, (and of course to information from proxy indicators such as moraines, erratic blocks, lake sediments, etc.), there might

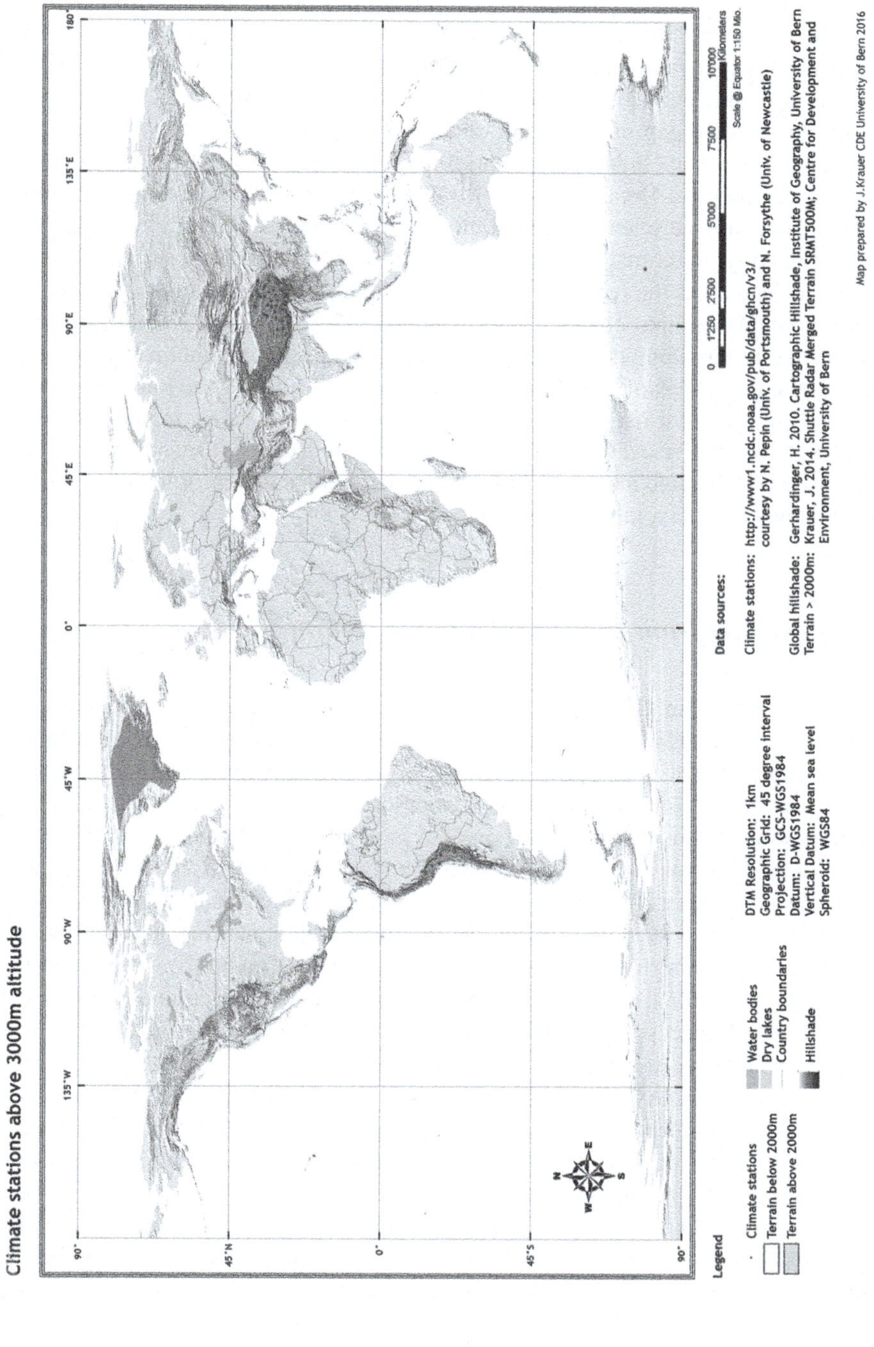

Fig. 5 Climate stations above 3000 m altitude. The map shows stations included in the Global Historical Climatology Network (GHCNv3) database (version 2011). Sources mentioned in the map

exist *forgotten archives*, including private records not incorporated into official weather and climate databases. A case in point is presented by the daily rainfall records kept by large scale ranches in the highlands of Kenya, which reach back into the 1920 and 1930s in some cases and have allowed new insights into regional climate change patterns (Schmocker et al. 2015). Efforts must be made to search for such archives, validate them, and include them in official records.

Closing the data gap on mountain climate change demands long term institutional anchorage. The Global Framework for Climate Services (GFCS), led by the World Meteorological Organization (WMO), set up for the development and application of standardised science-based climate information and services in support of decision-making in climate sensitive sectors, could provide such institutional anchorage, as well as the requisite expertise for improving the information base on global mountain climate change (www.wmo.int/gfcs/). The GFCS could also be instrumental in providing climate change models for specific mountain regions and sub-regions. It could also help in creating the necessary capacity and knowledge for the uptake of this information by regional and national authorities, mountain communities, and other relevant stakeholders, and for translating it into policies and action. The GFCS could also make sure that access to data and information is guaranteed between sectors, institutions, and countries.

3.3 Learning from Proven Adaptation Action

Closing the mountain climate data and information gap goes beyond improving climate records, though. As mentioned earlier, mountain communities have a unique and rich experience with a variety of climate regimes and with adaptation to these. This experience must be tapped for adaptation policy and action. Across the mountain world, there exists an extensive and growing body of knowledge on local adaptation and lessons learnt from it (Wymann von Dach et al. 2013; Kohler et al. 2014). What is lacking is a systematic global overview of such local adaptive action, to provide a detailed account of what was done, by whom, at what cost, and with what outcomes. This overview should also include information on the natural, societal and institutional environment in which the actions took place, and summarise main factors of success as well as issues addressed or circumvented. Such an inventory would need standardised approaches to allow cross-regional comparison, and regular updating. As adaptive action relates to local circumstances and differs with local development contexts, likely agents to establish and host such an inventory could be regional centres of competence with a proven track record in mountain research and development, which already exist in the major mountain regions of the world including the European Alps, the Hindu Kush Himalaya, the Andes, and the mountain countries of Central Asia. The important role of these centres is explicitly acknowledged in the Rio + 20 Outcome Document (UNCED Conference Rio 2012). Their competence is increasingly used to advance the mountain agenda at the global level, as shown by the SMD4GC[2] Programme of the Swiss Agency for Development and Cooperation (SDC), which links four of these regional centres with institutions acting at different levels, with the aim to promote knowledge sharing and policy dialogue, and create a stronger global support for mountain matters.

While collection of adaptive action is best done by region, access to the repository should be global. On this level, the partners of the Mountain Partnership (MP) together with its secretariat could support and complement the efforts done by regional centres by promoting use of the repository by its diversified membership. The MP could combine its efforts with The Nairobi Work Programme (NWP), established under the UNFCCC to promote the development and dissemination of information and knowledge that informs and supports climate change adaptation policies and practices (www3.unfccc.int/pls/apex/f?p=333:1:2454183892108257).

3.4 *Improving Climate Change Governance*

Barriers that hinder the formulation of effective climate change adaptation policy and action are not only a result of flaws in the science-policy dialogue or lack of data and information, but permeate climate governance as a whole (see also Clar et al. 2015, in this volume).

At the global level, for example, the procedures adopted for the UNFCCC climate negotiation process contain important barriers for policy formulation. Research has come up with suggestions on how these negotiations could be made more effective (Chasek et al. 2015). Propositions include the idea to work on a single negotiation text instead of compilation texts, which make consensus difficult and inflate text length. Moreover, the norm that "nothing is agreed before everything is agreed" should be eliminated. Finally, there is a need for broader competence in negotiations so as to fully represent national interests and expertise in climate change matters.

Barriers must also be removed at country level, the more so as nation states are in the driver seat when it comes to adaptation. For national climate change adaptation to become a reality on the ground, national strategies need to be worked out and implemented, which address critical climate relevant sectors of administration, economy and society, and secure coordination and linkage between sector policies and activities. This entails addressing sector activities that propel climate change adaptation, capacity building for sector adaptation, and cross-sector collaboration. Tailored policy and action are needed for subnational development contexts with specific adaptation issues and potentials such as mountain areas. All these processes come at a cost. Funding is thus a critical element as are awareness, acceptance and ownership. The higher the hierarchical level that is in charge of adaptation policy, the greater the chance that the key stakeholders accept and endorse the policy, and

[2]Accronym for: Sustainable Mountain Development for Global Change

take an active part in its implementation. Conversely, the better local specificities are reflected and local adaptation issues addressed, the higher is the likeliness for effective adaptive action.

As climate change does not consider political boundaries, including national ones, adaptation policies and measures must therefore also address regional challenges including transboundary issues. This is particularly important for water resources management. As water flows may change in amount and seasonality, upstream and downstream countries need to find ways and means to agree on how best to adapt to new hydrological regimes in order to secure safe water supplies for domestic use, irrigation, hydropower generation, and for preventing water-related hazards such as floods (Price 2015).

4 Conclusions

To conclude, we put three key aspects of climate change adaptation in mountain regions in the foreground:

First, how can the data gap relating to mountain climate observation and climate change be closed? Long term site-based records are few and far between in many mountain areas and are typically at lower elevations and in valleys. This stands in sharp contrast to the great variation of climate and its effects in mountains at the local scale as compared to lowland areas and the arctic. Still, and despite the fact that mountains are crucial contexts for sustainable development, they are still marginalized and only poorly reflected in global overviews such as the assessment reports of the International Panel on Climate Change IPCC. We think that a special report on climate change in mountains could help provide a sound overview on the gaps and needs of climate data and observation in mountain regions. Under the lead of Switzerland, a consortium has therefore submitted a proposal for a request of a special IPCC report on mountains in December 2015.

Closing the identified data gaps and the establishment of long-term climate observation will allow more accurate projections of climate change to support the formulation of adaptive policy and guide action in mountain regions. A strategy towards this end should be based on the combined use of on-site monitoring, remote sensing, and higher-resolution climate models. Data collection and analyses should involve standardised procedures and an open access policy so as to allow comparison within and across countries and mountain ranges, and between mountains and other regions. We hold that long term climate observation is a national task. It should thus be carried out by a national authority or a trusted institution such as an institution of higher learning or a regional centre of competence. In order to guarantee standardised procedures an international organisation such as the WMO could be mandated for advisory services and backstopping.

While site-based scientific data on climate are crucial, promising practices may already be in place, or experimented with through local initiative. It is important to collect such practices and make them available to potential users in the form of an inventory that is regularly updated. Given the wide range of development contexts of mountain regions and the great variance of mountain climates, it is suggested to organise the inventory at national or regional levels, but to secure exchange and access worldwide. The task could be assigned to national authorities or centres of competence in mountain development, with a global linkage for sharing, for example through the Secretariat of the Mountain Partnership hosted by the FAO.

Second, how can the necessary capacity building for climate change adaptation in mountains be enhanced? In order to make full use of the information collected by climate observation and existing adaptive practices, capacity must be developed within countries for the uptake of this information and for sharing it between all actor groups with a stake in climate change adaptation. This could be achieved by platforms established for exchange and learning between science and policy *and beyond*, including, specifically, local communities that are exposed to climate change, and have to find answers to it and incorporate these in their livelihood strategies. Platforms for learning and exchange could be established under national line ministries, regional centres of competence in mountain research and development, institutions of higher learning, preferably under patronage of trusted high-ranking national institutions such as for example an office of the president or a prime minister's office. This will increase the chance that key stakeholders accept the platform, endorse the adaptation process and take an active part in its implementation. International organizations such as those within the UN system can play an important role in supporting such platforms and make the information collected available at the global level.

Third, who will pay? Implementing a strategy for climate change adaptation in mountains, including climate change observation, and capacity building including exchange platforms and global linkages involve considerable resources and long term financial commitment. Countries should be in the drivers' seat when it comes to devising policy and action for their mountain regions, but those in need of external support should have access to funding specifically designed for this purpose, open for individual countries or a group of countries that share the same mountain region. Institutional anchorage of such an opportunity could e.g. be a special mountain window within the Green Climate Fund, the donor base of which should be enlarged to include all countries in a position to provide support. Justification for the window can be derived from the crucial ecosystem goods and services that mountain regions provide for humankind and global sustainable development, and also by the fact that climate change is a huge externality for mountain regions, as they contribute relatively little to global greenhouse gas emissions, but are fully exposed to the effects of these emissions.

References

Adam, J. C., Hamlet, A. F., & Lettenmaier, D. P. (2009). Implications of global climate change for snowmelt hydrology in the twenty-first century. *Hydrological Processes, 23*, 962–972.

ADB. (2014). Climate change and sustainable water management in Central Asia, Asian Development Bank ADB Central and West Asia working paper series, NO. 5, May 2014, FCG International, Dr. Mikko Punkari, Dr. Peter Droogers, Dr. Walter Immerzeel, Natalia Korhonen, Arthur Lutz, and Dr. Ari Venäläinen, 2014, Asian Development Bank.

AGOCA. (2015). Rapid assessment survey findings for 32 AGOCA villages of 3 countries of Central Asia, Central Asia Mountain Hub at University of Central Asia (CAMH/UCA) and Alliance of Central Asian Mountain Villages (AGOCA), 2015. Bishkek.

Barnett, T. P., Adam, J. C., & Lettenmaier, D. P. (2005). Potential impacts of a warming climate on water availability in snow-dominated regions. *Nature, 438*, 303–309.

Brönnimann, S., Andrade, M., & Diaz, H. F. (2014/1). Climate change and mountains. In *Mountains and climate change: A global concern* (Sustainable mountain development series, pp. 8–13). Bern: Centre for Development and Environment, Swiss Agency for Development and Cooperation, and Geographica Bernensia.

Brönnimann, S., Appenzeller, C., Croci-Maspoli, M., Fuhrer, J., Grosjean, M., Hohmann, R., Ingold, K., Knutti, R., Liniger, M. A., Raible, C. C., Röthlisberger, R., Schär, C., Scherrer, S. C., Strassmann, K., & Thalmann, P. (2014/2). Climate change in Switzerland: A review of physical, institutional, and political aspects. *WIREs Climate Change, 5*, 461–481. doi:10.1002/wcc.280.

Chasek, P., Wagner, L., & Zartman, I. W. (2015). *Six ways to make climate negotiations more effective* (Policy Brief Fixing Climate Governance Series, 3/2015). Ontario: Centre for International Governance Innovation.

Clar, C., Prutsch, A., & Steurer, R. (2015). *Barriers and guidelines in adaptation policy making: Taking stock, analysing congruence and providing guidance*. Vienna: Institute of Forest, Environment and Natural Resource Policy, University of Natural Resources and Life Sciences.

Collins, M., Knutti, R., Arblaster, J. M., Dufresne, J. L., Fichefet, T., Friedlingstein, P., Gao, X., Gutowski, W. J., Johns, T., Krinner, G., Shongwe, M., Tebaldi, C., Weaver, A. J., & Wehner, M. (2013). Long-term climate change: Projections, commitments and irreversibility. In T. F. Stocker, D. Qin, G. K. Plattner, M. Tignor, S. K. Allen, J. Boschung, A. Nauels, Y. Xia, V. Bex, & P. M. Midgley (Eds.), *Climate change 2013: The physical science basis. Contribution of Working Group I to the Fifth Assessment Report of the Intergovernmental Panel on Climate Change* (pp. 1029–1136). Cambridge: Cambridge University Press.

Dee, D. P., Uppala, S. M., Simmons, A. J., Berrisford, P., Poli, P., Kobayashi, S., Andrae, U., Balmaseda, M. A., Balsamo, G., Bauer, P., Bechtold, P., Beljaars, A. C. M., van de Berg, L., Bidlot, J., Bormann, N., Delsol, C., Dragani, R., Fuentes, M., Geer, A. J., Haimberger, L., Healy, S. B., Hersbach, H., Hólm, E. V., Isaksen, L., Kållberg, P., Köhler, M., Matricardi, M., McNally, A. P., Monge-Sanz, B. M., Morcrette, J.-J., Park, B.-K., Peubey, C., de Rosnay, P., Tavolato, C., Thépaut, J.-N., & Vitart, F. (2011). The ERA-interim reanalysis: Configuration and performance of the data assimilation system. *Quarterly Journal of the Royal Meteorological Society, 137*(656), 553–597. http://dx.doi.org/10.1002/qj.828.

European Climate Adaptation Platform. (2015). http://climate-adapt.eea.europa.eu/uncertainty-guidance/topic. Accessed 06 Apr 2015.

Federal Council. (2012). *Strategie des Bundesrates zur Anpassung an den Klimawandel in der Schweiz*. http://www.bafu.admin.ch/klima/13877/14401/index.html?lang=de. Accessed 08 Mar 2015.

Hansen, J., Ruedy, R., Sato, M., & Lo, K. (2010). Global surface temperature change. *Reviews Geophysics, 48*(4), 1–29. http://dx.doi.org/10.1029/2010RG000345.

IDB. (2012). Interamerican Development Bank. 2012. The challenge of the watershed integrated management: Analysis of the IDB actions in the watershed management programs. IDB Evaluation and Supervision Unit.

Immerzeel, W. W., Pellicciotti, F., & Shrestha, A. B. (2012). Glaciers as a proxy to quantify the spatial distribution of precipitation in the Hunza Basin. *Mountain Research and Development, 32*(1), 30–38.

Kohler, T., Wehrli, A., & Jurek, M. (2014). *Mountains and climate change: A global concern* (Sustainable mountain development series). Bern: Centre for Development and Environment, Swiss Agency for Development and Cooperation, and Geographica Bernensia.

Martin-Ortega, J., Ojea, E., & Roux, C. (2013). Payments for ecosystem services in Latin America: A literature review and conceptual model. *Ecosystem Services, 6*, 122–132.

MoH Kyrgyzstan. (2011). Health sector programme for the Kyrgyz Republic on Climate Change Adaptation for 2011–2015, Ministry of Health of the Kyrgyz Republic (MoH Kyrgyzstan), 2011. Bishkek.

MRI. (2015). Mountain Research Initiative EDW Working Group. 2015. Elevation-dependent warming in mountain regions of the world. *Nature Climate Change, 5*, 424–430.

Myhren, S. M. (2007). *Rural livelihood and forest management by Mount Elgon*. Kenya: Noragric.

Neu, U. (2009). Climate change in mountains. In T. Kohler & D. Maselli (Eds.), *Mountains and climate change: From understanding to action* (pp. 6–9). Bern: Geographica Bernensia, with the support of the Swiss Agency for Development and Cooperation.

OcCC. (2015). *Organe consultative sur les changements climatiques*. www.OcCC.ch. Accessed 06 Sept 2015.

OcCC and ProClim. (2007). *Climate change and Switzerland 2050. Expected impacts on environment*. Bern: Society and Economy. OcCC/ProClim – Forum for Climate and Global Change.

Price, M. F. (2015). Transnational governance in mountain regions: Progress and prospects. *Environmental Science & Policy, 49*, 95–105. doi:10.1016/j.envsci.2014.09.009.

Schmocker, J., Liniger, H. P., Ngeru, J. N., Brugnara, Y., Auchmann, R., & Brönnimann, S. (2015). Trends in mean and extreme precipitation in the Mount Kenya region from observations and reanalyses. *International Journal of Climatology*. doi:10.1002/joc.4438. Published online in Wiley Online Library (wileyonlinelibrary.com).

UNCED Conference Rio. (2012). Outcome document. http://www.un.org/disabilities/documents/rio20_outcome_document_complete.pdf. Accessed on 07 July 2015.

Viviroli, D., Dürr, H. H., Messerli, B., Meybeck, M., & Weingartner, R. (2007). Mountains of the world, water towers for humanity: Typology, mapping, and global significance. *Water Resources Research, 43*(7), W07447.

von Wymann, D. S., Romeo, R. L., Vita, A., Wurzinger, M., & Kohler, T. (Eds.). (2013). *Mountain farming is family farming. A contribution from mountain areas to the International Year of Family Farming 2014*. Rome: FAO; Centre for Development and Environment University of Bern; Centre for Development Research of the University of Natural Resources and Life Sciences, Vienna.

Weingartner, R. (2014). Mountain waters and climate change from a socio-economic perspective. In T. Kohler, A. Wehrli, & M. Jurek (Eds.), *Mountains and climate change: A global concern* (Sustainable mountain development series). Bern: Centre for Development and Environment, Swiss Agency for Development and Cooperation, and Geographica Bernensia.

Wiesmann, U., Kiteme, B., & Mwangi, Z. (2014). *Socio-economic Atlas of Kenya: Depicting the National Population Census by county and sub-location*. Nairobi: KNBS, Nanyuki: CETRAD, Bern: CDE.

Further References (Referring to Table 2, Central Asia)

Tajikistan. The third national communication of the Republic of Tajikistan under the United Nations Framework Convention on Climate Change, State Agency for Hydrometeorology of the Committee for Environmental Protection. Available at: www.unfccc.int/essential_background/library/items/3599.php?rec=j&priref=7785#beg.

The Kyrgyz Republic's Second National Communication to the United Nations Framework Convention on Climate Change, Ministry of Ecology and Emergencies of the Kyrgyz Republic. Available at: www.unfccc.int/resource/docs/natc/kyrnc2e.pdf.

Towards Paris. (2015). What does a new global climate policy mean for Central Asia? Regional Environmental Center for Central Asia (CAREC) and Asia Pacific Adaptation Network (APAN), August 2015, Almaty.

The manufacturer's authorised representative in the EU is Springer Nature Customer Service Centre GmbH, Europaplatz 3, 69115 Heidelberg, Germany. If you have any concerns regarding our products, please contact ProductSafety@springernature.com

Printed and bound by CPI Group (UK) Ltd, Croydon, CR0 4YY
30/09/2026
02225238-0004